D1560874

The Food Safety Hazard Guidebook

The Food Safety Hazard Guidebook

Richard Lawley, Laurie Curtis & Judy Davis
Food Safety Info, London, UK

RSC Publishing

ISBN: 978-0-85404-460-3

A catalogue record for this book is available from the British Library

Published by The Royal Society of Chemistry,
Thomas Graham House, Science Park, Milton Road,
Cambridge CB4 0WF, UK

Registered Charity Number 207890

For further information see our web site at www.rsc.org

Preface

Food safety is important. Consumers have a right to expect that those who supply the food that they buy have taken every care to manufacture products that will do them no harm. Those with a responsibility for the regulation of the global food industry recognise this principle and legislate accordingly. This confers a legal and a moral duty, as well as an economic incentive, on all food businesses to ensure that the food they supply is as free from food safety hazards as is practically possible. The food business that tries to evade its responsibilities in this regard will not remain in business for very long.

The business of managing and regulating the safety of the food supply chain has come a long way in the last 25 years or so. Prompted by the emergence of new food safety hazards, such as the bacterial pathogens *Listeria monocytogenes* and *E. coli* O157, powerful new techniques for evaluating and managing the risks presented by these threats have been developed. For example, hazard-analysis critical control point, or HACCP, has now become the food safety management system of choice worldwide. Similarly, the technique of risk assessment has been developed to the point where it can be applied to almost anything. There now exists a comprehensive toolbox of techniques for managing the safety of food, and a plethora of training and guidance options for learning how to use the tools. As a result, there is now little to excuse any food business that fails to protect its customers from known food safety hazards.

Although the food safety management tools are now widely available, they are still virtually useless unless they are supported by adequate and accurate information. HACCP does not work unless its practitioners have access to enough data and scientific knowledge to enable them to understand hazards and how to control them effectively. For example, there is little point in deciding that pasteurisation is the best way to control a bacterial pathogen unless its heat resistance is known. There is plenty of information available, in countless excellent books and other publications, and increasingly on-line.

The Food Safety Hazard Guidebook
By Richard Lawley, Laurie Curtis & Judy Davis

Unfortunately, accessing that information can be problematic, especially for smaller food businesses.

The *Food Safety Hazard Guidebook* is an attempt to address that problem by distilling the key facts about a wide range of individual food safety hazards into a single text. We have tried to adopt a clear format and to keep the information included as concise as possible so that it is easy to find the important facts. We would not claim for one moment that the book is a comprehensive or exhaustive reference work on food safety hazards, and it is not meant to be. As the title suggests, it is intended as a guidebook rather than an encyclopaedia, and has been conceived as a portal for the immense and ever-expanding body of scientific knowledge that exists for food safety. To that end, we have included "sources of further information" in every chapter for those needing more detail. As authors, we have drawn on our experience of supplying the technical and scientific information that food safety professionals require to address a real need for accessible knowledge. We have tried to produce a book that is accurate and reliable, as up to date as possible, and above all, useful.

Disclaimer

The material contained in this book is presented after the exercise of every possible care in its compilation, preparation and issue. However, the authors can accept no liability whatsoever in connection with its application and use.

Contents

The Food Safety Hazard Guidebook
By Richard Lawley, Laurie Curtis & Judy Davis

Introduction

Food Safety

The term food safety has no universally accepted definition. In fact, it is sometimes used, wrongly, in relation to defects in food commodities that are much more to do with food quality than with safety. For example, microbial spoilage of food may make it unattractive, or even inedible, but if neither the micro-organisms concerned, nor the by-products of their growth and metabolism have any adverse effect on health, then it is not strictly a food safety issue, but one of acceptability. For the purposes of this book, food safety can usefully be defined as the practice of ensuring that foods cause no harm to the consumer. This simple definition covers a broad range of topics, from basic domestic and personal hygiene, to highly complex technical procedures designed to remove contaminants from sophisticated processed foods and ingredients.

Essentially, the practice of food safety can be distilled down to three basic operations:

- protection of the food supply from harmful contamination;
- prevention of the development and spread of harmful contamination;
- effective removal of contamination and contaminants.

Most food safety procedures fall into one, or more than one, of these categories. For example, good food-hygiene practice is concerned with the protection of food against contamination, effective temperature control is designed to prevent the development and spread of contamination, and pasteurisation is a measure developed to remove contaminants.

Food Safety Hazards

A food safety hazard can be defined as any factor present in food that has the potential to cause harm to the consumer, either by causing illness or injury.

The Food Safety Hazard Guidebook
By Richard Lawley, Laurie Curtis & Judy Davis

Food safety hazards may be biological, such as pathogenic bacteria, chemical, such as a toxin produced during processing, or a physical object, like a stone or piece of metal. In other words, hazards are the factors that food safety practice seeks to protect against, contain and eliminate from foods. In order to be effective, food safety practice must be informed about the nature of these hazards, and food safety procedures must be science based. A thorough understanding of biological and chemical hazards is the first essential step in their control. This is less important for physical hazards, which also tend to have a much lower potential impact on public health. Physical hazards are not considered further here.

Biological Hazards

It is generally biological hazards that pose the greatest immediate food safety threat to the consumer. For example, the ability of food-poisoning bacteria to cause large outbreaks of acute illness within a short time is a threat with which most food businesses are likely to have to contend. There are few foods that are not vulnerable to biological hazards at some point in their manufacture, storage and distribution.

Technically, biological hazards may include larger organisms, such as insects and rodents. However, these rarely present a direct threat to health and are not considered further here. It is micro-organisms and certain foodborne parasites that are of most concern as food safety hazards.

Bacteria

A significant number of bacterial species can be classified as food safety hazards. Some of these, such as *Salmonella* and *Listeria monocytogenes*, are very well known and familiar to consumers, whereas others are much less common and less well understood. Examples include *Vibrio parahaemolyticus*, a comparatively rare cause of food poisoning associated with seafood, and *Yersinia enterocolitica*, a cause of gastroenteritis that predominantly affects young children. *Campylobacter* is another example of a less well known cause of foodborne illness. Few consumers have heard of this organism, yet it is now the cause of more reported cases of food poisoning than any other agent, including *Salmonella*. *Campylobacter* is also less familiar to the food industry and there are still many unknowns surrounding its transmission to humans. This underlines the importance of continued research and scientific investigation for increasing our understanding of biological hazards.

Bacterial food safety hazards fall into one of two categories according to the mechanism by which they cause illness.

Infection

Most foodborne bacterial pathogens cause illness by multiplying in the gut after ingestion of contaminated food. They may then provoke symptoms by invading the cells lining the intestine, or in some cases, invading other parts of the body and causing more serious illnesses. *Salmonella, Campylobacter* and *E. coli* O157 are all examples of bacteria that cause infective food poisoning. This type of food poisoning is usually characterised by a delay, or incubation time, of at least 8–12 h (sometimes much longer) before symptoms develop.

This category also includes some bacteria that produce symptoms by multiplying in the gut and producing toxins, rather than by actively invading the tissues. An example of this type is *Clostridium perfringens*, a food-poisoning bacterium usually associated with cooked-meat products.

Intoxication

There are a few foodborne pathogenic bacteria that produce illness not by infection, but by intoxication. These organisms are able to grow in certain foods under favourable conditions and produce toxins as a by-product of growth. The toxin is thus pre-formed in the food before ingestion and in some cases toxin may still be present even after all the bacterial cells have been destroyed by cooking. *Bacillus cereus* and *Staphylococcus aureus* are examples of bacteria able to cause intoxication, but the most important and potentially serious cause of intoxication is *Clostridium botulinum*. Intoxications usually have much shorter incubations times than infections, because the toxins are pre-formed in the food.

Viruses

Viral gastroenteritis is very common worldwide. There are a number of viruses that are capable of causing foodborne infections, although in most cases, other forms of transmission are more common. Perhaps the best known are noroviruses and hepatitis A, which has been responsible for a number of serious foodborne disease outbreaks, often as a result of poor personal hygiene by infected food handlers.

"New" viruses may also pose a threat to food safety. For example, highly pathogenic avian influenza viruses primarily affect birds, but in some cases may be transmitted to humans and cause serious disease. So far, there is no direct evidence that this transmission can be foodborne, but these viruses are a source of great concern to the poultry industry and there is still much to learn about them.

Parasites

A wide range of intestinal parasites can be transmitted to humans via contaminated foods, although for most, faecal–oral, or waterborne transmission are more common. These organisms are much more prevalent in developing countries with poor sanitation, but the increasingly global nature of the food supply chain may increase their importance in the developed world. Currently, protozoan parasites are the most important, but other types also need to be considered as food safety hazards.

Protozoans

The protozoan parasites that can cause foodborne illness in humans include several well known species, such as *Entamoeba histolytica*, the cause of amoebic dysentery, and *Cryptosporidium parvum*. However, in recent years, some unfamiliar species have emerged as threats to food safety, especially as contaminants in imported produce. An example is *Cyclospora cayetanensis*, the cause of several outbreaks of gastroenteritis in the USA associated with imported fruit.

Other Types of Parasite

Other types of foodborne parasite include nematode worms, such as *Trichinella spiralis* and the anisakid worms found in fish, and cestodes (tapeworms), such as *Taenia solium*. Although many of these are far less prevalent in developed countries than was once the case, thanks to improved sanitation, they are still significant causes of illness worldwide.

Prions

Prions are a relatively recent threat to food safety and are still not fully understood, but their probable involvement in potentially foodborne new variant Creutzfeldt–Jakob disease (vCJD), an invariably fatal brain disease, has lead to considerable concern.

Chemical Hazards

The presence of chemical hazards in food is usually less immediately apparent than that of bacteria and other biological hazards. Acute toxicity caused by foodborne chemical contaminants is now very rare in developed countries. Of much more concern is the potentially insidious effect of exposure to low levels of toxic chemicals in the diet over long periods. In some cases this can lead to chronic illness and there is also the risk that some contaminants may be carcinogenic.

There is potential for an enormous range of chemical contaminants to enter the food chain at any stage in production. For example, agricultural chemicals, such as herbicides and insecticides, may contaminate fresh produce during primary production, some commodities may contain "natural" biological toxins, and chemicals such as detergents and lubricants may enter food during processing. It is also possible for chemical contaminants to leach out of packaging into foods during storage.

Some of the main classes of chemical contaminant important in food safety are as follows:

- agricultural chemicals, pesticides, *etc.*;
- veterinary drugs;
- natural biological toxins;
  - fungal toxins;
  - plant toxins;
  - fish toxins;
- environmental contaminants (*e.g.* dioxins and heavy metals);
- contaminants produced during processing (*e.g.* acrylamide);
- contaminants from food-contact materials (*e.g.* plasticisers);
- cleaning and sanitising chemicals;
- adulterants (*e.g.* illegal food dyes).

The total number of potentially harmful chemicals that may contaminate food is very large. For example, UK legislation contains maximum residue levels (MRLs) for over 28 000 pesticide/commodity combinations. It is therefore not practical to cover pesticides here in anything but the most general terms. Fortunately, the use of pesticides is very strictly controlled in many countries and residues in imported foods are regularly monitored. Links are provided in the "Sources of Further Information" section for readers needing specific information on pesticides.

The list of potential adulterants is also an extensive one. Almost by definition, adulterants are often compounds that would not be expected to be present in foods and little may be known about their health significance if present in the diet. Recent examples include synthetic Sudan dyes found in imported spices and other commodities in the EU. These are illegal for food use, but the health effects of low levels in foods are uncertain, and there has been some discussion over their food safety significance. For these reasons, it is not practical to cover adulterants here.

The chemical-hazards section focuses on contaminants that are known to be food safety hazards, and that have received some attention from food safety researchers and regulators to establish the level of risk they carry.

Allergens

In recent years, the problem of food allergy has been growing in importance for the food industry as the number of people, particularly children, affected by allergy symptoms has increased. Food manufacturers have been encouraged to respond to this development, particularly in terms of labelling foods clearly. Along with clear allergen labelling comes a responsibility to ensure that such labels are accurate. When foods are labelled as not containing specific allergens, it is extremely important that they do not become contaminated with those allergens during production. This is vital for allergens such as peanuts, which may cause life-threatening anaphylactic reactions in sensitive individuals. The presence of undeclared allergens in foods is a growing cause of product recalls in Europe, North America and elsewhere.

The control of allergens in food is now a rapidly developing aspect of food safety, which many manufacturers will need to be concerned with. Twelve specific major food allergens are currently recognised by EU legislation, although many more foods are likely to be capable of causing allergic reactions in sensitive individuals.

These are:

- cereals containing gluten (*i.e.* wheat, rye, barley, oats, spelt or their hybridised strains);
- crustaceans;
- fish;
- egg;
- peanuts;
- soya beans;
- milk;
- tree nuts;
- celery;
- mustard;
- sesame seeds;
- sulfur dioxide and sulfites.

It is probable that food allergy will continue to grow in importance in the coming years, and that further allergens will be recognised in legislation.

The Obligations of Food Businesses

In most countries, the safety of the food supply is regulated by national and by local authorities. Food businesses are required to meet the demands of food safety regulations, at the very least, in order to protect consumers from hazards in food. These are likely to include the setting up of an effective food safety management system, such as hazard-analysis critical control point (HACCP). In addition, many food businesses will need to meet the requirements of their customers, such as large retail chains, or will need to comply with the food safety provisions of third-party audit schemes. Most of these will expect more extensive food safety measures than are required by relevant legislation.

Most businesses will find it necessary to adopt a risk assessment and HACCP-based approach to addressing food safety, and there is considerable assistance and support available to help with this. Nevertheless, it is important that every food business develops at least a basic understanding of the specific food safety hazards that may be relevant to their products and processes. Only then can food safety management systems operate effectively. The following pages are designed to help provide that basic understanding.

Section 1: Biological Hazards

CHAPTER 1.1

Bacteria

1.1.1 *AEROMONAS* SPECIES

Hazard Identification

What are Aeromonas *Species?*

Aeromonas species are gram-negative, non-spore-forming, bacteria, many of which are psychrotrophic (*i.e.* able to grow at low temperatures). Older references may state that these organisms are in the family *Vibrionaceae*, but they have recently been classified in a new family, the *Aeromonadaceae*, and this family now includes at least 14 described *Aeromonas* species.

Although a number of these species have been associated with human disease, the role of *Aeromonas* species as foodborne pathogens has yet to be confirmed. *Aeromonas hydrophila*, *Aeromonas caviae* and *Aeromonas sobria* are the main species that are thought to cause gastrointestinal disease in man and it is considered that the main vehicle for these organisms is drinking water. Many *Aeromonas* species can be divided into two groups based on the temperature range at which strains are able to grow and within a specific species some strains are psychrotophic, while others are mesophilic (not able to grow below 10 °C). For *A. hydrophila*, evidence suggests that those strains that are pathogenic to humans are mesophilic, whereas psychrotrophic strains are pathogenic to fish.

Occurrence in Foods

Aeromonas species are common contaminants in unprocessed foods and on occasions numbers can be high, exceeding 10^6 CFU/g (CFU = colony forming unit). Because of their widespread occurrence it is thought likely that not all strains of *Aeromonas* species are pathogenic. *Aeromonas* species have been isolated from the following food commodities: fresh vegetables; salads; fish; seafood; raw meats including beef, lamb, pork and poultry; and raw milk as well as

The Food Safety Hazard Guidebook
By Richard Lawley, Laurie Curtis & Judy Davis

high-pH cheeses produced from raw milk. *Aeromonas* spp. have also, on occasions, been isolated from some processed foods including pasteurised milk, whipped cream, ice cream and ready-to-eat animal products.

Possible gastroenteritis-causing species have been isolated from most of the above food groups. However, *A. caviae* is more commonly isolated from vegetables and salad, while *A. hydrophilia* is more commonly isolated from meat, fish and poultry.

Hazard Characterisation

Effects on Health

Although there is increasing evidence to suggest that *A. hydrophila, A. caviae* and *A. sobria* are causative agents of foodborne gastroenteritis in humans, this is still the subject of debate. However, aeromonads are often detected in gastrointestinal infections.

The infectious dose is unknown, although data suggests that it is high, probably $>10^6$ cells. Volunteer feeding studies involving ingesting high numbers of *A. hydrophila* cells ($>10^7$) have been inconclusive, whereas the organism has been isolated from the stools of divers who became ill after taking in small amounts of contaminated water. Gastroenteritis associated with *Aeromonas* species is most frequently reported in young children, although it can occur in individuals of any age with the number of cases peaking in the summer months.

It is thought that when ingested, these organisms can cause gastrointestinal disease in healthy individuals, and septicaemia in the immunocompromised. Symptoms are thought to start to occur within 24–48 h of ingestion of cells. Infection can manifest itself in one of two distinct forms. The more common form is a cholera-like illness (watery diarrhoea accompanied by a mild fever), sometimes accompanied by vomiting in children less than 2 years old. The less common form is a dysentery-like illness (diarrhoea with blood and mucus in the stools). The disease is usually self-limiting, lasting between 1–7 days. Occasionally however, the diarrhoea can last for several months, or even longer (12 months plus).

Incidence and Outbreaks

Most *Aeromonas* infections are thought to be caused by contaminated water and there are few reported outbreaks of *Aeromonas*-associated gastroenteritis where food is the suspected vehicle of infection. These few incidents are mostly associated with seafood products such as oysters, sashimi, cooked prawns, shrimp cocktail and raw fermented fish. The literature suggests that other food groups such as edible land snails, egg salad and smorgasbord (comprising shrimp and various ready-to-eat meat products) have also been involved.

Sources

Aeromonas species are ubiquitous, although the main source of the organisms is generally accepted as water. The organisms are found in flowing and stagnant fresh water, in water supplies (including chlorinated water), sewage and in marine waters, particularly those that border with fresh water such as in estuaries. *Aeromonas* species are also often found in household environments such as drains and sinks, and can be isolated from soil.

Aeromonads are found in aquatic animals such as frogs, fish and leeches, in reptiles and in domestic animals such as pigs, sheep, poultry and cows. They can also be carried by humans without symptoms on occasion, although carriage rates are higher in tropical or developing regions.

Growth and Survival Characteristics

The growth temperature range for *Aeromonas* species is variable, but is reported to be between <5 °C and 45 °C. Within a particular species there can be psychrotrophic strains (capable of growth at chill temperatures) and mesophilic strains (cannot grow below 10 °C). Although the optimum temperature for growth is generally reported as 28 °C, this figure is likely to vary depending on strain. Although environmental strains may not grow at 37 °C, many clinical strains can grow at 5–7 °C. *A. hydrophila* is reported to grow between 1–42 °C, with an optimum of 28 °C.

Aeromonads are reported to survive freezing temperatures and have been isolated from frozen foods after storage for approximately two years.

The optimum pH range for the growth of aeromonads is between 6.5 and 7.5. The organisms are tolerant of pH values of up to 10 and many strains will grow down to pH 5.5 or less (under otherwise ideal conditions), but this characteristic is uncommon at chill temperatures.

Many aeromonads will not grow at salt levels >4%, although there are reports of some strains growing at concentrations of 6%. Studies have shown that when foods are stored at chill temperatures *Aeromonas* species are unlikely to grow when the salt levels are more than 3–3.5% and pH values are below 6.0.

Aeromonas species are facultative anaerobes (capable of growth with or without oxygen). At chilled temperatures however, it has been reported that growth rate is either unaffected, or possibly reduced, when fish is modified-atmosphere/vacuum packaged. Modified atmospheres containing high levels of oxygen (>70%) have been shown to retard the growth of *A. caviae* on ready-to-eat vegetables at refrigeration temperatures.

Aeromonas species are not notably resistant to preservatives or sanitisers. It is thought that their presence in chlorinated water is the result of post-treatment contamination or inefficiencies in the chlorination process.

Thermal Resistance

Aeromonads are not heat-resistant organisms and are readily inactivated by pasteurisation or equivalent processes. *D*-values (decimal reduction time) of 3.20 – 6.23 min at 48 °C in raw milk have been recorded.

Control Options

Processing

At present, research suggests that if some *Aeromonas* spp strains are indeed foodborne pathogens, it is foods containing high numbers of the organisms that pose the greatest health risk.

Measures to reduce the likelihood of high numbers occurring should include: using treated water supplies in food processing; keeping foods chilled; and the thorough, frequent cleaning of equipment used to process foods, especially those that are not later cooked by the consumer, *e.g.* salads and vegetables.

Aeromonas species are easily inactivated by pasteurisation, or equivalent processes used by the food industry. Preventing the recontamination of heat-processed products, particularly those with a high water activity and neutral pH that are to be stored chilled, should ensure that aeromonads are not a potential health risk in these foods. Measures to reduce the risk of recontamination include keeping raw and cooked foods separate and implementing good handling and packaging practices.

Product Use

Aeromonas species should be considered as possible pathogens and it has been suggested that very young children, the elderly and the immunocompromised should avoid foods that could be contaminated with high numbers of these organisms.

Legislation

There is no specific legislation in the EU and the US on levels of *Aeromonas* species in foods.

Sources of Further Information

Published

Kirov, S.M. *Aeromonas* species in "Foodborne micro-organisms of public-health significance." ed. Hocking, A.D. & Australian Institute of Food Science and Technology, 6th edn. Waterloo DC. AIFST, 2003, 553–75.

Isonhood, J.H. and Drake, M. *Aeromonas* species in foods. *Journal of Food Protection*, 2002, 65(3), 575–82.

On the Web

Guidelines for drinking water quality. Addendum: Microbiological agent in drinking water. 2nd edn. *Aeromonas*. World Health Organization. (2002). http://www.who.int/water_sanitation_health/dwq/en/admicrob2.pdf

1.1.2 *ARCOBACTER*

Hazard Identification

What is Arcobacter*?*

Arcobacter species are potentially pathogenic bacteria and are closely related to *Campylobacter*. Species in both genera share some similar morphological and metabolic characteristics. Arcobacters are gram-negative, non-spore-forming bacteria, and are often described as aerotolerant *Campylobacter*-like organisms. Both genera belong to the family *Campylobacteraceae*. There are currently six described *Arcobacter* species, but it is *Arcobacter butzleri*, and more rarely *Arcobacter cryaerophilus*, that have been implicated in cases of human illness. On one occasion however, *Arcobacter skirrowi* was isolated from an individual suffering from chronic diarrhoea.

It is thought that the consumption of food contaminated with *Arcobacter* species may play a role in the transmission of these pathogens, although this has not yet been conclusively demonstrated. However, the most significant source of the organisms is thought to be contaminated-water sources.

Occurrence in Foods

Arcobacters are associated with foods of animal origin and have been detected in beef, poultry, pork and lamb, but are most frequently found in poultry and pork products. Chicken carcasses and poultry-processing plants are often contaminated with *Arcobacter* spp. and the organisms have been isolated from retail chicken and turkey products. However, evidence suggests that eggs are not usually contaminated with these bacteria. Arcobacters are not routinely examined for in foods, and so their prevalence in other food types is unknown.

Hazard Characterisation

Effects on Health

Arcobacter butzleri is the most common *Arcobacter* species implicated in human disease. Those most at risk from developing the symptoms associated with *Arcobacter* infection are very young children, although any age group is susceptible. Asymptomatic infections are reported to occur.

The infective dose and incubation time is unknown. Clinical symptoms include abdominal pain, nausea and acute watery diarrhoea, typically lasting between 3–15 days, although this can persist or reoccur on occasions for up to 2 months. Occasionally, vomiting, fever and chills are reported. Extra-intestinal disease such as septicaemia has also been documented occasionally.

Incidence and Outbreaks

The incidence of *Arcobacter* enteritis is unknown, and outbreaks caused by *Arcobacter* species have rarely been reported. One reason for this may be because these organisms are not routinely included in clinical screening.

Sources

Humans suffering from *Arcobacter* infections can be a source of *Arcobacter* species and the faecal–oral route is one probable route of transmission.

Arcobacters are also a cause of enteritis and abortion in animals, although the organisms can also be isolated from apparently healthy animals. Cattle, pigs, sheep, poultry and even horses are thought to be reservoirs for these bacteria. Although meat and associated products from all these animals could be contaminated with *Arcobacter* species, the organisms are most frequently associated with poultry and pork products. Unlike campylobacters, arcobacters are not considered to be normal inhabitants of the poultry intestine, and it is thought that poultry carcasses become contaminated with the organism after slaughter.

Animal faeces can lead to the contamination of soil and water with *Arcobacter* species. Arcobacters have been isolated from water sources, such as drinking-water reservoirs, as well as from canal and river waters. They have also been found in raw sewage and disinfected effluent.

Growth and Survival in Foods

Arcobacters can be differentiated from the Campylobacters in that they are aerotolerant and are able to grow at lower temperatures.

The temperature range within which Arcobacters are able to grow is between 15 and 37 °C (although some isolates are reported to grow up to 42 °C). The organisms are tolerant of refrigerated storage, although numbers do decrease very gradually over time. Arcobacters survive well when frozen at –20 °C.

Arcobacters can grow or survive in both aerobic and microaerophilic atmospheres and can grow over a pH range of 5.5–8.5, possibly up to pH 9.0. Arcobacters do not grow at water activities below 0.980.

Thermal Resistance

Arcobacters are relatively heat sensitive and are readily inactivated at temperatures of 55 °C and above. For *A. butzleri*, *D*-values in phosphate-buffered saline at pH 7.3 have been reported as 0.07 to 0.12 min at 60 °C, 0.38 to 0.76 min at 55 °C, and 5.12 to 5.81 min at 50 °C. Reducing pH has been found to increase the heat sensitivity of the organism. *D*-values in pork have been reported as 18.51 min and 2.18 min, at 50 °C and 55 °C, respectively.

Control Options

Although there is no direct evidence linking *Arcobacter* to foodborne disease in humans, the presence of the organisms in foods suggests that contaminated foods may play a role in their transmission. Effective controls should therefore focus on prevention of contamination.

Processing

Research suggests that these organisms are not normal contaminants of the poultry gastrointestinal tract, and so concentrating on the prevention of contamination in the poultry processing environment and ensuring the rapid chilling of carcasses could reduce the prevalence of these organisms in associated products.

Product Use

Arcobacters are easily inactivated during normal cooking processes. Consumers should be advised to avoid the consumption of inadequately cooked-meat products, and to avoid cross-contamination between raw and ready-to-eat foods.

Legislation

There are no specific requirements for levels of *Arcobacter* species in foods under EU legislation or in the US Food Code.

Sources of Further Information

Published

Forsythe, S.J. *Arcobacter* in "Emerging Foodborne Pathogens." ed. Motarjemi, Y. and Adams, M., Cambridge, UK. Woodhead Publishing Ltd, 2006, 181–221.

1.1.3 *BACILLUS* SPECIES

Hazard Identification

What are Bacillus *Species?*

The *Bacillus* genus is a group of gram-positive, spore-forming bacteria some of which, notably *Bacillus cereus* and more rarely *Bacillus licheniformis, Bacillus subtilis* and *Bacillus pumilus*, have been implicated in foodborne disease. Of the *Bacillus* species, *B. cereus* is recognised as most frequently causing food poisoning and therefore much of this section will focus on this pathogen. It is important to note that not all strains of *B. cereus* are capable of causing food-borne illness.

Occurrence in Foods

Bacillus species are found in many raw and unprocessed foods. However, *B. cereus* is commonly associated with dried foods, spices, cereals (particularly rice and pasta), as well as milk and dairy products. The presence of low numbers of *B. cereus* in raw foods is of little concern because large numbers of the bacteria (usually $>10^5$ CFU/g) are required to cause illness. However, *B. cereus* spores can survive cooking processes and high numbers of *B. cereus* spores in herbs and spices can be a problem if these seasonings are used in processed foods where conditions permit the growth of the vegetative cells.

Hazard Characterisation

Effects on Health

Bacillus cereus food poisoning is caused by toxins produced during the growth of the bacteria and these toxins cause two distinctly different forms of food poisoning – the emetic or vomiting type, and the diarrhoeal type. Both forms of food poisoning require the bacteria to reach high numbers in the food (usually $>10^5$ CFU/g) before sufficient toxin to cause illness can be produced.

The more common emetic type is caused by the presence of a pre-formed toxin (a heat- and acid-stable, ring-form peptide called "cereulide") in the food. It is important to note that live cells of *B. cereus* do not need to be ingested for this form of *B. cereus* food poisoning to occur and foods containing toxin, but no viable cells, can still cause illness. This form of intoxication is characterised by rapid (between 0.5–6 h) onset of symptoms, which include nausea, vomiting and sometimes abdominal cramps and/or diarrhoea. Symptoms usually last less than 24 h.

The less common diarrhoeal type is caused by the formation and release of heat- and acid-labile enterotoxins in the small intestine, although enterotoxin can also be pre-formed in food. This "intermediate" form of food poisoning has an incubation time of between 6–24 h (typically 10–12 h). Typical symptoms, which last for between 12–24 h, are primarily watery diarrhoea, abdominal cramps and pain, with occasional nausea and vomiting.

Some strains of *B. subtilis* and *B. licheniformis* linked to outbreaks of food-borne illness also produce heat-stable toxins similar to cereulide. Rapid onset of vomiting is the main feature of *B. subtilis* food poisoning, usually followed by diarrhea. However, in outbreaks linked to *B. licheniformis*, diarrhoea is usually the main feature of illness, with vomiting occurring in half of the cases.

Recovery from food poisoning caused by *Bacillus* species is usually within 24 h with no complications. Fatalities have rarely been reported.

Incidence and Outbreaks

Most outbreaks of *Bacillus* food poisoning are associated with the consumption of cooked food that has been cooled too slowly and/or incorrectly stored, providing conditions for the micro-organism to increase to significant numbers. Outbreaks caused by the *B. cereus* emetic toxin are most frequently linked to starchy foods such as boiled or fried rice, as well as pasta, potato and noodle dishes. The diarrhoeal form of *B. cereus* food poisoning has been linked to a wide variety of foods but is most commonly associated with meat and vegetable dishes, soups, sauces and puddings.

Food poisoning caused by other *Bacillus* species has also been linked to a wide variety of foods including cooked meat and vegetable dishes, cooked reheated rice, "ropy" bakery products, custard powder, pastries, infant formula, synthetic fruit drinks, mayonnaise, canned tomato juice, sandwiches and pizza.

Sources

Bacillus species are ubiquitous and are widespread in the environment, being found in dust, soil, water, air and vegetable matter. It is thought that climate can influence *B. cereus* populations in soil with surveys indicating that psychrotrophic strains are more dominant in samples from cold regions. *Bacillus* species are also often present in low numbers in human stools, reflecting dietary intake. During bouts of *Bacillus* food poisoning fairly high numbers of the organism will be excreted for up to 48 h after onset.

Growth and Survival in Foods

The optimum growth temperature range for *B. cereus* is around 30–35 °C with an upper limit of up to 55 °C. Some strains, particularly from milk and dairy sources, are reported as being able to grow at chill temperatures, having a minimum temperature for growth of 4 °C (these are described as psychrotrophic). These psychrotrophic strains usually have a maximum temperature for growth of 37 °C. Psychrotrophic *B. cereus* strains have been shown to produce enterotoxins and research suggests that this may occur at temperatures of 7 °C. Emetic toxin production at refrigeration temperatures is thought not to occur.

Although growth of *B. cereus* can occur at less than 10 °C, both lag time and growth rate are significantly increased at these temperatures.

Data on growth temperature ranges for other *Bacillus* species associated with food poisoning is limited. Although there have been occasional reports of some strains of *B. subtilis* and *B. pumilus* growing at 5 °C, these organisms are not generally considered psychrotrophic.

Bacillus cereus can grow under otherwise ideal conditions at pH values between 4.3 and 9.3. The emetic toxin is stable over the pH range 2–11, but the diarrhoeal enterotoxin is less stable at acid pH values.

Some strains of *B. subtilis* and *B. pumilus* can grow at relatively low pH values, and have been implicated in the spoilage of canned tomato products. Whether these strains are capable of causing food poisoning is not known.

The minimum water activity for the growth of *B. cereus* is generally considered as 0.93 but may be as low as 0.91. *Bacillus* spores can survive for extended periods of time in low water activity conditions and are resistant to desiccation.

Bacillus cereus and *B. licheniformis* are facultative anaerobes, being able to grow either aerobically or anaerobically, although studies have shown that both growth and toxin production by *B. cereus* are reduced under anaerobic conditions. *B. subtilis* and *B. pumilus* are obligate aerobes. The growth of *B. cereus* is adversely affected by increasing concentration of carbon dioxide, and the use of appropriate gas mixtures in modified-atmosphere packaged products can extend safe shelf life.

The vegetative cells of *Bacillus* species are not notable resistant to commonly used preservatives and sanitisers, but the spores are much more difficult to destroy. The "natural preservative" nisin, which prevents spore germination, has been shown to be effective at preventing the growth of *Bacillus* species in various food commodities.

Thermal Resistance

The vegetative cells of *B. cereus* are fairly heat sensitive, being readily destroyed by typical pasteurisation processes; however, *Bacillus* spores are moderately heat resistant and can survive quite harsh heat treatments. *B. cereus* spores can vary in their resistance to heat with D_{85}-values of 33.8–106 min and D_{95}-values of between 1.2–36 min being described. Spores are more heat resistant in high-fat or low water activity products.

The *B. cereus* emetic toxin is heat stable (withstanding 126 °C for 90 min), whereas the diarrhoeal enterotoxins are heat sensitive, being inactivated at 56 °C for 5 min.

Control Options

Processing

The risk from *Bacillus* species in foods is usually highest where the pH and/or water activity of the product will permit the growth of the pathogen. The risk also applies for products designed to be rehydrated by the consumer prior to consumption, such as infant formula and soup mixes. For these foods control is achieved by ensuring a low initial level of the micro-organism in the product. This can be done by using ingredients with low levels of *Bacillus*, as well as by using well-designed equipment with effective cleaning regimes to prevent biofilm formation.

Further control of *Bacillus* numbers is achieved by the appropriate use of temperature, either to destroy spores (sterilisation temperatures used for many low-acid canned products are effective), or to minimise the germination and outgrowth of spores during the manufacture of chilled foods. Heat processes sufficient to inactivate the very heat stable emetic toxin are not practical, and the preferred approach is to prevent its formation before heat is applied. For many refrigerated products, heating processes should be devised so that foods reach processing temperatures quickly, and are cooled rapidly, particularly over the temperature range 10–55 °C. The cooling of small portions is easier to control than large volumes of product. Published cooling processes devised to control *Cl. perfringens* will usually also control the growth of *B. cereus* and other *Bacillus* species.

Product Use

Manufacturers should ensure that *B. cereus* levels do not reach hazardous levels ($>10^3$ CFU/g) during the shelf life of the food. Cooked foods should be held hot (minimum 63 °C) prior to consumption, and refrigerated foods should be held at chill temperatures (ideally 4 °C or below) throughout the shelf life of the product.

Legislation

There are no specific requirements for *B. cereus* and other species in foods under European Community (EC) legislation. EC legislation does require, however, that foodstuffs should not contain micro-organisms or their toxins in quantities that present an unacceptable risk for human health.

Both the UK Health Protection Agency (HPA) and the Food Safety Authority Ireland (FSAI) have published guidelines on acceptable levels of micro-organisms in various ready-to eat foods (see links below). These state that the acceptable level of *B. cereus* and other pathogenic *Bacillus* species in these products is $<10^4$ CFU/g.

Sources of Further Information

Published

Granum, P.E. and Baird-Parker, T.C. *Bacillus* Species in Microbiological Safety and Quality of Food, Volume 2. ed. Lund, B.M., Baird-Parker, T.C and Gould, G.W. Gaithersburg. Aspen Publishers, 2000, 1029–39.

International Commission on Microbiological Specifications for Foods. *Bacillus cereus,* in Microorganisms in Foods, Volume 5: Microbiological Specifications of Food Pathogens. ed. International Commission on Microbiological Specifications for Foods. London. Blackie, 1996, 20–35.

On the Web

Opinion of the Scientific Panel on Biological Hazards on *Bacillus cereus* and other *Bacillus* spp in foodstuffs. European Food Safety Authority. (January 2005). http://www.efsa.europa.eu/etc/medialib/efsa/science/biohaz/biohaz_opinions/839.Par.0001.File.dat/biohaz_ej175_op_bacillus_enfinal1.pdf

Risk profile: *Bacillus* spp in rice. Institute of Environmental Science and Research Limited. (February 2004). http://www.nzfsa.govt.nz/science/risk-profiles/bacillus-in-rice-1.pdf

Guidelines for the microbiological quality of some ready-to-eat foods sampled at the point of sale. Health Protection Agency. (September 2000). http://www.hpa.org.uk/cdph/issues/CDPHvol3/No3/guides_micro.pdf

Guidelines for the interpretation of results of microbiological analysis of some ready-to-eat foods sampled at point of sale. Food Safety Authority of Ireland. (2001). http://www.fsai.ie/publications/guidance_notes/gn3.pdf

1.1.4 *CAMPYLOBACTER*

Hazard Identification

What is Campylobacter*?*

Campylobacter spp. are gram-negative, non-spore-forming bacteria, some of which (*C. jejuni, C. coli, C. lari* and *C. upsaliensis*) are associated with gastroenteritis, although most cases of human campylobacteriosis are caused by *C. jejuni. Campylobacter* is now the leading cause of bacterial gastroenteritis in many developed countries.

Campylobacter is unique amongst food-poisoning bacteria in that it is not normally able to grow in foods. This is because it has specific atmospheric requirements (microaerophilic conditions) for growth and can only grow at temperatures above ambient.

Occurrence in Foods

Campylobacter is most often associated with fresh poultry meat and related products. A UK Food Standards Agency study has found that the level of poultry carcass contamination in the UK is 50%, but elsewhere studies have found contamination rates of at least 60%, with up to 10^7 *Campylobacter* cells per carcass being recorded. Fresh poultry is more frequently and more heavily contaminated than frozen.

Campylobacter species have also been isolated from other fresh meats such as beef, lamb, pork and offal, but at lower frequencies than in poultry. *Campylobacter* can also be found in raw milk, shellfish, mushrooms and salads.

Hazard Characterisation

Effects on Health

The infective dose for *Campylobacter* may be less than 500 cells. Symptoms associated with *Campylobacter* infections appear between 1 to 11 days (typically 2–5 days) after infection. Symptoms can vary widely and usually start with muscle pain, headache and fever. Most cases involve diarrhoea, and both blood and mucus may be present in stools. Nausea occurs, but vomiting is uncommon. Symptoms can last from 1 to 7 days (typically 5 days). The infection is usually self-limiting. *Campylobacter* enteritis is most commonly associated with children (less than 5 years) and young adults. Death rarely occurs, particularly in healthy individuals. However, mortality rates associated with *C. jejuni* in the US have been estimated at 1 per 1000 cases.

Although complications of campylobacteriosis are rare, arthritis (*e.g.* Reiter's syndrome) can occur and severe abdominal pain can be confused with appendicitis. Reactive arthritis occurs in 1% of cases and 0.1% can suffer Guillain–Barré syndrome (a severe nerve disorder, which can lead to paralysis). Around

15% of those affected recover from Guillain–Barré syndrome, 3–8% die and the remainder suffer from some degree of disability. Bacteraemia can also occur, particularly in the elderly.

Incidence and Outbreaks

Campylobacter has recently been recognised as the principal cause of bacterial gastroenteritis in Europe and nearly 200 000 cases were reported in the EU in 2005. The majority of these are thought to be foodborne. A similar situation exists in North America and other countries. In 2004, New Zealand was reported to have the highest incidence of *Campylobacter* infection in the developed world.

Most cases of *Campylobacter* enteritis are sporadic, so definitive sources of infection are difficult to establish. However, most cases are thought to be associated with undercooked, or recontaminated, poultry meat. Documented outbreaks are relatively rare, but have been linked to raw and inadequately pasteurised milk, raw clams, garlic butter, fruits and contaminated water supplies. In one recorded incident in 2005, at least 80 people at offices in Copenhagen were made ill by contaminated chicken salad in canteen meals.

Sources

Campylobacters are found in the intestinal tract of many warm-blooded animals, such as cattle, sheep, pigs, goats, dogs and cats, although they are especially common in birds, including poultry. Wild birds are thought to be a reservoir for domestic and food animals.

If hygiene is poor, infected humans can transfer *Campylobacter* to food via the faecal–oral route and asymptomatic carriers have also been reported. Excreta from infected animals can contaminate water and mud, and *Campylobacter* can survive for some time in these environments, particularly when temperatures are low.

Growth and Survival in Foods

As previously stated, *Campylobacter* is unable to grow at temperatures normally used to store food. The temperature range for growth is 30–45 °C, with an optimum of 42 °C. Although survival at room temperature is poor, *Campylobacter* can survive for a short time at refrigeration temperatures – up to 15 times longer at 2 °C than at 20 °C. The organism dies out slowly at freezing temperatures.

The optimum pH for growth is 6.5–7.5, and the organism does not grow below pH 4.9. Survival at acid pH values is temperature dependent, but inactivation is rapid at pH values less than 4.0, especially above refrigeration temperatures.

The minimum water activity for growth is $\geq$0.987 (2% sodium chloride). The organism is sensitive to salt and, depending on temperature, levels of 1% or more can be bactericidal (less effect being observed with decreasing temperature).

Although *Campylobacter* is sensitive to desiccation, there are reports of survival for some time on wooden cutting boards.

Campylobacter is microaerophilic, requiring reduced levels of oxygen (5–6%) to grow. The cells usually die out quickly in air but survive well in modified or vacuum packaging.

Thermal Resistance

Campylobacter is heat sensitive and the cells are destroyed at temperatures above 48 °C. They do not therefore survive normal pasteurisation processes applied to milk. Heat processes targeted at other poultry pathogens (*e.g. Salmonella*) will easily inactivate *Campylobacter*.

Control Options

Processing

Poultry and poultry products are considered to be the main source of *Campylobacter* food poisoning and controls focus on measures to minimise the level of contamination during primary production and processing of poultry meat.

In many European countries measures are in place to encourage effective biosecurity and hygiene strategies to prevent the introduction of *Campylobacter* to flocks and reduce the incidence of infection. For example, in Denmark, "Campylobacter-free" chicken meat can be marketed at a premium price, providing that it comes from flocks that meet required monitoring standards.

Much attention has also been given to measures designed to reduce high rates of cross-contamination during the processing of poultry, particularly chicken, by improving the hygienic design and operation of equipment such as defeathering machines and immersion chiller tanks.

Product Use

As previously discussed, *Campylobacter* is unable to grow in foods stored at normal temperatures. However, the potentially low infective dose means that undercooking of raw foods and/or cross-contamination from raw to ready-to eat foods is a major risk factor for human campylobacteriosis.

Clear and effective cooking instructions can help to ensure that the pathogen is destroyed during the cooking stage. Undercooking and/or cross-contamination at barbeques are thought to be linked to an increase in reported *Campylobacter* infections during summer months.

Consumer education and domestic hygiene training can help prevent the transfer of *Campylobacter* from raw to ready-to-eat foods. Consumers should be advised not to wash meat and poultry carcasses prior to cooking to help prevent water splashes and aerosols from contaminating kitchen surfaces. Any surfaces that could be potentially contaminated, such as in meat-preparation areas, as well as chopping boards, should be thoroughly disinfected after use.

Legislation

No specific requirement is made under European Commission legislation with regard to levels of *Campylobacter* species in food. Requirements for their control are covered under EU general food safety requirements.

The UK Health Protection Agency (HPA) has published guidelines on acceptable levels of micro-organisms in various ready-to eat foods (see link below). These state that ready-to-eat foods should be free from *Campylobacter* spp. and that, even in small numbers, their presence in processed, ready-to-eat foods, "results in such foods being of unacceptable quality/potentially hazardous."

Sources of Further Information

Published

Nachamkin, I. *Campylobacter jejuni*, in Food Microbiology: Fundamental and Frontiers. 2nd edn. ed. Doyle, M.P., Beuchat, L.R and Monteville, T.J. Washington D.C. ASM Press. 2001. 179–192.

International Commission on Microbiological Specifications for Foods. *Campylobacter*. Microorganisms in foods, volume 5: microbiological specifications of food pathogens. Edited by International Commission on Microbiological Specifications for Foods. London. Blackie. 1996, 45–65.

On the Web

Risk profile: *Campylobacter jejui/coli* in mammalian and poultry offals. Institute of Environmental Science and Research Limited. (January 2007). http://www.nzfsa.govt.nz/science/risk-profiles/FW0465_Campy_in_Offal_PVDL_final_comments_Mar_2007.pdf

Risk profile: *Campylobacter jejuni/coli* in red meat. Institute of Environmental Science and Research Limited. (January 2007). http://www.nzfsa.govt.nz/science/risk-profiles/FW0485_Campy_in_red_meat_Final_sent_to_NZFSA_Jan_07.pdf

Advisory Committee on the Microbiological Safety of Food. Second Report on *Campylobacter*. (2005). http://www.food.gov.uk/multimedia/pdfs/acmsfcampylobacter.pdf

Risk Profile: *Campylobacter jejuni/coli* in poultry (whole and pieces). Institute of Environmental Science and Research Limited. (June 2003). http://www.nzfsa.govt.nz/science/risk-profiles/campylobacter.pdf

Risk assessment of *Campylobacter* spp. in broiler chickens and *Vibrio* spp. in seafood. World Health Organization. (2002). http://www.who.int/foodsafety/publications/micro/aug2002.pdf

Control of *Campylobacter* species in the food chain. Food Safety Authority Ireland. (2002). http://www.fsai.ie/publications/reports/campylobacter_report.pdf

Guidelines for the microbiological quality of some ready-to-eat foods sampled at the point of sale. Health Protection Agency. (September 2000). http://www.hpa.org.uk/cdph/issues/CDPHvol3/No3/guides_micro.pdf

1.1.5 *CLOSTRIDIUM BOTULINUM*

Hazard Identification

What is Clostridium botulinum*?*

Clostridium botulinum is a gram-positive, spore-forming bacterium that produces neurotoxins. It is these toxins (the most potent natural toxins known) that cause the severe illness known as botulism. In recent years, some strains of *Clostridium butyricum* and *Clostridium baratii* have also been found to produce botulinum neurotoxins and there have been outbreaks of foodborne illness associated with these species.

There are at least two types of foodborne botulism:

Classic botulism – an intoxication caused by the ingestion of pre-formed toxins in food.
Infant botulism (also known as floppy baby syndrome) – a condition arising from toxin produced when *Cl. botulinum* grows in the intestines of unweaned infants.

Seven different types of *Cl. botulinum* (A to G) are recognised and are typed by the toxin they produce. These seven types are divided into four groups based on physiological differences. When assessing risk, food safety professionals should consider two of these groups:

Group I – proteolytic, mesophilic (comprising types A, B & F)
Group II – non-proteolytic, psychrotrophic (comprising types B, E & F)

Occurrence in Foods

Clostridium botulinum spores are present at low levels in a wide variety of foods. However, surveys to determine levels in foods have concentrated on fish, meat and honey. The highest incidence is in fish, with *Cl. botulinum* type E commonly associated with farmed trout, Pacific salmon and Baltic herring. Types A and B have been isolated in very low numbers from meats such as pork, bacon and liver sausage as well as fruit and vegetables, including mushrooms. *Cl. botulinum* has also been isolated, usually at low levels, from some honey samples. However, levels as high as 60 CFU/g have occasionally been reported, and 80 spores/g of types A and B were found in a sample of honey linked to a case of infant botulism.

It is important to remember that most low acid (pH >4.6) foods stored in conditions that permit the growth of *Cl. botulinum* have the potential to be associated with botulism unless sufficient thermal processing to inactivate spores has been applied.

Hazard Characterisation

Effects on Health

Botulinum toxins are neurotoxins that affect the neuro-muscular junction, leading to muscle paralysis. Botulism is the most severe form of food poisoning and unless it is recognised and treated promptly, it carries a high risk of mortality (35–40%). Prompt treatment can reduce this mortality rate to below 10%. The presence of live organisms is unnecessary for "classic" foodborne botulism to occur and very small concentrations of pre-formed toxin (possibly as low as a few nanograms) in food can cause illness. The ingestion of viable *Cl. botulinum* spores, at levels as low as 10 to 100 spores, is required for infant botulism to occur.

All individuals are susceptible to classic foodborne botulism and onset times and the severity of symptoms depend on the amount of toxin ingested. Typically, the onset of symptoms occurs within 12–36 h, although the recorded range is 4 h–8 days. Early symptoms may include abdominal distension, mild diarrhea and vomiting, before more severe neurological symptoms develop. These include blurred or "double" vision, dryness of mouth, weakness, and difficulties in talking, swallowing and breathing. Death is usually the result of respiratory paralysis. General paralysis may also develop in some cases.

Infant botulism is associated with babies under a year old and symptoms include constipation, poor feeding, lethargy, and an unusual cry as well as a loss of head control.

Incidence and Outbreaks

The incidence of botulism around the world reflects regional eating patterns and outbreaks are relatively rare. The highest nationally reported incidence of botulism in the world is in the Republic of Georgia. However, the highest incidence in the EU is in Poland, where a large number of "high-risk" home-preserved (bottled/canned) foods are consumed. In the USA, infant botulism is the most common form of botulism.

Notable outbreaks in the UK linked to commercially produced foods have been associated with canned salmon, hazelnut conserve used as a flavouring in yogurt and duck paste. Elsewhere "unusual" foods causing botulism have been baked potatoes, potato salad made from baked potatoes, uneviscerated dry salted fish, vegetable-in oil products (such as garlic and aubergines), Brie and Mascarpone cheeses, cheese containing onion and hot and cold smoked fish. A large outbreak in Thailand linked to dishes containing preserved bamboo shoots occurred during the Spring of 2006, when at least 143 individuals were taken ill, although there were no fatalities. More recently, an outbreak in the US in late 2006 is thought to have been caused by temperature abused, commercially produced carrot juice.

Honey and possibly glucose syrup, are the only food vehicles known to cause infant botulism. However, infant milk powder may have caused a case in the UK.

Sources

Clostridium botulinum is widely distributed in nature, being found in soil and marine environments throughout the world, as well as in the intestinal tracts of animals (including fish). The frequency of isolation and variation of type varies with geographical region. Type A dominates in the western US, South America and China, type B in the eastern US and Europe and type E in northern areas and in temperate aquatic environments.

Growth and Survival in Foods

Cl. botulinum is an obligate anaerobe (only grows in the absence of oxygen), but the risk from the pathogen is not limited to products packaged in obviously anaerobic conditions such as canned, bottled or vacuum/modified-atmosphere packaging. Conditions in products packed in air can be anaerobic beneath the surface of the food providing a suitable growth environment for the pathogen.

In other respects Group I (proteolytic) and Group II (psychrotrophic, or non-proteolytic) *Cl. botulinum* differ significantly in their growth and survival characteristics.

Group I

The minimum temperature for growth is 10 °C, with a maximum of 45–50 °C and an optimum of 35–40 °C. Both toxins and spores will survive freezing.

The minimum pH for growth is generally accepted as 4.6. This value is important in defining which foods will receive a botulinum cook (see below). For example, in the UK a low-acid food (*i.e.* low in acid and not low pH) is defined as having a pH value equal to or greater than 4.5. *Cl. botulinum* toxin is stable at low pH but is quickly inactivated at pH 11.

Although the minimum water activity for growth can be affected by solutes in the product it is accepted that 10% sodium chloride (salt), or a water activity of 0.94 is required to inhibit the growth of Group I *Cl. botulinum*.

Group II

The minimum temperature for growth is 3 °C, with a maximum of 40–45 °C and an optimum of 18–25 °C. The ability of Group II *Cl. botulinum* to grow at refrigerated temperatures has raised concerns over products that receive a mild heat treatment and are given an extended shelf life at chilled temperature, particularly if the products are modified-atmosphere/vacuum packaged.

The minimum pH for growth is 5.0.

The salt concentration and water activity value required to inhibit the growth of Group II *Cl. botulinum* are 3.5%, and 0.97, respectively.

Other Toxin-producing Species

The minimum temperature for growth for *Cl. butyricum* and *Cl. baratii* is 7–8 °C, although for *Cl. butyricum* strains known to produce toxins, it is around 10–11 °C.

A recent study has found that the minimum pH for the growth of other *Clostridium* species that may produce botulinum toxins is 4.1, although minimum pH values are influenced by the type of acid in the product.

Cl. butyricum and *Cl. baratii* have minimum water activities for growth of 0.95.

Thermal Resistance

Vegetative cells of *Cl. botulinum* are not particularly heat resistant. Heat processes designed to inactivate *Cl. botulinum* target the much more heat resistant spores of this pathogen.

Group I

Although heat resistance of spores varies between different strains the most heat resistant spores are found from *Cl. botulinum* types in Group I ($D_{121\,°C}$= ca 0.21 min). Consequently foods that will be stored at temperatures at 10 °C or above and where conditions can support the growth of *Cl. botulinum* are usually given a heat process (known as a "botulinum cook") designed to inactivate Group I spores. This encompasses many canned or bottled products with a pH > 4.6. For commercial food processing purposes a botulinum cook is a process equivalent to 121 °C for at least 3 min at the slowest heating point in the container (an F_o 3 process).

Group II

Group II (psychrotrophic) *Cl. botulinum* spores are not as heat resistant at Group I spores. For refrigerated foods where psychrotophic *Cl. botulinum* can grow (generally pH > 4.9 and a_w (water activity) > 0.96), heat processes to inactivate the pathogen need to be applied to the product when it is in its final packaging and should be the equivalent of a minimum of 90 °C for 10 min. "At-risk" products – especially, but not exclusively, those that are modified-atmosphere or vacuum packed – receiving a lesser heat treatment should have a very limited refrigerated shelf life to prevent the outgrowth of any viable *Cl. botulinum* spores.

All toxins produced by *Cl. botulinum* are heat labile and can be inactivated by heating at 80 °C for at least 10 min. However, toxins may be more heat stable at lower pH values.

Control Options

The ubiquitous nature of *Cl. botulinum* means it must be assumed that spores could be present in all raw food. It should be remembered that the growth of Group II, non-proteolytic (psychrotophic) *Cl. botulinum* does not cause obvious spoilage and can easily go undetected in foods. Group I *Cl. botulinum* growth is proteolytic and usually causes detectable spoilage.

Processing

Prevention of spore outgrowth and subsequent toxin production in foods can be achieved both by applying an effective thermal process as described above and by careful product formulation. For "at-risk" foods, there are a number of published processing guidelines and codes of practice, and these should be strictly followed where applicable.

Any change to a process or product formulation should be carefully evaluated using a HACCP approach and adequate controls implemented to ensure either the destruction or control of the growth of *Cl. botulinum*.

Factors that can be used to control the growth of *Cl. botulinum*

	Group I (proteolytic)	*Group II (non proteolytic, psychrotrophic)*
pH	<4.6	<5.0
Water activity	<0.94	<0.97
Temperature	<10 °C	<3.3
Heat processes (in sealed final container)	121 °C for 3 min, or equivalent	90 °C for 10 min, or equivalent

Although Group II *Cl. botulinum* strains will not grow below 3.0 °C, refrigeration alone should not be used to prevent growth for extended periods, except under very controlled and monitored conditions, because of the difficulty in maintaining the very low temperatures required. It is usually recommended that refrigerated processed foods with extended durability (REPFEDS), or sous-vide products, are heated to 90 °C for 10 min, or equivalent, to ensure safety with regard to Group II *Cl. botulinum*.

It is also important to note that *Cl. butyricum* can grow at lower pH values than Group I *Cl. botulinum* strains and this should be considered in acid products with a pH >4.0.

Preservatives can effectively control the growth of *Cl. botulinum* in foods. For example, nitrite is used, in combination with other factors (often referred to as hurdles) in cured-meat products. Sorbates, parabens, polyphospates, phenolic antioxidants, ascorbates, EDTA, metabisulfite, n-monoalkyl maleates and fumarates, lactate salts and liquid smoke (in fish) can all be used as additional hurdles in the control of *Cl. botulinum* under certain circumstances, although specific use should always be validated. The natural bacteriocin nisin, is sometimes used to prevent the germination of *Cl. botulinum* spores in products such as canned vegetables and processed cheese.

Product Storage and Use

Foods stored at ambient temperatures should never rely on shelf life as a control for *Cl. botulinum*. These products should be formulated, and/or heat processed, to ensure the prevention of growth of the pathogen, or the destruction of spores. For chilled foods where the pH and water activity could potentially permit the growth of psychrotrophic *Cl. botulinum* (*e.g.* many ready meals, chilled low-acid sauces and cooked-meat products) and where a 90 °C for 10 min or equivalent process in the final packaging has not been implemented, the UK Advisory Committee on the Microbiological Safety of Food (ACMSF) has given advice on restricting shelf life to control the growth of *Cl. botulinum*.

Well-chosen food packaging can play in role in reducing the risk from botulism. An outbreak of botulism associated with film-wrapped mushrooms in the US, where product respiration had quickly provided an anaerobic environment permitting growth and toxin production by *Cl. botulinum* naturally present on the produce, led to advice to suppliers to ensure holes at the bottom of containers of pre-packed mushrooms.

Infant botulism is controlled by advice to parents not to give their infants "at-risk" foods. These foods, notably honey, are recommended to carry warnings on their labels that they are not suitable for infants under 12 months of age.

Legislation

No specific requirement is made under European Commission legislation with regard to levels of *Cl. botulinum* in food. Requirements for its control is covered under EC general food safety requirements in which food should not be sold if it is unsafe.

Sources of Further Information

Published

Bell., C. and Kyriakides, A. *Clostridium botulinum*: A Practical Approach to the Organism and its Control in Foods. Oxford. Blackwell Science. 2000.

Advisory Committee on the Microbiological Safety of Food. Report on vacuum packaging and associated processes. London. HMSO, 1992.

International commission on microbiological specifications for foods. *Clostridium botulinum*, in micro-organisms in foods, volume 5: microbiological specifications of food pathogens. ed. International Commission on Microbiological Specifications for Foods. London. Blackie, 1996, 66–111.

Department of Health and Social Security, Ministry of Agriculture Fisheries and Food, Scottish Home and Health Department, Department of Health and Social Services Northern Ireland, Welsh Office. Food hygiene codes of practice no. 10. The canning of low acid foods: a guide to good manufacturing practice. London. HMSO, 1981.

On the Web

Risk profile: *Clostridium botulinum* in honey. Institute of Environmental Science and Research Limited. (April 2006). http://www.nzfsa.govt.nz/science/risk-profiles/oct-2006-c-bot-honey.pdf

Risk profile: *Clostridium botulinum* in ready-to-eat seafood in sealed packaging. Institute of Environmental Science and Research Limited. (April 2006). http://www.nzfsa.govt.nz/science/risk-profiles/c-bot-seafood-sept.pdf

Foodborne clostridia and sporulation in "Food Microbiology and sporulation of the genus Clostridia". (2004). http://www.genusclostridium.net/scbooklet4.pdf

Opinion of the scientific committee on veterinary measures relating to public health on honey and microbiological hazards. European Commission. (June 2002). http://ec.europa.eu/food/fs/sc/scv/out53_en.pdf

Food safety implications of potentially pathogenic clostridia. UK Food Standards Agency. (June 2006). http://www.food.gov.uk/science/research/researchinfo/foodborneillness/foodbornediseaseresearch/b14programme/b14projlist/b14007proj/b14007results

1.1.6 *CLOSTRIDIUM PERFRINGENS*

Hazard Identification

What is Clostridium perfringens*?*

Clostridium perfringens is a gram-positive spore-forming bacterium and is a relatively common cause of food poisoning. It is an anaerobe, although it can also grow in the presence of very low levels of oxygen. *Clostridium perfringens* was previously known as *Clostridium welchii*, and the organism may be referred to by this name in older references. *Cl. perfringens* strains are classified by the types of exotoxin they produce (types A, B, C, D & E). Most cases of *Cl. perfringens* food poisoning are caused by type-A strains, although type-C strains can also produce the enterotoxins that cause *Cl. perfringens* food poisoning.

Occurrence in Foods

Clostridium perfringens can be found in low numbers in many raw foods, especially meat and poultry, as the result of soil or faecal contamination. Spores of *Cl. perfringens* will survive many heating and drying processes, and the presence of low numbers of the spores in raw, cooked and dehydrated products is not necessarily a cause for concern because high numbers of vegetative cells are required to cause illness. In addition, research has suggested that only strains of *Cl. perfringens* repeatedly exposed to heating are able to cause food poisoning and that strains freshly isolated from the environment do not.

Hazard Characterisation

Effects on Health

Clostridium perfringens food poisoning is a relatively mild form of food poisoning and is caused by strains that produce enterotoxins (it is important to note that not all strains of *Cl. perfringens* are enterotoxin producers). The enterotoxins are produced when vegetative cells of the bacterium start to multiply in the human intestine and then sporulate. During sporulation, the organism also releases the enterotoxin that causes the symptoms associated with food poisoning. Some cases of *Cl. perfringens* food poisoning have reported very rapid onset of illness suggesting that toxin was pre-formed in food. However, toxin pre-formed in food is not usually at sufficient levels to cause illness, although low levels may contribute to a rapid onset of symptoms.

High numbers ($>10^5$/g, usually 10^6–10^8/g) of viable vegetative cells of enterotoxin producing *Cl. perfringens* are necessary to cause food poisoning. Symptoms generally appear 8–22 h (typically 12–18 h) after ingestion of contaminated food and usually comprise profuse watery diarrhea and severe abdominal pain. Vomiting and nausea occur only rarely. The duration of illness is short, usually lasting for 24 h and not exceeding 48 h. In the majority of cases there is a full recovery, although occasional deaths do occur in elderly and debilitated individuals.

Incidence and Outbreaks

There is little information on the incidence of *Cl. perfringens* food poisoning. However, because of its mild nature, it is probable that it is grossly under-reported, even in countries with well-developed disease-reporting systems. In the UK, multiple outbreaks are reported each year with 300–500 people being affected on average.

Outbreaks of *Cl. perfringens* are usually associated with meat dishes and are frequently linked to facilities or events catering for large numbers of people, such as institutions, restaurants or receptions. Dishes prepared and cooked in large quantities can be difficult to cool quickly to refrigeration temperatures, or can be held at improper temperatures and served warm instead of piping hot. Slow cooling or holding of food at incorrect temperatures can result in the germination of surviving *Cl. perfringens* spores and the rapid multiplication of vegetative cells. The organism can grow extremely rapidly and relatively short times at abuse temperatures can give rise to high enough numbers of vegetative cells to cause illness.

Cooked-meat and poultry products are often associated with *Cl. perfringens* food poisoning because spores of *Cl. perfringens* are likely to be present and protected from extreme heat at the centre of stuffed poultry, rolled meats and meat pies. Cooling at the centre of these products can be slow, oxygen levels are low and the food is protein-rich, providing ideal conditions for the outgrowth of surviving spores. Anaerobic conditions are also created during the rapid boiling of gravies, casseroles and stews, and if improperly cooled or held at inappropriate temperatures, these products too may be the cause of *Cl. perfringens* food poisoning.

Non-meat-derived foods such as vegetable curries and soups have also been associated with outbreaks of *Cl. perfringens* food poisoning, although fish and fish products are rarely implicated.

Sources

Cl. perfringens is ubiquitious and spores of *Cl. perfringens* type A are widely distributed in the environment. *Cl. perfringens* spores are found in soil and dust, as well as in the faeces of many animals. Well-manured soil can have high numbers (10^3–10^4/g) of the spores present. *Cl. perfringens* spores are also present as part of the faecal flora in healthy humans (typically 10^3–10^4/g) and it has been reported that healthy humans can serve as a reservoir for *Cl. perfringens* type-A strains carrying enterotoxin genes. Contaminated food handlers could therefore potentially play a role in the spread of *Cl. perfringens* type-A food poisoning.

Growth and Survival in Foods

Cl. perfringens can grow over the temperature range 15–55 °C, and growth does not occur below 10–12 °C. The optimum temperature for growth is 43–47 °C, and at these temperatures *Cl. perfringens* has the fastest recorded growth rate (shortest generation time) of any bacterium. Generation times (time for a defined population to double in size) of around 7 min at 41 °C have been recorded although 10 min is more typical. The optimum temperature for enterotoxin production is 35–40 °C.

Cl. perfringens vegetative cells die out relatively rapidly (93.5% were killed after 30 days at –17.7 °C) at freezing temperatures. However, they die out less quickly during storage at chill temperatures. Spores survive both refrigeration and freezing.

The optimum pH for the growth of *Cl. perfringens* is 6.0–7.0, the pH of most cooked meat and poultry products. The organism is able to grow over the pH range 5.0–8.3 under otherwise ideal conditions. The spores can survive more extreme pH values.

The minimum water activity for spore germination and growth of *Cl. perfringens* is between 0.94–0.95, but minimum values will be affected by the nature of the solute. Vegetative cells are not very tolerant of low water activity but spores are very resistant to desiccation.

Cl. perfringens in an anaerobe and grows best when oxygen is absent or present at very low levels, such as at the centre of cooked meat and poultry dishes. It is unable to grow on the surface of foods unless they are vacuum or modified-atmosphere packaged.

The vegetative cells are not especially resistant to preservatives and sanitisers, but the spores are much more resistant. Curing salts used in meat products can be effective at controlling *Cl. perfringens*, but unacceptably high levels of sodium nitrite are required to inhibit the growth of the pathogen when it is used as the sole preservative. However, in combination with sodium chloride, sodium nitrite can be effective at preventing its growth.

Thermal Resistance

Vegetative cells of *Cl. perfringens* are not very heat resistant and will usually be inactivated at temperatures exceeding 60 °C. The heat resistance of *Cl. perfringens* spores has been shown to vary significantly. *D*-values at 90 °C of between 0.015 and 8.7 min, and at 110 °C of between 0.5 and 1.29 min have been recorded. Some spores have been shown to survive boiling for one hour.

The enterotoxins are heat labile and heating food to >70 °C throughout will inactivate enterotoxin.

Control Options

A HACCP approach to the control of *Cl. perfringens* in food is preferred and control measures focus on effective temperature control.

Processing

The key control for *Cl. perfringens* during processing is the rapid cooling of "high-risk" product after cooking, especially through the temperature range from 55–15 °C, followed by storage at temperatures below 4 °C (although below 10 °C ensures no growth of *Cl. perfringens*, refrigeration below 5 °C is essential to control other pathogens). In some countries there is legislation, and in others published guidelines, for the rapid cooling of various products (see legislation section).

When developing a product, food processors should ensure that the intended use of the product should not pose a risk to the consumer. For example, in a dehydrated soup product, acceptable levels of *Cl. perfringens* in the dried ingredients should take into consideration that the consumer will be rehydrating the product by adding hot water and that it will later be consumed warm.

Product Use

Food to be served hot should either be freshly cooked and kept hot at temperatures not permitting the growth of *Cl. perfringens* (>63 °C), or if cooked product is reheated, it should reach temperatures that inactivate vegetative cells and enterotoxin (at least 72 °C throughout the product).

Legislation

EU regulations and the US Food code do not have specific requirements relating to levels of *Cl. perfringens* in foods.

However, the European Food Safety Authority's (EFSA) Scientific Panel on Biological Hazards has recommended that, "when new or modified products are developed, that might support the growth of *Cl. perfringens* and/or enterotoxin production, processors should ensure that target levels of 10^5/g are not exceeded under the anticipated conditions of storage and handling." In addition, the UK Health Protection Agency (HPA) has issued guidelines on the microbiological quality of some ready-to-eat foods at the point of sale. These state that levels of *Cl. perfringens* of 100/g to <10^4/g in these products is unsatisfactory, and levels >10^4/g are unacceptable/potentially hazardous.

In the UK and the US there are requirements for the control of temperature aimed at limiting the growth of *Cl. perfringens* in "at-risk" foods.

The US has a mandatory requirement for the times and temperatures used during the cooling of large joints of meat. These are the same as guidelines published in the UK by Campden & Chorleywood Food Research Association for the chilling of a large piece of meat and can be summarised as follows:

	Good practice (h)	*Maximum (h)*
>50 °C	1	2.5
50–12 °C	6	6
12–5 °C	1	1.5
Total cooling time	8	10

For chilled prepared foods (excluding cook-chill foods used within integrated catering systems) a guideline by the UK Chilled Food Association advises that a heated product should be cooled as quickly as possible through the temperature range 63 °C to 5 °C or less to minimise the risk of spore germination and outgrowth. The time taken for cooling will vary from product to product, but as a guideline, should be no more than 4 h. Rapid cooling for these products can be facilitated by preparing product in relatively small portions/packages and ensuring their separation during the cooling process.

There are also requirements in the US under the US Food Code (2005) for the cooling of potentially hazardous cooked food. The code requires that these should be cooled:

1. Within 2 h from 57 °C (135 °F) to 21 °C (70 °F); and
2. Within a total of 6 h from 57 °C (135 °F) to 5 °C (41 °F) or less, or to 7 °C (45 °F) or less (under certain conditions).
3. Product prepared from ingredients at ambient temperatures, such as reconstituted foods and canned tuna, and that are potentially hazardous food, should be cooled within 4 h to 5 °C (41 °F) or less, or to 7 °C (45 °F) under certain circumstances.

For the hot holding of foods, UK legislation requires food served hot to be held at temperatures >63 °C. In the US, food that "is received hot" should be at a temperature of 57 °C (135 °F) or above.

Sources of Further Information

Published

Bates, J.R. and Bodnaruk P.W. *Clostridium perfringens*, in Foodborne Microorganisms of Public Health Significance. ed. Australian Institute of Food Science and Technology. 6th edn. Waterloo, DC. AIFST, 2003, 479–504.

McClane, B.A. *Clostridium perfringens*, in Food Microbiology: Fundamentals and Fronteirs. ed. Doyle, M.P., Beuchat, L.R. and Monteville, T.J. 2nd edn. Washington DC. ASM Press, 2001, 351–72.

International Commission on Microbiological Specifications for Foods. *Clostridium perfringens*, in Microorganisms in Foods, Vol. 5. Microbiological Specifications of Food Pathogens. ed. International Commission on Microbiological Specifications for Foods. London. Blackie, 1996, 112–25.

On the Web

A Risk Assessment for *Clostridium perfringens* in Ready-to-Eat and Partially Cooked Meat and Poultry Products. United States Department of Agriculture's Food Safety and Inspection Service (FSIS) (September 2005) http://www.fsis.usda.gov/PDF/CPerfringens_Risk_Assess_Sep2005.pdf

Opinion of the Scientific Panel on Biological Hazards on the request from the Commission related to *Clostridium* spp in foodstuffs. European Food Safety Authority. (March 2005). http://www.efsa.europa.eu/etc/medialib/efsa/science/biohaz/biohaz_opinions/885.Par.0005.File.dat/biohaz_op_ej199_clostridium_en1.pdf

Guidelines for the microbiological quality of some ready-to-eat foods sampled at the point of sale. Health Protection Agency. (September 2000). http://www.hpa.org.uk/cdph/issues/CDPHvol3/No3/guides_micro.pdf

1.1.7 *ENTEROBACTER SAKAZAKII*

Hazard Identification

What is Enterobacter sakazakii?

Enterobacter sakazakii is a gram-negative bacterium belonging to the family *Enterobacteriaceae*. Microbiologists recognise a small number of genera within the *Enterobacteriaceae*, including *Enterobacter*, as the coliform group.

Occurrence in Foods

Although of most concern in infant formula, where it has been isolated from both powdered and rehydrated product, *E. sakazakii* has also been found in a number of other foods, including dried milk powders, other dried infant foods, dried herbs and spices, lettuce, mung bean sprouts, vegetables, rice flour, cheese, eggs, minced beef and sausages.

Hazard Characterisation

Effects on Health

Enterobacter sakazakii has most frequently been associated with illness in infants, particularly those in intensive-care units fed with contaminated rehydrated powdered milk-based formula. Symptoms of infection are bloody diarrhoea, and in rare cases sepsis and meningitis resulting in high death rates. Infants at greatest risk of infection appear to be premature babies who are immunosuppressed and those of low birth weight. However, cases have also been reported in full-term newborns.

There is little reported evidence of any risk to older children and adults from *E. sakazakii* in food. In assessing the risk of *E. sakazakii* in dairy products to the general (non-infant) population a report published by the New Zealand Food Safety Authority in May 2004 concluded that "There is no evidence however that *E. sakazakii* poses any significant risk to general populations consuming food products that comply with recognised international food processing or public health standards. While there have been eight cases of *E. sakazakii* infection reported in adults suffering from underlying health problems no connections to food could be made in any of these episodes."

Incidence and Outbreaks

Reported *E. sakazakii* infections in infants are rare. However, cases have occurred in a number of countries including England, Canada, the Netherlands, Belgium, Israel, the United States and France. Although cases of what is now thought to be *E. sakazakii* infections (initially documented as being caused by atypical yellow-pigmented *Enterobacter cloacae*) have been described since

1958, it was not until 2001 that an outbreak was linked to product, when the micro-organism was isolated from an unopened can of powdered infant formula. Previously, it had been difficult to ascertain if outbreaks were caused by product contaminated with the pathogen after opening, or if the outbreak strain was present as part of the manufacturing process.

During the 1990s and the first part of the 21st century a number of outbreaks of *E. sakazakii* infection were reported, the most recent in France during late 2004. In this outbreak all the affected cases, 9 infants with 2 deaths, had been fed with a particular brand of powdered infant formula, which was subsequently recalled from the market. *E. sakazakii* was later isolated from 31 unopened cans of the implicated product.

Sources

Studies have isolated *E. sakazakii* from a range of sources and it is thought that the micro-organism is ubiquitous, being widespread in both the environment and in plant material.

Growth and Survival in Foods

Strains of *E. sakazakii* have been reported to grow over a range of temperatures from 5.5 °C to 47 °C, and induced acid resistance at pH 3 has also been demonstrated. *E. sakazakii* has been shown to form biofilms on latex, polycarbonate, silicon rubber and glass. The organism also appears to be able to survive well in dry conditions and is reported as being atypical, when compared to other members of the *Enterobacteriaceae*, in its ability to survive desiccation. The micro-organism's ability to survive in dry conditions for extended periods of time (possibly up to 2 years) has been attributed to capsule formation. *E. sakazakii* has also recently been reported as being more resistant to osmotic stress than a number of other members of the *Enterobacteriaceae*, including strains of *E. coli* and *Salmonella* serotypes.

Thermal Resistance

Originally it was thought that the presence of an unusually high number of *Enterobacter* spp. in dried infant formula indicated that some species in the genus must be relatively heat resistant; however, there has been no conclusive evidence to suggest that *E. sakazakii* can survive typical milk-pasteurisation treatments. Thermal inactivation studies have indicated that the organism cannot survive commercial pasteurisation processes, with studies in rehydrated infant formula indicating a D-value of 2.50 min at 60 °C with a z-value (the temperature change resulting in a tenfold change in D) of 5.82 °C. However, there is some evidence to suggest that when rehydrated, previously desiccated *E. sakazakii* cells may have different thermal inactivation characteristics to those not exposed to dry conditions. There are also indications of a degree of variability in thermal inactivation characteristics between different strains.

Reheating infant formula by applying microwave heating until the first signs of boiling has been shown to be particularly effective in inactivating *E. sakazakii*. However, due to the potential scalding hazard this may pose to infants, many authors have suggested that rehydration of infant formula at 70 °C is sufficient to minimise the risk from the pathogen.

Control Options

Processing

In an opinion published in September 2004, the European Food Safety Authority (EFSA) concluded that "*E. sakazakii* is inactivated by the pasteurisation processes used in the manufacture of infant formula. However, due to the widespread occurrence of the micro-organism it appears very difficult to control it in the processing environment, and as a consequence the recontamination of product does occur during handling and filling processes." EFSA has advised that measures to reduce the risk of *E. sakazakii* recontaminating product during manufacture include: using ingredients of good microbiological quality; closely monitoring and controlling levels of *Enterobacteriaceae* in the production environment using the results to indicate the likely presence of pathogens such as *E. sakazakii*; and imposing strict hygiene measures such as the control of movement of personnel, the separation of wet and dry processes, and avoiding condensation and water ingress in dry areas.

Product Use

At present the dose–response relationship of *E. sakazakii* infections in humans is unknown, however the widespread occurrence of the micro-organism in the environment would indicate that the consumption of low numbers of the pathogen in infant formula is unlikely to cause illness in healthy infants. To protect the most "at-risk" infants, food safety experts in many countries have advised that where possible commercially sterile ready-to-use infant formula should be used in neonatal intensive-care settings.

The storage of reconstituted product at temperatures in excess of 5 °C can lead to the rapid increase of numbers of pathogens such as *E. sakazakii* and EFSA has advised on the safe preparation, handling, storage and use of infant formula in the home and in hospitals. A joint FAO/WHO workshop meeting in early 2004 advised that care-givers should be alerted by health-care providers that dried infant formula products are not sterile. At a joint FAO/WHO expert meeting held in January 2006 it was recommended that product labels should be revised so that safe preparation instructions and other safety information is included on the packaging of dried infant formula products.

This recent FAO/WHO report also includes an evaluation of a quantitative risk assessment model for *E. sakazakii* in powdered infant formula. The risk assessment can be accessed at the web link below and is to be published at a later date in a user-friendly format.

Legislation

The European Union (EU) regulation for microbiological criteria for foodstuffs that came into force in January 2006 has specific requirements with regards to limits for *Enterobacteriaceae* and *E. sakazakii* in dried infant formula and dried dietary foods for special medical purposes intended for infants below 6 months of age. These state that *E. sakazakii* should be absent in 10 g of product.

Sources of Further Information

Published

Friedemann, M. *Enterobacter sakazakii* in food and beverages (other than infant formula and milk powder). *International Journal of Food Microbiology*, 2007, 116, 1–10.

Drudy, D., Mullane, N.R., Quinn, T., Wall, P.G. and Fanning, S. *Enterobacter sakazakii*: an emerging pathogen in powdered infant formula. *Clinical Infectious Diseases*, 2006, 42, 996–1002.

Gurtler, J.B., Kornacki, J.L. and Beuchat, L.R. *Enterobacter sakazakii*: A coliform of increased concern to infant health. *International Journal of Food Microbiology*, 2005, 104, 1–34.

On the Web

Enterobacter sakazakii and *Salmonella* in powdered infant formula: Meeting report. Microbiological risk assessment series 10. The Food and Agriculture Organization of the United Nations/World Health Organization. (2006). http://www.fao.org/ag/agn/jemra/enterobacter_en.stm

Overview of a risk assessment model for *Enterobacter sakazakii* in powdered infant formula. The Food and Agriculture Organization of the United Nations/World Health Organization. (2006). http://www.who.int/foodsafety/micro/jemra/r_a_overview.pdf

Hazards associated with *Enterobacter sakazakii* in the consumption of dairy foods by the general population. Draft report to the New Zealand Technical Consultation Committee. (May 2004). http://www.nzfsa.govt.nz/dairy/publications/information-pamphlets/enterobacter-sakazakii/index.htm

Report of the Joint FAO/WHO Workshop on *Enterobacter Sakazakii* and Other Microorganisms in Powdered Infant Formula. World Health Organization. (April 2004). http://www.who.int/foodsafety/micro/meetings/feb2004/en/

Opinion adopted by the BIOHAZ Panel related to the microbiological risks in infant formulae and follow-on formulae. European Food Safety Authority. (September 2004). http://www.efsa.eu.int/science/biohaz/biohaz_opinions/691_en.html

1.1.8 ENTEROCOCCI

Hazard Identification

What are the Enterococci?

The enterococci belong to a genus of gram-positive, non-spore-forming bacteria previously known as Lancefield's group D streptococci, or faecal streptococci. At least 20 *Enterococcus* species have been described, but the most common species associated with foods and human disease are *Enterococcus faecium* and *Enterococcus faecalis*. The enterococci are recognised as the causative agents of a number of non-foodborne clinical infections, such as bacteraemia and endocarditis, and in recent years there has been increasing concern over the number of emerging vancomycin-resistant enterococci strains (VREs).

However, the enterococci are also important in food microbiology for a number of seemingly opposing reasons. When present in food they can be viewed as potential pathogens very occasionally associated with outbreaks of foodborne disease, as important spoilage micro-organisms of dairy and meat products, as starter micro-organisms used in the production of various traditional fermented foods, or even as probiotic micro-organisms. It is important to remember that the possession of "virulence" factors (*i.e.* the ability to cause disease) and resistance to antibiotics are strain specific and that many strains are entirely non-pathogenic.

Occurrence in Foods

Enterococci are found in a wide variety of foods. They are common contaminants of milk and meat products and are used as starter cultures in some traditional European cheeses. They are also found on plant materials such as olives and vegetables.

Hazard Characterisation

Effects on Health

Symptoms associated with foodborne outbreaks associated with *Enterococcus* species have been described as "milder than *Staphylococcus* food poisoning". Human volunteer feeding studies have been conflicting and so the description of *Enterococcus* food poisoning is vague and variable. All individuals are thought to be susceptible to food poisoning caused by enterococci.

The infectious dose for foodborne outbreaks is thought to be high ($>10^7$ cells), and the incubation period is reported to vary widely (between 2–60 h). Symptoms described include abdominal cramps, diarrhoea, nausea, vomiting and dizziness. The disease is thought to be typically of short duration and self-limiting.

Incidence and Outbreaks

There is very little data on the incidence of foodborne enterococcal infections. It is the presence of high numbers of *Enterococcus* species and the absence of other foodborne pathogens that has caused some outbreaks of foodborne disease to be linked with the enterococci. However, it is important to remember that many foods (*e.g.* cheeses) can contain high numbers of these bacteria without causing illness.

Foodborne outbreaks have been associated with sausages, ham, evaporated milk, cheeses and chocolate pudding.

Sources

Enterococcus species are found in the intestine of most animals, including man. They are excreted in the faeces of animals leading to the contamination of the environment. *E. faecalis* is the species found most frequently in human faeces (10^5–10^7 cells/g of faeces) whereas *E. faecium* is the most common species found in the faeces of cattle. Dairy-processing equipment can become contaminated with enterococci and surveys have frequently isolated them from pig, poultry and beef carcasses.

Although associated with faeces, the presence of enterococci in foods is not always related to direct faecal contamination. Due to environmental contamination, the enterococci are also found in soil, insects, water and plant materials such as vegetables.

Growth and Survival Characteristics

The enterococci can grow over a wide growth range; some strains can grow at temperatures as low as 1 °C. The maximum reported growth temperature is 50 °C, but the optimum for most strains is 37 °C. The enterococci are resistant to freezing and are reported to survive storage at –70 °C for several years.

Growth can occur over the pH range 4.4–10.6. Although the minimum water activity for growth is dependent on solute present, *E. faecalis* is reported to grow at 0.93. The enterococci are generally able to tolerate salt concentrations of 10%. In addition, these organisms are resistant to drying and are extremely persistent in the environment. *E. faecalis* and *E. faecium* are reported to survive for weeks on environmental surfaces, in soil for up to 77 days and in cheese for up to 180 days.

Although the enterococci are generally persistent in the environment they are not particularly resistant to sanitisers (including sodium hypochlorite) or preservatives. There is concern, however, that some enterococci strains isolated from food have demonstrated multiple antibiotic resistance, including resistance to vancomycin.

Thermal Resistance

The enterococci are relatively heat resistant and are able to survive many mild pasteurisation processes. This is often why they are present in, and associated with, the spoilage of some heat-processed foods such as pasteurised milk and cooked meats.

E. faecium ($D_{70\,°C}$-values of 1.4–3.4 min) is more heat resistant than *E. faecalis* ($D_{70\,°C}$-values of 0.02–0.6 min).

Control Options

Processing

The enterococci can survive mild pasteurisation treatments and can be present in mildly heat processed, or undercooked foods. Strict adherence to heat-processing regimes and the subsequent control of the chill chain assists in minimising the numbers of any enterococci present in pasteurised foods.

The presence and persistence of enterococci in the environment and on raw materials means that processing equipment and establishments can become sources of the organisms. Strict adherence to cleaning regimes and the use of appropriate sanitisers can control the organisms in food-processing establishments.

The use of enterococci as starter organisms or as probiotics in foods has been a cause for concern even though there is a history of safe use for the organisms in both these roles. Antibiotic resistance and the ability to cause disease appear to be strain dependent. However, both of these factors should be carefully considered in any risk assessment when selecting an *Enterococcus* strain for use in the food industry.

Legislation

European legislation has requirements for levels of enterococci in drinking water and for water used in the food industry unless it can be demonstrated that the use of the water does not affect the wholesomeness of the food. These requirements are a level of 0/100 ml. For water on sale in bottles or containers there is a more stringent requirement of 0/250 ml.

Sources of Further Information

Published

Franz, C.M.A.P. and Holzapfel, W.H. Enterococci, in Emerging foodborne pathogens. ed. Motarjemi, Y. and Adams, M. Cambridge. Woodhead Publishing, 2006, 557–613.

Franz, C.M.A.P., Stiles, M.E., Schleifer, K.H. and Holzapfel, W.H. Enterococci in foods – a conundrum for food safety. *International Journal of Food Microbiology*, 2003, 88 (2–3), 105–122.

1.1.9 *LISTERIA*

Hazard Identification

What is Listeria?

The genus *Listeria* are gram-positive, non-spore-forming rod-shaped bacteria. The genus contains a number of species including *L. monocytogenes, L. innocua, L. welshimeri, L. seeligeri, L. ivanovii* and *L. grayi.* Although the first four of these have all been implicated in human infection nearly all cases of *Listeria* infection are caused by *L. monocytogenes.*

At least 13 different serotypes of *L. monocytogenes* are known. All can cause human listeriosis, but most cases are caused by serotypes 1/2a, 1/2b and 4b. The majority of significant reported foodborne outbreaks have been caused by serotype 4b.

Occurrence in Foods

Listeria monocytogenes has the potential to be present in all raw foods. Cooked foods can also be contaminated, usually as the result of post-process contamination. *L. monocytogenes* has been isolated from a very wide range of processed foods including pâtés, milk, soft cheeses, ice cream, ready-to-eat cooked and fermented meats, smoked and lightly processed fish products and other seafood products. *L. monocytogenes* is usually found only in low numbers (<10 CFU/g) in foods. However, some product such as pâtés and soft cheese have occasionally been found to contain populations of >10 000 CFU/g.

Hazard Characterisation

Effects on Health

Listeria monocytogenes causes one of the most severe forms of foodborne infection and it is fortunate that listeriosis is a relatively rare disease. The overall mortality rate associated with the disease is 30%, although it can be as high as 40% in susceptible individuals. Those most at risk at acquiring the disease are pregnant women (20 times greater risk than healthy individuals), the elderly and the immunocompromised, although healthy individuals can develop listeriosis, particularly if the food is heavily contaminated. Monitoring in the USA suggests that *Listeria* infections are more likely to result in the hospitalisation of affected individuals in comparison with those affected by other foodborne pathogens such as *Salmonella* (hospitalisation rate 95% for *Listeria* compared with 21% for *Salmonella*).

The incubation period is 1 to 90 days (mean 30 days). The onset of illness is typically marked by flu-like symptoms (fever and headache), and sometimes by nausea, vomiting and diarrhoea. In some cases these symptoms can lead on to meningitis and septicaemia. Symptoms in pregnant women can lead to infection of the foetus, which can result in miscarriage, stillbirth, or the birth of an infected infant, although the mother usually survives.

The infective dose is unknown, although it is generally considered to be $>10^3$ CFU/g for healthy individuals. Due to the length of the incubation period, it can be difficult to determine the numbers of organisms in foodstuffs at the time of consumption and an outbreak associated with frankfurters in the USA in 1998 is thought to have been caused by product containing less than 0.3 CFU/g, although it is thought that the causative strain may have carried enhanced virulence.

Incidence and Outbreaks

The first outbreak of *L. monocytogenes* that could be definitely linked to food was caused by commercially prepared coleslaw in Canada in 1981 (at least 41 cases with 7 deaths). Manure from *Listeria*-infected sheep had been used as a fertiliser when growing the cabbages used to prepare the salad.

The incidence of reported *Listeria* infections increased dramatically during the 1980s as did the number of food-related outbreaks. An outbreak in Los Angeles County during 1985 was caused by Mexican-style cheese (142 cases with 48 deaths) and during the late 1980s an outbreak in the UK was associated with pâté (>350 cases with >90 deaths).

Notable outbreaks occurring in the 1990s were linked to smoked mussels (1992; New Zealand); "rillettes" or potted pork (1993, France); pasteurised chocolate milk (1994, USA); raw milk soft cheese (1995, France); frankfurters (1998–9, USA); butter (1998–9, Finland) and pork tongue in jelly (1999–2000, France).

During the first few years of the 21st century there have been a number of large *Listeria* outbreaks caused by ready-to-eat (deli) poultry products in the USA. In 2000 a multistate outbreak (29 cases, with 7 deaths) was linked to turkey deli meat, and during 2002 another outbreak (at least 46 cases with 11 deaths) was linked to poultry deli products produced by the Pilgrims Pride Corporation. This outbreak resulted in the recall of 27.4 million pounds of product, the largest meat recall in US history.

Strategies to reduce the incidence of *Listeria* infections were implemented in many countries during the 1990s resulting in a reduction in the incidence of the disease. However, outbreaks have continued to occur and incidence has again risen in some countries in recent years. For example, in the UK 278 cases were reported in 1988 and this number fell to only 87 in 1995. But by 2003 the number of reported cases had risen to 237. A similar pattern has been reported in the EU, but the USA has reported a downward trend in recent years and the incidence of infection in 2005 was 0.27 cases per 100 000 people, despite some significant outbreaks.

Sources

Listeria is ubiquitous in the environment. It is found in soil, where it can survive for extended periods and leads to the contamination of plant material. *Listeria* has been isolated from a wide variety of fresh produce. It is also found in marine environments and the organism is often associated with fish and seafood products. Animals such as sheep, goats and cattle are recognised carriers of the organism, often acquired from the consumption of contaminated (usually poor quality) silage. Healthy humans can also be carriers of the organism.

Kitchen and food-processing environments, particularly those that are cold and wet or moist, can be reservoirs for *Listeria*. The organism can be particularly persistent and difficult to control because of its psychrotrophic nature and resistance to environmental conditions. The efficacy of hygiene standards in food-production facilities producing ready-to-eat products is usually monitored and this can include environmental swabbing for *L. monocytogenes*. Although other *Listeria* species are not normally associated with human disease, a positive test for *Listeria* species other than *L. monocytogenes* can be a useful indicator that there is the potential for *L. monocytogenes* to be present.

Growth and Survival in Foods

Listeria monocytogenes is psychrotrophic and the ability to grow at chill temperatures is why it is a particular risk in extended-shelf-life chilled foods that can support its growth. Extremely slow growth of *L. monocytogenes* has been recorded at temperatures as low as –1.5 °C and the maximum temperature for growth is generally accepted as 45 °C. The organism survives well in frozen foods, but survival times can be adversely affected under acid conditions.

The pH range for the growth of *L. monocytogenes* is 4.3–9.4 under otherwise ideal conditions. These values are affected by the specific acid in the product, and the minimum pH is likely to be higher in real foods and at low temperatures. However, *L. monocytogenes* can survive for extended periods in acid conditions, particularly at chilled temperatures.

The minimum water activity for the growth of *L. monocytogenes* is 0.92. The organism is tolerant of high sodium chloride levels and is able to grow in environments of up to 10% salt, and to survive in concentrations of 20–30%. *L. monocytogenes* is also able to survive for some time in low water activity environments, and may survive drying processes. Survival times are extended at chilled temperatures.

L. monocytogenes grows well in aerobic and anaerobic conditions. Its growth is unaffected by many modified atmospheres even at low temperatures. High concentrations of carbon dioxide are necessary to inhibit growth.

Although *L. monocytogenes* is not especially resistant to antimicrobials, it can prove difficult to control on food-contact surfaces such as stainless steel because the bacteria can form persistent biofilms. It is important to clean equipment prior to using sanitisers because organic matter can affect their efficacy at inactivating the pathogen.

Thermal Resistance

Although *L. monocytogenes* is not particularly heat resistant it is more heat resistant than some other foodborne pathogens, such as *Salmonella* and *E. coli* O157:H7. It is readily inactivated at temperatures above 70 °C and heat processes such as normal commercial milk pasteurisation will destroy numbers typically found in milk. Typical *D*-values in food substrates are: between 5–8 min at 60 °C, and 0.1–0.3 min at 70 °C. Concern about the pathogen in particular food-product categories has led to heating guidelines being issued by various heath authorities. The UK Department of Health advised that ready meals or similar products should receive a heat treatment of at least 2 min at 70 °C, or equivalent, to ensure the destruction of *L. monocytogenes*. For consumers, terms such as heating till "piping" hot in the UK, and "steaming" hot in the USA are used to describe heat processes required to ensure the safety of foods identified as being a potential risk of causing *Listeria* food poisoning.

Control Options

The control of *Listeria* in foods relies largely on a HACCP approach and the establishment of effective critical control points in the process.

Processing

The careful design and layout of processing equipment in conjunction with the implementation of regular, thorough cleaning regimes of the processing environment can significantly reduce the level of *Listeria* contamination in many processed foods. However, because of its ubiquitous nature it is virtually impossible to totally eliminate the pathogen from many food products. The organism should be inactivated by heat applied during the cooking process and the presence of *Listeria* in cooked products can indicate poor hygiene, either during manufacture, distribution or at retail.

Other critical controls include strict temperature control, the prevention of cross-contamination between raw and processed foods and between the processing environment and processed foods, as well as the use of a restricted shelf life for potentially contaminated products that could support the growth of the pathogen.

Product Use

Appropriate scientifically based methods should be used to devise safe shelf lives for "at-risk" chilled foods and these restricted shelf lives should be rigorously implemented and adhered to in order to reduce the risk from *L. monocytogenes*. Clear cooking instructions are needed on the packaging of many chilled foods requiring reheating prior to consumption, to ensure that all parts of the product receive a listericidial process.

Vulnerable individuals, especially pregnant women, the elderly and the immunosuppressed are advised to avoid eating specific foods to reduce the risk from listeriosis. Health authorities in the UK advise these groups not to eat soft mould-ripened or blue-veined cheeses, pâté and unpasteurised dairy products. These groups are advised that they may also choose to avoid cold (pre-cooked) meats and smoked salmon, and that they should thoroughly wash pre-packed salads and heat chilled meals and ready-to-eat chicken adequately before eating.

In the US the FDA also includes hot dogs, luncheon meats, cold cuts and smoked seafood (unless thoroughly reheated) to the list of foods that at-risk consumers should definitely avoid.

Legislation

Countries differ in their regulatory approach to the presence of *L. monocytogenes* in RTE food. In the USA a "zero tolerance policy" is taken on the presence of *L. monocytogenes* in any RTE food. Recent European Union regulations generally permit a count of up to 100 CFU/g at the end of shelf life for RTE foods, except those intended for infants and for special medical purposes.

Specific regulatory guidance on *Listeria* for food manufacturers is also available in a number of countries.

In July 2004, Canadian authorities published a regulatory policy on *L. monocytogenes* in ready-to-eat (RTE) foods, which included the following guidelines:

A refrigerated RTE food not supporting the growth of L. monocytogenes includes the following:

1. pH 5.0–5.5 and $a_w < 0.95$
2. pH < 5.0 regardless of a_w
3. $a_w \leq 0.92$ regardless of pH, or
4. frozen foods

In the USA, regulations encourage producers of ready-to-eat (RTE) cold meat or "deli" products to use HACCP or similar programs to control *L. monocytogenes* and require these companies to give authorities access to data used to verify procedures, such as environmental and finished product testing for *L. monocytogenes*. These establishments are encouraged to make food safety enhancement claims on their RTE product labels that describe the processes used to eliminate or reduce *L. monocytogenes*, or suppress its growth in products.

Various countries have standards/legislation for the pasteurisation of ice cream/frozen desserts; these heat processes are more severe than high temperature short-time milk (HTST) pasteurisation (which is at least 15 s at 72 °C) because ingredients such as sugars, fat, emulsifiers and stabilisers in these products protect *L. monocytogenes* from heat, resulting in an increase in *D*-value. In New Zealand a heat process of at least 15 s at 79.5 °C (or equivalent) is required for ice cream, and in the US standards require a process of 30 min at 68.3 °C or 25 s for 79.4 °C.

Sources of Further Information

Published

Bell, C. and Kyriakides A. *Listeria*: a practical approach to the organism and its control in foods. 2nd edn. Oxford. Blackwell Publishing. 2005.

Ryser, E.T. and Marth, E.H. *Listeria*, listeriosis and food safety. 2nd edn. New York. Marcel Dekker. 1999.

International Commission on Microbiological Specifications for Foods. *Listeria monocytogenes,* in Microorganisms in Foods, Volume 5, Microbiological Specifications of Food Pathogens, ed. International Commission on Microbiological Specifications for Foods. Blackie. London, 1996, 141–82.

On the Web

Risk profile: *Listeria monocytogenes* in soft cheeses. Institute of Environmental Science and Research Limited. (November 2005). http://www.nzfsa.govt.nz/science/risk-profiles/FW0382_L_Mono_in_soft_cheese_November_2005.pdf

Risk profile: *Listeria monocytogenes* in low moisture cheeses. Institute of Environmental Science and Research Limited. (July 2005). http://www.nzfsa.govt.nz/science/risk-profiles/FW0440_L_mono_in_low_moisture_cheese_Final_Mar_2007.pdf

Risk profile: *Listeria monocytogenes* in ready-to-eat salads. Institute of Environmental Science and Research Limited. (July 2005). http://www.nzfsa.govt.nz/science/risk-profiles/FW0446_L_mono_in_RTE_salads_2005.pdf

Risk assessment of *Listeria monocytogenes* in ready-to-eat foods, MRA Series 4 & 5 World Health Organization (2004). http://www.who.int/foodsafety/publications/micro/mra_listeria/en/index.html

Policy on *Listeria monocytogenes* in Ready-to-Eat Foods. Health Canada (July 2004). http://www.hc-sc.gc.ca/fn-an/legislation/pol/policy_listeria_monocytogenes_politique_toc_e.html

Risk profile: *Listeria monocytogenes* in ice cream. Institute of Environmental Science and Research Limited. (October 2003). http://www.nzfsa.govt.nz/science/risk-profiles/lmono-in-ice-cream.pdf

Quantitative Assessment of Relative Risk to Public Health from Foodborne *Listeria monocytogenes* Among Selected Categories of Ready-to-Eat Foods (US FDA: September 2003). http://www.foodsafety.gov/~dms/lmr2-toc.html

Risk profile: *Listeria monocytogenes* in processed ready-to-eat meats. Institute of Environmental Science and Research Limited. (October 2002). http://www.nzfsa.govt.nz/science/risk-profiles/listeria-in-rte-meat.pdf

1.1.10 *MYCOBACTERIUM AVIUM* SUBSP *PARATUBERCULOSIS*

Hazard Identification

What is Mycobacterium avium *Subsp* paratuberculosis*?*

Mycobacterium avium subsp *paratuberculosis*, often referred to as *Mycobacterium paratuberculosis* or MAP, is a gram-positive, strictly aerobic bacterium belonging to the family *Mycobacteriaceae*. It is a slow-growing organism and is difficult to cultivate in laboratory conditions.

It is known to be the causative agent of Johne's disease, a widespread chronic condition in ruminants, particularly cattle. However, there is some evidence that it may also have a role in the development of a chronic inflammatory bowel condition in humans called Crohn's disease.

Occurrence in Foods

MAP can be isolated from the raw milk of clinically infected cattle and from the milk of subclinically infected, apparently healthy cattle. A survey in the UK also found MAP in about 2% of pasteurised milk, giving rise to concerns that the organism is able to survive standard high temperature/short time milk-pasteurisation treatments (72 °C for 15 s). In addition, MAP has been isolated from commercial milk in the US and Switzerland. The bacterium is generally acid-resistant and may survive the low-pH conditions in cheese making. It could therefore be present in cheeses made from raw milk, or milk subjected to less severe pasteurisation processes. Sheep and goat's milk and associated dairy products may also be potential sources of MAP.

Infected ruminants excrete MAP into the environment where the organism is known to persist for sometime. It is likely that MAP enters the water supply and is present on raw vegetables and raw meats from ruminants. However, data on its prevalence from these sources is very limited.

Hazard Characterisation

Effects on Health

The evidence to link MAP as the causative agent of Crohn's disease is not conclusive and claims that the two are linked are not widely accepted by gastroenterologists. There is evidence that hereditary and environmental factors play in role in the development of Crohn's disease, suggesting that if MAP is involved, it is not the sole aetiological agent. However, while research is ongoing to determine any implications for the presence of MAP in foods, it has been suggested that the food industry adopt a precautionary approach.

Crohns's disease is a chronic inflammatory disease in humans, which can occur in any part of the gastrointestinal tract, although it usually affects the small intestine. Symptoms, which include loss of weight, abdominal pain and cramps, diarrhoea, fatigue, muscle and joint pains, usually first occur when individuals are 14 to 24 years of age. There is no known cure, and it is a life-long debilitating illness. The disease is managed by the use of drugs, although surgical intervention is also often necessary. Crohn's disease is rarely fatal, though life expectancy is often reduced.

Incidence and Outbreaks

Cases of Crohn's disease have not yet been conclusively linked to MAP-contaminated food. The incidence of Crohn's disease is higher in developed countries than in the developing world, although some of this difference could be because diagnosis is more likely where there is a higher standard of health care. In Europe and North America, Crohn's disease is estimated to occur with an overall incidence of 5.6 cases per 100 000 individuals per year.

If MAP is involved in the development of Crohn's disease, food is likely to be an important vehicle for transmission. An epidemiological study has reported a statistical link between the consumption of beef and Crohn's disease. However, if MAP is a zoonosis, it is thought that the most likely source for humans is cow's milk.

Sources

Infected ruminants are the major source of MAP and transmission of the bacterium is mainly via the faecal–oral route. Infected ruminants such as cows, sheep, goats, deer and rabbits excrete MAP into the environment where the organism is known to persist in pastures for sometime. This can also lead to water supplies becoming contaminated with the organism. MAP may survive processes used to produce potable water and could be present in drinking water.

Breast milk samples from Crohn's disease patients have been found to contain MAP.

Growth and Survival Characteristics

MAP can grow at temperatures between 25–45 °C, with an optimum of 37 °C. It can grow at salt concentrations below 5% and can grow at pH values ≥ 5.5. It is a very slow growing bacterium, and on laboratory media incubated at 37 °C it can take many weeks for colonies to be visible to the naked eye.

Although MAP is unlikely to increase in numbers in food, the organism can survive for extended periods depending on conditions. It can survive for some time under acid conditions and studies have recorded *D*-values of approximately 10 and 19 days at pH 4.0 and pH 5.0, respectively, when stored at 20 °C. Salt concentrations of between 2 and 6% had little effect on the survival of the organism regardless of pH. It is therefore possible that MAP may survive some cheese-making processes. MAP can survive outdoors in pastures and the environment for up to 9 months although exact survival times are dependent on conditions.

Evidence suggests that standard water treatments such as slow sand filtration and chlorination may not be sufficient to remove MAP from drinking water. MAP can survive chlorination at 2 ppm and this resistance is increased in the low-nutrient, low-temperature conditions found in many water systems.

MAP in not usually inactivated by food preservatives.

Thermal Resistance

MAP is more heat resistant that other Mycobacteria of concern in milk, notably *M. bovis* (which can cause tuberculosis in humans). Following extensive studies on the thermal inactivation of MAP in milk, coupled with the fact that MAP can be isolated from commercially pasteurised milk, it has been concluded that the organism may occasionally survive standard commercial milk-pasteurisation processes. This has led to recommendations in the UK for extended high-temperature/short-time milk-pasteurisation treatments of 72 °C/25 s, although there is evidence to suggest that even this extended heat process is insufficient to ensure that the organism is absent in pasteurised milk.

Control Options

Processing

Strategies to control MAP in milk focus on reducing or even eliminating Johne's disease in dairy cattle on the farm. There are difficulties with this approach, such as the possibility of reinfection of MAP-negative herds from infected wild-animal reservoirs. Nevertheless, initiatives such as cattle-health schemes, vaccination and veterinary advice to farmers on husbandry, basic hygiene and biosecurity measures are in place in many countries.

Other measures to lessen the risk of MAP-contaminated milk reaching consumers include, minimising faecal contamination of raw milk during the milking process to reduce initial MAP numbers in milk, therefore lessening the chance of the organism being present after pasteurisation, and ensuring that dairies carry out pasteurisation correctly and that cross-contamination between raw and pasteurised milk does not occur.

Product Use

Consumers can reduce the possible risk of MAP by only using correctly pasteurised milk and other dairy products.

Legislation

There is no specific legislation in the EU and the US on levels of MAP in foods.

There are food-hygiene requirements in many countries, which include controls on hygiene standards for the production and distribution of milk and dairy products. In addition, there are recommendations on steps to reduce the prevalence of MAP in dairy herds.

The UK Food Standards Agency has recommended taking a precautionary approach with respect to MAP and has said that steps should be taken to reduce human exposure to the organism. In the UK it is recommended that the minimum holding time for high temperature/short time milk pasteurisation at 72 °C should be increased from 15 to 25 s.

Sources of Further Information

Published

International Life Sciences Institute, Gould G. *Mycobacterium avium* subsp. *paratuberculosis* (MAP) and the food chain, Brussels, ILSI Europe. 2004. 32pp.

Mycobacterium paratuberculosis, IDF bulletin 362/2001, Brussels, IDF. 2001. 61pp.

On the Web

Strategy for the control of *Mycobacterium avium* subspecies *paratuberculosis* (MAP) in cows milk. UK Food Standard Agency. (2003). http://www.food.gov.uk/multimedia/pdfs/map_strategy.pdf

Possible links between Crohn's disease and Paratuberculosis. Report of the Scientific Committee on Animal Health and Animal Welfare. European Commission. (March 2000). http://ec.europa.eu/food/fs/sc/scah/out38_en.pdf

1.1.11 *PLESIOMONAS SHIGELLOIDES*

Hazard Identification

What is Plesiomonas shigelloides*?*

Plesiomonas shigelloides is a gram-negative, non-spore-forming bacterium that has, on occasions, been thought to have caused foodborne disease. Although the role of the organism in causing enteric disease has yet to be conclusively established, it is strongly implicated as a cause of human diarrhoea by a number of factors.

In many ways the organism is very similar to *Aeromonas*. Indeed the organism was called *Aeromonas shigelloides* for a short time, and for years both genera were included in the family *Vibrionaceae*. Recently, however, *P. shigelloides* has been classified into a different family, the *Plesiomonadaceae*. In addition, unlike *Aeromonas* species, *P. shigelloides* is not regarded as a psychrotroph (it is unable to grow at refrigeration temperatures).

Occurrence in Foods

Plesiomonas shigelloides is primarily an aquatic organism and most infections are thought to be caused by the ingestion of contaminated water. The few studies identifying foods contaminated with the organism have mostly isolated *P. shigelloides* from fish and seafoods, and many foodborne infections are associated with the consumption of raw oysters.

Hazard Characterisation

Effects on Health

All individuals are susceptible to *P. shigelloides* infections, although the organism is likely to cause more severe disease in children and the immunocompromised. Infections peak in the summer months and are more often reported in the tropical and subtropical regions.

P. shigelloides causes gastroenteritis and in rare cases extraintestinal infections. Although the infective dose is unknown, it is thought to be high ($>10^6$ organisms). The incubation period is not well defined and symptoms may begin between 20 to 50 h after ingesting the contaminated water or food.

Symptoms of gastroenteritis last from 1–9 days and can include diarrhoea, nausea, vomiting, abdominal pain, chills, fever and headaches. The diarrhoea is usually characterised by watery stools although in severe cases the stools have been described as greenish-yellow, mucoid and blood tinged. This form of the disease is usually self-limiting.

Occasionally, extraintestinal infections, such as meningitis and septicaemia, can occur, particularly in immunocompromised individuals. These infections can be very severe and are associated with a high mortality rate.

Incidence and Outbreaks

There is very little reported information on the incidence of foodborne *P. shigelloides* infections and foodborne outbreaks caused by the organism are not often reported. However, foodborne disease caused by the organism has been associated mostly with raw oysters. Other foods thought to have caused outbreaks of *P. shigelloides* gastroenteritis are chicken, fish, shrimp and temperature-abused buffet food comprising cold fish and egg with mayonnaise.

Sources

Plesiomonas shigelloides is regarded as an aquatic micro-organism, and is found in fresh and marine waters, especially during warm weather. The organism is unable to grow below 8 °C, so is more often found in tropical and subtropical waters and in river water from temperate climates during the summer months.

The organism is naturally found in finfish and shellfish, again more often in those originating from warmer waters. During the warmer months samples can be heavily contaminated. *P. shigelloides* has also been isolated from snakes, toads, dogs, cats, cattle, pigs, goats and birds.

P. shigelloides has been found in healthy humans at very low rates in Japan (<1%) but at higher rates in developing countries such as Thailand (23–24%).

Growth and Survival Characteristics

P. shigelloides is not regarded as being psychrotrophic and most strains will not grow below 8 °C. At least one strain, however, has been reported to grow at 0 °C and the organism can be isolated from waters in cold climates such as those in Northern Europe. The maximum temperature for growth is around 45 °C.

P. shigelloides can survive freezing temperatures and the organism has been isolated from foods stored at –20 °C for some years.

P. shigelloides has been shown to grow in salt concentrations up to 5% and the pH range for growth is generally 4.5–8.5. However, a few isolates have been shown to grow at low pH values of 3.5, and some at high pH values of 9.0.

P. shigelloides is a facultative anaerobe (it is able to grow in the presence or absence of oxygen). Studies using vacuum/modified-atmosphere (80% CO_2) packaged cooked crayfish tails have shown some inhibition in the growth of *P. shigelloides* compared to product stored in air.

Thermal Resistance

The organism is not particularly heat resistant and pasteurisation process of 60 °C for 30 min or equivalent heat processes will ensure its inactivation.

Control Options

P. shigelloides is primarily a risk when contaminated water and raw seafoods are ingested. It is easily inactivated by heat and normal cooking processes should ensure its destruction.

Processing

Using water from a potable source in food-processing establishments, and ensuring that cross-contamination between raw and cooked foods does not occur reduces the risk of infection. High numbers of cells are thought to be necessary to cause illness, so ensuring adequate refrigeration of raw and cooked foods will limit the growth of any *P. shigelloides* present.

Product Use

Consumers can reduce the risk from *P. shigelloides* infections by avoiding the consumption of raw shellfish and contaminated water.

Legislation

There is no specific legislation in the EU and the US on levels of *P. shigelloides* in foods.

Sources of Further Information

Published

Krovacek, K. *Plesiomonas shigelloides* in International Handbook of Foodborne pathogens. ed. Miliotis, M.D. and Bier, J.W. New York. Marcel Dekker. 2003, 369–73.

International Commission on Microbiological Specifications for Foods. *Plesiomonas* in Microorganisms in Foods, Volume 5. Microbiological Specifications of Food Pathogens. ed. International Commission on Microbiological Specifications for Foods. London. Blackie, 1996, 208–133.

1.1.12 *PSEUDOMONAS AERUGINOSA*

Hazard Identification

What is **Pseudomonas aeruginosa***?*

Pseudomonas aeruginosa is a gram-negative, non-spore-forming, strictly aerobic bacterium. The organism has on rare occasions been implicated in cases of food poisoning, but is more often associated with disease in the immunocompromised, in hospital patients and in infants. It is very rarely a problem for healthy individuals and is generally regarded as an opportunistic pathogen.

Occurrence in Foods

Pseudomonads are ubiquitous and are normal contaminants of vegetables, meats, milk and water. In many foods, these organisms, including *Ps. aeruginosa*, are regarded as potential spoilage micro-organisms. *Ps. aeruginosa* contamination is of particular concern in potable water supplies and bottled water.

Hazard Characterisation

Effects on Health

Pseudomonas aeruginosa can cause a range of infections, such as soft tissue, respiratory tract, urinary tract and systemic infections in "at-risk" individuals. However, it can also invade the intestinal tract, sometimes leading to acute gastroenteritis.

Ps. aeruginosa infection in healthy individuals can on occasions lead to mild gastroenteritis, whereas in susceptible individuals, in particular infants, it can lead to serious diarrhoea sometimes resulting in death.

Incidence and Outbreaks

There is little information on the incidence of foodborne *Ps. aeruginosa* infections, but outbreaks of infections in hospital caused by the bacterium being introduced from water or food sources have been documented. These outbreaks are not necessarily associated with gastroenteritis. For example, an outbreak of pneumonia in an intensive-care unit was traced back to patients drinking *Ps. aeruginosa*-contaminated bottled mineral water.

Sources

Pseudomonads are ubiquitous and are commonly present in environmental sources such as soil and water. They are frequently found on plant surfaces, and occasionally on the skin of animals. *Ps. aeruginosa* can be found on the skin or in the throat of some healthy human individuals.

It is thought that *Ps. aeruginosa* may enter hospital environments on foods such as fruits and vegetables and surveys have found that *Ps. aeruginosa* is present on vegetables and meats as well as in frozen foods.

Growth and Survival Characteristics

Pseudomonads are sensitive to heat and are readily inactivated by normal cooking processes. They are sensitive to desiccation and are not tolerant of acid pH.

However pseudomonads, including *Ps. aeruginosa*, are notable for their relative resistance to many disinfectants, and they can form biofilms on surfaces, making them very difficult to remove.

Control Options

Processing

Low levels of *Ps. aeruginosa* are not usually a concern in foods destined for consumption by healthy individuals. However, the organism should be considered when designing and preparing foods intended for consumption by the immunocompromised such as those found in intensive-care units. Mild heat processes readily inactivate the micro-organism, but high hygiene standards need to be implemented to prevent post-process contamination.

Product Use

Healthy consumers need not be unduly concerned about the presence of low levels of *Ps. aeruginosa* in foods.

Legislation

There are requirements within European Community legislation for *Ps. aeruginosa* in water offered for sale in bottles or containers. These require that no *Ps. aeruginosa* cells can be detected in 250 ml of water.

Sources of Further Information

Published

Pitt, T.L. *Pseudomonas, Burkholderia* and related genera, in "Topley and Wilson's microbiology and microbial infections, volume 2: systematic bacteriology." ed. Balows, A. and Duerden, B.I. 9th edn, London. Arnold Publishers, 1998, 1109–1138.

International Commission on Microbiological Specifications for Foods. *Pseudomonas cocovenenans*, in "Microorganisms in Foods, volume 5. Microbiological Specifications of Food Pathogens." ed. International Commission on Microbiological Specifications for Foods. London. Blackie, 1996, 214–6.

1.1.13 *SALMONELLA*

Hazard Identification

What is Salmonella*?*

The Salmonellae are gram-negative, non-spore-forming rod-shaped bacteria belonging to the family *Enterobacteriaceae*. However, *Salmonella* is not included in the group of organisms referred to as coliforms. *Salmonella* is one of the principal causes of foodborne gastroenteritis worldwide and is also an important pathogen of livestock. Salmonellosis is a zoonotic infection (can be transmitted to humans from animals).

Salmonella nomenclature has been revised over the years and is based on biochemical and serological characteristics. Many microbiologists now use a classification that recognises only two species of *Salmonella*. These are *S. enterica* (which includes 6 subspecies) and *S. bongori*. The subspecies most important in foodborne disease is *S. enterica* subspecies *enterica*.

The genus *Salmonella* can be further divided into serotypes, of which there are a great many (>2500). Most serotypes (sometimes referred to as serovars) belong to the species *S. enterica* and only 20 belong to *S. bongori*. *Salmonella enterica* subspecies *enterica* contains nearly 1500 serotypes, including many of the serotypes that are known to cause foodborne disease. Under the currently accepted classification, an example of the correct way to denote a serotype would be *Salmonella enterica* subspecies *enterica* serotype Enteritidis, although fortunately convention allows this to be abbreviated to *Salmonella* Enteritidis (*S.* Enteritidis). In addition, each *Salmonella* serotype can be divided further by phage typing. A particular phage type can be denoted using the term PT. For example, *Salmonella* Enteritidis PT4 is an organism commonly associated with eggs and human illness. Other common serotypes involved in human illness are *S.* Typhimurium and *S.* Virchow.

Occurrence in Foods

Food animals can become infected with *Salmonella* from feed and from the environment, and many foods of animal origin such as meat, poultry, eggs and raw milk can be contaminated with the pathogen. Many studies to determine *Salmonella* contamination rates in food commodities have been conducted. For example, in 2005 a Europe-wide study found that about one in five large-scale commercial egg-producing facilities had hens infected with *Salmonella*, with the lowest levels of infection being found in Sweden and Luxembourg, and the highest levels in Portugal, Poland and the Czech Republic. A UK study reported contamination levels in poultry of 5.7% in 2001, and a 2003 study of UK produced shell eggs found contamination levels of 0.34%. In the US, testing during 2003 found that 3.6% of raw meat and poultry samples were contaminated with *Salmonella*.

Fresh produce may also become contaminated with *Salmonella* from animals and environmental sources. The pathogen has been isolated from tomatoes, lettuce and salad greens, sprouting seeds, fruit juice, cantaloupe melons and nuts.

Cooked ready-to-eat foods can become contaminated as the result of cross-contamination from raw foods. Although contamination can occur as the result of direct contact, it can also occur via food-preparation surfaces or equipment used for both raw and cooked foods. A wide variety of processed foods have been found to be contaminated with *Salmonella*, including chocolate, breakfast cereal, flavoured potato crisps, peanut butter, fermented meats, cheeses, milk powder and ice cream.

Hazard Characterisation

Effects on Health

Some *Salmonella* serotypes have a limited host spectrum (*i.e.* they cause specific and often serious clinical disease in one or a few animal species), such as *S.* Typhi and *S.* Paratyphi in humans (causing typhoid fever), *S.* Dublin in cattle, and *S.* Choleraesuis in pigs. These are not considered further here.

The more usual foodborne form of the illness is caused by non-typhoid salmonellae, which invade the cells lining the small intestine. These organisms cause gastroenteritis lasting between 1–7 days, with symptoms that include diarrhoea, abdominal pains, nausea, vomiting, and chills, leading to dehydration and headaches. Susceptible individuals, such as the young, the elderly and those who are immunocompromised can sometimes develop more severe symptoms from non-typhoid salmonellae such as septicaemia, or chronic conditions, such as reactive arthritis. The death rate for infection by non-typhoid salmonellosis is $<1\%$ although this figure is higher amongst some groups, particularly the elderly.

The incubation time is between 6 and 48 (usually 12–36) h. The infective dose is thought to vary widely and can depend on the individual consuming the infected food, the type of food involved and possibly the serotype involved. Small numbers (between 10–100) of cells can cause illness if consumed by the young or the elderly, or if the food consumed has a high fat content (*e.g.* chocolate, cheese or peanut butter) because the fat is thought to protect the cells from the gastric acids. In general however, it is thought that high numbers (between 10^5–10^6 cells) of salmonellae need to be consumed to cause illness.

Individuals recovering from salmonellosis can continue to shed *Salmonella* in their stools for some time. Food handlers reporting *Salmonella* gastroenteritis should be excluded from work until shedding has stopped.

Incidence and Outbreaks

The incidence of human salmonellosis in Europe has been declining steadily since 1995. In 2005, just over 181 000 cases were reported in 27 countries, but this is likely to represent considerable under-reporting. The decline is thought to be mainly due to the success of measures taken to reduce *S*. Enteritidis contamination in hen's eggs. Similar trends have been observed in other developed countries, including the USA, where the incidence of salmonellosis fell sharply between 1996 and 2001, but has since remained at approximately 15 cases per 100 000 of the population.

Foodborne *Salmonella* outbreaks are commonly associated with inadequately cooked eggs and poultry, or products containing these ingredients, such as egg mayonnaise. However, many other food types have been linked with outbreaks. These include dairy products (such as milk, cheese and ice cream), fruit juice, tomatoes, melons, lettuce and other salad leaves, sprouted seeds, cereals, potato crisps, coconut, black pepper, chocolate, almonds, products containing sesame seed paste (tahini), peanut butter, herbal infusions, cooked meats, fermented meats such as salami, bottled water and reconstituted dried infant formula. Outbreaks involving processed foods can be very large. For example, an outbreak of *S*. Enteritidis associated with ice cream that occurred in the USA in 1994 may have affected as many as 224 000 people.

Sources

Salmonella can be shed in the faeces of infected humans. Shedding can occur for some time after symptoms have subsided and some individuals become chronic carriers. However, foodborne illness caused by an infected food handler is rare and is the result of poor personal hygiene.

Many *Salmonella* infections in animals are asymptomatic, and many animals such as birds, rodents, reptiles, frogs, fish and snails can be infected with *Salmonella*. This can result in contamination of the soil and surface waters, leading to the infection of food animals and contamination of fruits and vegetables, herbs, spices, seeds, nuts and shellfish. In addition, food animals can also become infected via their feed or from other infected food animals. Although some *Salmonella* serotypes are species specific, many are able to cross between species and cause disease in man (zoonoses). Both poultry and pigs are considered to be significant reservoirs of *Salmonella* but many foods of animal origin, such as raw meats and unpasteurised milk are also important sources of the pathogen.

Growth and Survival in Foods

Most *Salmonella* serotypes can grow over the temperature range 7–48 °C, although growth is reduced at temperatures below 10 °C. Reports in the literature suggest that some serotypes can grow at temperatures as low as 4 °C, but this is not universally accepted.

Although most *Salmonella* serotypes are unable to grow at refrigeration temperatures, the organism is able to survive for extended periods at chill temperatures, particularly under freezing conditions.

A few *Salmonella* serotypes can grow over a range of pH values from 3.7–9.5 under otherwise ideal conditions, but the optimum is 6.5–7.5. Other factors such as temperature, the type of acid present and the presence of antimicrobials can affect the minimum pH for growth. Although *Salmonella* cannot grow under very acid conditions, the organism is able to survive for some time in acid environments. Survival times are dependent on type of acid present and temperature (chilled temperatures favour survival).

Salmonellae are able to grow at water activities down to 0.94 (and possibly 0.93), lower values are dependent on serotype, food sources, temperature and pH. *Salmonella* will die out at water activities below that permitting growth, but inactivation can be extremely slow in some products (measured in years), particularly those with very low moisture and high fat content, such as chocolate. *Salmonella* is relatively resistant to drying and can survive on food-production surfaces for some time.

Salmonellae are facultative anaerobes (can grow with or without oxygen) and growth is only slightly reduced under nitrogen. The organism is able to grow in atmospheres containing high levels of carbon dioxide (possibly up to 80% in some conditions).

Salmonella is not especially resistant to sanitisers used in the food industry, but is able to form biofilms that may reduce the efficacy of a sanitiser if cleaning is inadequate.

Thermal Resistance

The majority of *Salmonella* serotypes are not particularly heat resistant and are usually inactivated by pasteurisation or equivalent heat processes. *D*-values are typically between 1–10 min at 60 °C and < 1 min at 70 °C, with typical *z*-values of 4–5 °C. However, there are some important exceptions. Some rare serotypes such as *S.* Senftenberg are much more heat resistant (approximately 10–20 times) than other *Salmonella* serotypes at high water activities, and some foods such as those with high fat content or with low water activities reduce the effectiveness of heat treatments normally expected to inactivate the organism.

Control Options

A HACCP approach is essential for the effective control of *Salmonella* in food production.

Processing

The control of *Salmonella* in food should start on the farm with the careful production of fresh produce and animal-derived raw materials, such as eggs, poultry and pork. Many countries have policies that encourage measures to reduce the levels of *Salmonella* in egg-production units, in poultry houses, during the growing of fresh produce and also during transport of raw commodities. Such measures are especially important for products that will not receive a heat treatment prior to consumption. Food manufacturers should carefully source their ingredients and supplies from producers implementing such measures, or purchase pasteurised products (such as milk or egg) to reduce the risk of *Salmonella* entering their facilities or reaching the consumer.

Salmonella can be effectively controlled by relatively mild heat processing (*e.g.* milk pasteurisation), but it is essential that adequate measures are in place to avoid cross-contamination between raw and cooked product. HACCP should be used to identify and implement adequate controls for *Salmonella* (ensuring the organism is absent) in all foods that will be supplied to the consumer as ready-to-eat (or drink). The HACCP plan should be rigorously reviewed when product is reformulated as such exercises can affect the efficacy of heat treatments, or the use of acid or solute as a control for *Salmonella*. General good hygiene procedures and effective temperature controls are also very important.

Product Use

To ensure that ready-to-eat foods remain free from *Salmonella*, careful handling and storage of product should be encouraged at the retail stage and in the consumer's home. Avoidance of cross-contamination is particularly important in this respect.

In the UK, consumers and caterers are encouraged to refrigerate eggs once purchased and to adhere to the "use by date" stamped on the egg, which should mean it is consumed within 3 weeks of date of laying.

Careful labeling and cooking instructions for raw product is very important, especially when it may appear cooked. Raw chicken entrée products have caused illnesses in the US because they were not clearly labeled and appeared ready-to-eat. Consumers should also be advised to wash fresh produce, such as bagged lettuce, even when it appears ready prepared.

Consumers should be advised of "high-risk" foods. These include raw or partly cooked egg products, such as home-made mayonnaise and ice cream, undercooked meat and meat products, unpasteurised dairy products, unpasteurised fruit juices and raw or lightly cooked seed sprouts.

Legislation

There are codes of practice in many countries around the world for the production of various food commodities that include measures to control *Salmonella*. Although it is unacceptable for any ready-to-eat product to contain viable Salmonellae, there are regulations in many countries enforcing requirements in specified products.

European Union regulations have specific requirements pertaining to *Salmonella* in a wide range of products, including meat and meat products, cheese, butter and cream that have not undergone standard pasteurisation processes, milk powder, whey powder, some ice cream and egg products, various shellfish products, ready-to-eat sprouted seeds, ready-to-eat fruit and vegetables, unpasteurised fruit and vegetable juices and infant formula and dried dietary foods. Sampling plans and absence requirements vary depending on product. There are also EU requirements for *Salmonella* testing of cattle, sheep, goats, horses, poultry and pig carcasses.

US food law also requires *Salmonella* to be absent from ready-to-eat food products that are not intended to be heated before being consumed. There are also specific requirements for the labeling of eggs not treated to inactivate the pathogen and for control of *Salmonella* in foods prepared for vulnerable populations.

Some countries have specific storage, labeling requirements and heat treatments for foods that are aimed at controlling foodborne salmonellosis. In the US these include mandatory refrigerated storage of eggs (from farm to the consumer) and labeling requirements for the inside of egg boxes advising of safe egg-handling practices. In the EU, legislation requires many eggs to be stamped with a distinguishing mark and country of origin to help trace the farm of origin in case of an outbreak.

Sources of Further Information

Published

Jay, L.S., Davos, D., Dundas, M., Frankish, E. and Lightfoot, D. *Salmonella*. in *Foodborne Microorganisms of Public Health Significance*. ed. Australian Institute of Food Science and Technology. 6th edn. Waterloo DC, AIFST, 2003, 207–266.

Bell, C. and Kyriakides, A. *Salmonella* in Foodborne pathogens: Hazards, Risk Analysis and Control. ed. Blackburn C de, W. and McClure, P.J. Cambridge. Woodhead Publishing Ltd, 2002, 307–5.

D'Aoust, J.-Y., Maurer, J. and Bailey, J.S. *Salmonella* species, in *Food Microbiology: fundamentals and frontiers*. ed. Doyle, M.P., Beuchat, L.R. and Monteville T.J. 2nd edn, Washington DC, ASM Press, 2001, 141–178.

On the Web

Risk assessment of *Salmonella* Enteritidis in shell eggs and *Salmonella* spp. in egg products. United States Department of Agriculture's Food Safety and Inspection Service (FSIS) (October 2005). http://www.fsis.usda.gov/PDF/SE_Risk_Assess_Oct2005.pdf

Risk assessment of the impact of lethality standards on salmonellosis from ready-to-eat meat and poultry products. United States Department of Agriculture's Food Safety and Inspection Service (September 2005) http://www.fsis.usda.gov/PDF/Salm_RTE_Risk_Assess_Sep2005.pdf

Risk profile *Salmonella* (non typhoidal) in poultry (whole and pieces). Institute of Environmental Science and Research Limited. (June 2004). http://www.nzfsa.govt.nz/science/data-sheets/salmonella-poultry-update.pdf

Risk profile *Salmonella* (non typhoidal) in and on eggs. Institute of Environmental Science and Research Limited. (May 2004). http://www.nzfsa.govt.nz/science/data-sheets/salmonella-eggs.pdf

Risk assessments of *Salmonella* in eggs and broiler chickens, MRA Series 1 & 2. World Health Organization. (2002). http://www.who.int/foodsafety/publications/micro/salmonella/en/index.html

1.1.14 *SHIGELLA*

Hazard Identification

What is Shigella*?*

Shigella species are gram-negative, non-spore-forming bacteria belonging to the group *Enterobacteriaceae*. They have many similarities with *E. coli*, but are not included in the group microbiologists refer to as coliforms. There are four *Shigella* species, *S. sonnei, S. dysenteriae, S. flexneri* and *S. boydii*, which cause the disease known as shigellosis (also called bacillary dysentery). Although the most common route of transmission is from person-to-person via the faecal–oral route, all have been linked to foodborne outbreaks. *Shigella* infections can also occur as the result of drinking, or swimming in, contaminated water.

S. sonnei is the leading cause of shigellosis from food as well as being the leading cause of shigellosis in industrialised countries. The other three species are largely associated with contaminated water. *S. dysenteriae* is the cause of epidemic dysentery and *S. flexneri* is largely sexually transmitted.

Occurrence in Foods

Humans are the main reservoir for *Shigella* and almost any food can become infected if it is contaminated with faecal material from infected individuals, or with sewage-contaminated water. Foods that require a lot of handling during preparation and are not subsequently cooked, such as salads and sandwiches, are at particular risk of contamination from infected food handlers.

Hazard Characterisation

Effects on Health

Shigellae usually only infect humans and some other primates. In humans, all individuals are susceptible to *Shigella* infections but infants, the immunocompromised and the elderly are at risk of developing the severest form of the disease.

The infective dose can be very low – as few as 10 cells can cause illness. The incubation time for illness ranges from 12 h to 7 days (usually 1–3 days). *Shigella* species can cause an asymptomatic infection, mild diarrhoea, or can cause acute dysentery. Typical symptoms are abdominal pain and cramps, fatigue, fever and diarrhoea with mucus and sometimes blood occurring in the faeces. Frequent bowel movements can lead to dehydration. Typically, symptoms last for between 3–14 days although longer-term complications such as Reiter's disease, reactive arthritis and haemolytic uraemic syndrome can occur as a result of infection.

The estimated fatality rate is 0.16%, although this can increase to 10–15% with some particularly virulent strains.

Incidence and Outbreaks

In countries where hygiene standards are good, the incidence of shigellosis is low. In the US there are about 18 000 cases reported each year, although the actual figure is thought to be considerably greater because of incorrect diagnosis and under-reporting. In Europe in 2005, just less than 7500 cases of shigellosis were reported across 26 countries and in England and Wales approximately 1000 cases are reported each year. In developing countries, where hygiene standards are low, shigellosis is much more common, and each year an estimated 1.1 million people die from *Shigella* infections.

A wide variety of foods have been implicated in foodborne shigellosis. These include various salads, lettuce, green onions, uncooked baby maize, milk, soft cheese, cooked rice, spaghetti, prawn cocktail, orange juice, mashed potato, chocolate pudding and stewed apples.

Notable recent foodborne outbreaks include an outbreak of *S. sonnei* infections in 1994 affecting several Northern European countries, which was associated with imported Spanish Iceberg lettuce. In 1998, chopped parsley used as garnish was implicated in a number of outbreaks, involving 493 confirmed and probable cases of *S. sonnei* infection in the US and Canada. *S. flexneri* caused an outbreak in the UK during 1998 with 46 cases linked to fruit salad purchased from a supermarket.

Sources

Humans and higher primates are the main reservoir for *Shigella* species. Individuals recovering from infection can continue to shed the pathogen for weeks after the symptoms have ceased and the organism can survive for some time in faeces.

The organism is not normally found free living in the environment and is only present in food as the result of faecal contamination.

Sewage-contaminated water can be a source of *Shigella* contamination. Although it is commonly thought that water, rather than food, is the more important vehicle for *Shigella*, public-health data suggests that the reverse may be the case. Food can become contaminated from soiled hands, from contaminated water, from the use of nightsoil as manure and from flies that have been feeding on human faeces.

Growth and Survival Characteristics

Shigella spp have a minimum temperature for growth of 6.1 °C, and a maximum of 47 °C. Although little is known about the growth of the organism in foods it has been shown to grow on parsley, as well as on sliced fruit at ambient temperatures. However, *Shigella* does not need to grow in food to cause illness, as the very low infective dose means that the presence of the organism in food is sufficient to cause infection. *Shigella* spp survive at frozen and chill temperatures, although the time of survival depends on the type of food environment as well as the temperature.

The reported pH range allowing growth of *Shigella* spp is 4.8–9.3, although actual values will depend on acid type. *Shigella* spp are gradually inactivated at pH values <4.0, but the organism can survive for some time in acid conditions. Fresh orange juice has been linked to a *S. flexneri* outbreak in South Africa, and *Shigella* spp survived for up to 14 days in tomato and apple juice stored at 7 °C.

Shigella spp can grow at water activities down to 0.96 (maximum salt concentration 5.2% NaCl). The organism dies out slowly at low water activities. Even at high NaCl concentrations (10%) some strains can survive for 4 days.

Shigella spp are facultative anaerobes (can grow with or without oxygen). At room temperature *S. sonnei* rapidly increased in numbers in shredded cabbage stored in vacuum/modified-atmosphere (30% N_2, 70% CO_2) packaging, and *Shigella* numbers remained static when stored under similar conditions at chilled temperatures.

Shigella spp are not particularly resistant to commonly used preservatives and sanitisers and 200 ppm free chlorine has been shown to give a >6 $\log_{10}$ reduction of *S. sonnei* on parsley held at 21 °C for 5 min.

Thermal Resistance

Shigella spp are easily inactivated by heat and death is rapid at temperatures above 65 °C.

Control Options

Measures to prevent food becoming contaminated with *Shigella* spp should focus on preventing faecal contamination of raw and processed foods and using safe or treated water supplies for irrigation of crops and for food processing.

Processing

Washing of fresh produce, even in water containing a disinfectant does not ensure inactivation/removal of any *Shigella* present. Good hygiene standards in countries supplying salad crops and fruit are very important to prevent the import of contaminated produce. Minimising handling, and insisting on good levels of personal hygiene, both reduce the risk of food becoming infected from food handlers.

Food handlers suffering or suspected of suffering from *Shigella* infections or individuals who have been in contact with people suffering from shigellosis should be excluded from food-handling areas until it is ensured they are free from the pathogen (typically three consecutive negative stool samples are required).

Product Use

The importance of good hygiene should be emphasised to consumers. When traveling to developing countries where shigellosis is endemic, consumers should be advised to only drink treated or boiled water, only eat cooked foods and fruits that they have peeled themselves.

Legislation

No specific requirement is made under European Commission legislation, or in the US Food Code (2005) with regard to levels of *Shigella* in food.

Control of the pathogen is required under EC general food safety requirements in which food should not be sold if it is unsafe. The presence of *Shigella* spp in food indicates poor hygiene, is unacceptable and the food is unsafe.

Sources of Further Information

Published

Warren, B.R., Parish, M.E. and Schneider, K.R. *Shigella* as a foodborne pathogen and current methods for detection in food. Critical Reviews in Food Science and Nutrition. 2006. 46 (7), 551–567.

Lightfoot, D. *Shigella* in "Foodborne micro-organisms of public-health significance." ed. Hocking, A.D., Australian Institute of Food Science and Technology. AIFST, Waterloo DC, 2003, 543–552.

1.1.15 *STAPHYLOCOCCUS AUREUS*

Hazard Identification

What is Staphylococcus aureus?

Staphylococcus aureus is a gram-positive, non-spore-forming bacterium that is able to grow both aerobically or anaerobically (a facultative anaerobe). Some strains of the organism have the ability to produce toxins (enterotoxins) in food, and it is the ingestion of these pre-formed enterotoxins that causes the symptoms associated with staphylococcal food poisoning.

Although *Staph. aureus* is the principle *Staphylococcus* species to cause food poisoning, other staphylococci have also been shown to produce enterotoxins. These include *Staph. intermedius, Staph. hyicus, Staph. xylosus, Staph. cohnii, Staph. epidermis* and *Staph. haemolyticus*, although of these, *Staph. intermedius* is the only non-*Staph. aureus* species to be clearly implicated in foodborne outbreaks.

To date, 14 different staphylococcal enterotoxins have been described (known by letters of the alphabet, A–O, although a few letters are missing from the sequence). All are heat-stable, water-soluble proteins that resist most proteolytic enzymes, such as pepsin or trypsin, therefore retaining their activity in the digestive tract after ingestion. Most food-poisoning strains produce enterotoxin A.

It is important to note that not all enterotoxin-producing staphylococci strains are coagulase or thermonuclease positive (tests for these enzymes are commonly used to indicate potential food-poisoning strains). In addition commercial kits used to test for staphylococcal enterotoxins in foods usually test for the enterotoxins classically causing staphylococcal food poisoning (A–E) and do not test for all staphylococcal enterotoxins that have been described.

Occurrence in Foods

Foods that have caused outbreaks of staphylococcal food poisoning have usually been temperature abused either during processing, or refrigerated storage. Foods particularly "at risk" of causing staphylococcal food poisoning are those that are handled and where the competing microflora has either been destroyed, or inhibited, by cooking or salting.

Foods involved in outbreaks have included milk and milk-based products, such as chocolate milk, cream, custard or cream-filled pastries, butter, ham and other cured meats such as corned beef and bacon. Cooked meats and poultry products are also commonly implicated, as are cheeses – especially where there has been a slow start in the fermentation process leading to a delay in acid production. Other foods linked to outbreaks have included sausages, canned meat, salads, cooked meals (particularly pasta-based products) and sandwich fillings.

Hazard Characterisation

Effects on Health

Staphylococcal food poisoning is considered a mild form of foodborne disease, although all individuals are thought to be susceptible. The toxin is pre-formed in the food, so the onset of symptoms is rapid, 30 min–7 h (average 2–4 h). The severity of symptoms is related to the amount of enterotoxin ingested and the susceptibility of the individual to the particular enterotoxin.

No live *Staph. aureus* cells need to be ingested for staphylococcal food poisoning to occur. However, for sufficient quantities of enterotoxin to be produced to cause illness, the organism needs to reach levels of 10^5–10^6 CFU/g in food. It is thought that the amount of enterotoxin needed to cause illness is between 0.1–1 μg. In instances where lower levels appear to have been involved, it is possible that more than one toxin type may have been present, with one or more types going undetected (see below).

Symptoms are usually nausea and vomiting with abdominal cramps, sometimes followed by diarrhoea. In more severe cases, headache, muscle cramping and dehydration occur, but patients usually recover within 2 days. Although deaths have occurred amongst children and the elderly, these are rare.

Incidence and Outbreaks

The mild nature of staphylococcal food poisoning means that it is probably a very under-reported illness and its true incidence is uncertain. However, a number of significant outbreaks have been recorded.

A large outbreak linked to chocolate milk in the US affected schoolchildren and was estimated to have been caused by quantities of enterotoxin as low as 144 (± 50) ng. The toxin was apparently produced during a period of temperature abuse prior to pasteurisation.

A mass outbreak (> 10 000 cases) of staphylococcal food poisoning occurred in Japan during 2000 and was linked to milk from a single dairy. This outbreak was thought to have involved staphylococcal enterotoxin A (SEA) at a very low level (80 ng), but later research suggested that samples of implicated product may have contained other enterotoxins (SEH), which had been overlooked in the original testing (only "classical" staphylococcal enterotoxins (A–E) are picked up by most commercial kits).

Sources

Humans are a primary reservoir for staphylococci. *Staph. aureus* is carried in the throats and nasal cavities of around 40% of healthy humans and also in infected cuts and sores. Almost any foodstuff can potentially become contaminated with *Staph. aureus* during physical handling and food handlers play a major role in contaminating foods with the pathogen. It can be transmitted to foods via manual handling as well as by coughing and sneezing.

Animals are also a key source of *Staph. aureus*. Mastitis in cows can be caused by *Staph. aureus* resulting in the contamination of raw milk and raw-milk products, such as cheeses. Raw meat, particularly pork, can be contaminated with the organism, as can raw poultry and seafood.

The organism is also able to persist in the food-processing environment. It is quite resistant to desiccation and can survive on dry surfaces such as glass, metal and porcelain. It is often found in dust in ventilation systems.

Growth and Survival in Foods

Staph. aureus can grow over the growth range 7–46 °C and the optimum temperature for growth is 37 °C. Enterotoxin can be produced over the temperature range 10–45 °C, with an optimum temperature for production of around 40 °C. The cells survive frozen storage well.

The pH range for the growth of *Staph. aureus* is 4.5–9.3, and the optimum is around 7.0. Enterotoxin can be produced between pH 4.8–9.0, although production is usually inhibited below pH 5.0. The optimum pH for enterotoxin production depends on strain and type of toxin and is between pH 6.5 and 7.3.

Staph. aureus is noted amongst food-poisoning bacteria as being unusually tolerant of low water activities. It is also more tolerant of salt (NaCl) than many other organisms and is generally able to grow in 7–10% NaCl, although some strains can grow at levels as high as 20%. Enterotoxin production has also been shown at around 10% NaCl. The minimum water activity for growth is generally considered to be 0.86. The ability to grow at such low water activity values confers a competitive advantage to *Staph. aureus* in low water activity products. Enterotoxin can be produced at a_w values as low as 0.87, but the optimum is ≥ 0.90. *Staph. aureus* is very resistant to drying and can survive for extended periods in dried foods.

Staph. aureus is best able to grow and produce enterotoxin in the presence of oxygen, but it also able to grow and produce small quantities of enterotoxin under anaerobic conditions. High concentrations of carbon dioxide (80%) effectively inhibit *Staph. aureus* growth.

Thermal Resistance

Under normal circumstances *Staph. aureus* is not particularly heat resistant and cells are inactivated by normal pasteurisation temperatures. D_{60}-values of around 2 min are typical in high water activity substrates. However, at reduced water activities, such as in salty foods (cheese, ham and bacon), pasta, or high fat foods, heat resistance is enhanced and D_{60}-values of up to 50 min have been documented.

Staphylococcal enterotoxins are very heat resistant. Inactivation of enterotoxin is affected by the water activity and pH of the substrate. Although heating at 100 °C for a minimum of 30 min will generally inactivate enterotoxin, the time for inactivation will be extended at lower water activities. If enterotoxins are present in sufficient quantities, it is possible for them to survive heat processes used in the sterilisation of low acid products. Correctly processed canned mushrooms were implicated in an outbreak of staphylococcal food poisoning in the US.

It is important to remember that heating of product is likely to inactivate *Staph. aureus* cells, but may not inactivate enterotoxin. Temperature abuse of product prior to heat processing could result in staphylococcal food poisoning even though no viable *Staph. aureus* is detectable in the product.

Control Options

Processing

The presence of low levels of *Staph. aureus* in raw products is not necessarily a cause for concern – it is the prevention of staphylococcal enterotoxin production that should be considered in risk assessments. However, measures to reduce the risk of *Staph. aureus* food poisoning during processing should focus on keeping levels low. This can be achieved by minimising physical handling of product, keeping work-preparation areas clean and by the implementation of good temperature control.

Using utensils and disposable gloves can help reduce direct human contact with food products. Individuals suffering from infected cuts and sores and from colds should temporarily be excluded from dealing with ready-to-eat products.

Systems where rework is fed back into the process (*e.g.* pasta/batter production), and where temperatures may permit the growth of *Staph. aureus*, can lead to fresh product being seeded with increasing levels of the pathogen. Cooking processes applied to these products will not usually be sufficient to inactivate enterotoxin. In these circumstances, short run times, discarding any remaining unused product and good cleaning regimes are important factors for minimising the risk from *Staph. aureus*.

Product Use

After processing, the physical handling of "at-risk" processed foods or cured/salted products should be kept to a minimum to reduce the risk of contamination with *Staph. aureus*.

"At-risk" products should either be kept well refrigerated ($<5\,^{\circ}C$) or kept hot ($>63\,^{\circ}C$): under these conditions any contaminating *Staph. aureus* cells will be unable to grow.

Legislation

EU legislation has requirements governing sampling plans and limits for coagulase-positive staphylococci in various cheeses, milk powder and whey powder. For these foods levels of coagulase positive staphylococci below $10–10^4$ CFU/g (depending on product) at the time of removal from the premises are generally satisfactory. However, tests for staphylococcal enterotoxin are required where levels of coagulase-positive staphylococci are detected at $>10^5$ CFU/g, and these toxins should be absent in 25 g. If coagulase-positive staphylococci are found at levels $>10^3$ CFU/g in shelled and shucked products of cooked crustaceans and molluscan shellfish, EU regulations require improvements in production hygiene.

The US Food & Drug Administration's (FDA) food compliance program suggests that any cheese or fish product could be removed from the market place if it is found positive for staphylococcal enterotoxin or if levels of *Staph. aureus* are $\geq 10^4$ CFU/g.

The UK Health Protection Agency (HPA) has issued guidelines on the microbiological quality of some ready-to-eat foods at the point of sale. These state that levels of *Staph. aureus* of 100/g to $<10^4$/g in these products is unsatisfactory, and levels $>10^4$/g is unacceptable/potentially hazardous.

Sources of Further Information

Published

Jay, J.M., Loessner, M.J. and Golden, D.A. Staphylococcal gastroenteritis, in Modern Food Microbiology. 7th edn. New York. Springer Science, 2005, 545–66.

Bergdoll, M.S. and Lee Wong, A.C. Staphylococcal intoxications, in Foodborne infections and intoxications. 3rd edn. ed. Reimann, H.P. and Cliver, D.O. London. Academic Press, 2005, 523–62.

Stewart, C.M. *Staphylococcus aureus* and staphylococcal enterotoxins, in Foodborne Microorganisms of Public Health Significance. 6th edn. ed. Australian Institute of Food Science and Technology. Waterloo DC. AIFST, 2003, 359–79.

On the Web

Opinion of the Scientific Committee on Veterinary Measures relating to Public Health on staphylococcal enterotoxins in milk products, particularly cheeses. European Commission. (2003). http://ec.europa.eu/food/fs/sc/scv/out61_en.pdf

1.1.16 STREPTOCOCCI

Hazard Identification

What are Streptococci?

Streptococcus is a genus of gram-positive, non-spore-forming bacteria. Most species are facultative anaerobes, but some are strict anaerobes and will not grow in the presence of oxygen. Although some streptococci have been implicated in human disease, the majority of species are non-pathogenic.

Some of the streptococci implicated in human illness, notably but not exclusively *Streptococcus pyogenes* and *Streptococcus equi* subspecies *zooepidemicus*, may be transmitted by food and have been linked to foodborne outbreaks associated with salads, milk and dairy products.

Str. pyogenes is a member of the Lancefield Group A streptococci (often abbreviated to GAS). There are around 80 distinctly different serological types of *Str. pyogenes*. It is a facultative anaerobe and it displays beta-haemolysis on blood agar.

Str. zooepidemicus belongs to the Lancefield Group C streptococci, and it too is beta-haemolytic on blood agar. The organism is a cause of zoonotic disease (transmitted from animals to humans).

Occurrence in Foods

Str. pyogenes and *Str. zooepidemicus* can both be present in unpasteurised milk taken from cows suffering from mastitis. Either organism could therefore be present in dairy products made from raw or inadequately pasteurised milk. *Str. pyogenes* can also be present in foods as the result of poor hygienic practices by food handlers suffering from *Str. pyogenes* infections.

Hazard Characterisation

Effects on Health

The main mode of transmission for *Str. pyogenes* infections is person-to-person contact, or via airborne droplets, but the organism can also be foodborne. Typically, *Str. pyogenes* causes pharyngitis, but it can also cause tonsillitis, scarlet fever, septic sore throat and skin infections (such as impetigo). The organism is also occasionally associated with very severe skin/wound infections, sometimes leading to necrotising fasciitis – in these cases the organism is often described in the media as "flesh eating".

All individuals are susceptible to infection. Although unknown, the infectious dose is thought to be relatively low (<1000 organisms) and onset of symptoms is 12–72 h after infection. Typically, these include a sore throat, fever, headache, runny nose, nausea and vomiting. Occasionally a rash occurs. Complications very occasionally occur and the fatality rate is low. If untreated the condition can remain infective for around 10–21 days although if properly treated the infectious period can be reduced to 24 – 48 h.

Many *Str. zooepidemicus* infections in humans are linked to handling animals, but foodborne outbreaks have also been reported. Typically, foodborne infections of *Str. zooepidemicus* cause pharyngitis, but it has also been associated with acute post-streptococcal glomerulonephritis (an inflammation of the kidney tubules). In the US in 1983, a foodborne outbreak associated with *Str. zooepidemicus* reportedly caused a range of symptoms, from fever and chills to systemic infections, such as pneumonia, endocarditis and pericarditis.

Incidence and Outbreaks

There is little information on the incidence of foodborne streptococcal infections.

Foods associated with outbreaks of Group-A streptococci infections include milk, yoghurt, ice cream, custard, rice pudding, meats, seafood, devilled eggs, salads and sandwiches made from eggs or mayonnaise. In many cases the foods had been prepared by infected food handlers and then stored at room temperature for a few hours prior to consumption.

Foodborne outbreaks of *Str. zooepidemicus* infections have been associated with unpasteurised milk and dairy products. For example, an outbreak occurred in the US during 1983 caused by contaminated "queso blanco", a homemade white cheese made from raw milk. Unpasteurised milk contaminated with *Str. zooepidemicus* caused an outbreak involving 7 deaths in the UK in 1984. More recently, an outbreak of *Str. zooepidemicus* infections in Spain in 2006 was associated with inadequately pasteurised cheese and involved 15 cases resulting in 5 deaths.

Sources

The natural reservoir for *Str. pyogenes* is humans. However, humans can transmit the organism to cows on occasion, causing mastitis. The organism is found on human skin, mucous membranes (particularly in the respiratory tract) and can sometimes colonise the rectum.

Although *Str. zooepidemicus* has been isolated mainly from horses, it has also been found in a wide range of animals including sheep, cattle and pigs.

Growth and Survival Characteristics

Streptococci cannot grow at chilled temperatures, and although some species can grow at elevated temperatures (*Str. thermophilus* can grow at 52 °C), this is not typical of the genus. The minimum temperature for the growth of *Str. pyogenes* is around 20 °C, with a maximum of 40 °C.

Str. pyogenes has been shown to survive in various environments outside the host. It can survive in cheese for up to 126 days, on the rim of a drinking glass for 2 days, on blankets for up to 120 days, and in dust for up to 195 days.

Outbreaks of *Str. pyogenes* infections have been associated with food vehicles with relatively low pH, such as yoghurt and products containing mayonnaise.

Thermal Resistance

Streptococcus species are not heat-resistant bacteria and are inactivated by normal milk-pasteurisation processes.

Control Options

Processing

The control of foodborne *Streptococcus* infections relies upon the implementation of strict hygiene, ensuring the rapid cooling of foods to refrigerated temperatures, and avoiding the use of unpasteurised milk. Food handlers with skin lesions or symptoms of respiratory illness should be excluded from food-handling duties.

Product Use

Consumers should be advised to avoid the consumption of raw milk and associated dairy products.

Legislation

There are no specific requirements for levels of *Streptococcus* species in foods in European Community or US legislation.

Sources of Further Information

Published

Gray, B.M. and Arnavielhe, S.R. *Streptococcus* species, in "International handbook of foodborne pathogens". ed. Miliotis, M.D. and Bier, J.W. New York. Marcel Dekker, 2003, 375–405.

1.1.17 VEROCYTOTOXIN-PRODUCING *ESCHERICHIA COLI* (VTEC)

Hazard Identification

What are VTEC?

The verocytotoxin-producing *Escherichia coli* (VTEC) are a group of strains within the species *E. coli*, some of which are highly pathogenic and capable of causing potentially serious foodborne infections in humans. *E. coli* are gram-negative, non- spore-forming bacteria belonging to the family *Enterobacteriaceae*. Microbiologists recognise a small number of genera within the *Enterobacteriaceae*, including *Escherichia* species, as the coliform group.

E. coli are found as part of the normal human gut flora, as well as in the environment, and the presence of *E. coli* in processed product can indicate faecal contamination (the reason why *E. coli* is used as an "indicator" organism). Most strains of *E. coli* do not usually cause illness, but a minority have been associated with infections resulting in diarrhoea, or sometimes more severe illness.

There are four different groups of diarrhoea-causing *E. coli* grouped by virulence characteristics as follows:

Enteropathogenic (EPEC)	Causing infantile gastroenteritis or summer diarrhoea mostly in the developing world.
Enterotoxigenic (ETEC)	Causing traveller's diarrhoea
Enteroinvasive (EIEC)	Causing a form of bacillary dysentery
Verocytotoxin-producing (VTEC) – sometimes referred to as Shiga-like toxin-producing (STEC). This group includes a subset of serotypes often referred to as enterohaemorrhagic *E. coli* (EHEC)	Not all VTEC are associated with human disease, but those that are EHEC can cause haemorrhagic colitis (bloody diarrhoea).

The group of most concern in developed countries is the VTEC, so named because they produce one or more toxins that are toxic to vero cells (a tissue cell culture line derived from the kidneys of an African Green monkey). In excess of 200 VTEC have been described and some of these organisms have been associated with outbreaks of severe foodborne disease in many countries. The VTEC most frequently associated with causing foodborne illness is the serotype *Escherichia coli* O157:H7. Other important VTEC that have caused foodborne infections are O26, O103, O111 and O145.

Occurrence in Foods

VTEC are usually associated with foods derived from cattle such as beef products, particularly minced/ground beef, and dairy products derived from raw milk. Although VTEC could be present on any raw-beef product, minced-meat products are considered more of a risk because the pathogen is transferred from the surface to the centre of the product during the mincing process.

Studies in the USA and the UK have found that VTEC can be present, at least occasionally, on most farms. However, surveys of food commodities have found that the prevalence of the organism in beef and raw-milk products is generally low. VTEC have also been found on fruits, vegetables and seeds and associated products. Fresh produce can be contaminated at any stage during cultivation or handling, possibly via contaminated water supplies, or cattle manure used as a fertiliser.

Hazard Characterisation

Effects on Health

The incubation period for illness caused by VTEC can be between 1 and 14 days, although on average it is 3–4 days. The infective dose is thought to be very low, possibly just 10 cells. This is probably because these bacteria are unusually acid tolerant. Symptoms may be restricted to mild diarrhoea only, and some individuals may be asymptomatic.

However, VTEC infection can cause more serious symptoms in some 50% of those infected, especially in vulnerable groups. These symptoms include bloody diarrhoea, abdominal cramps, vomiting and very occasionally, fever. The illness typically resolves itself after 5–10 days, but in a small number of cases, particularly in young children under 5 years of age and the elderly, VTEC infection can lead to haemolytic uraemic syndrome (HUS), potentially resulting in kidney failure. HUS in children can also result in seizures, coma and sometimes death. Thrombotic thrombocytopaenic purpura (TTP) is a form of HUS typically developed by the elderly and includes fever, platelet loss and neurological symptoms. Around one third of individuals showing signs of VTEC infection are hospitalised and the average mortality rate from HUS caused by VTEC infections in the UK and in North America is 3–5%.

Incidence and Outbreaks

Fortunately, in view of its potentially serious symptoms, VTEC infections are comparatively rare. Nevertheless, in Europe between 1995 and 2002, the incidence of infection more than doubled to 3.2 cases per 100 000 of the population, before leveling off. In 2005, just over 5200 cases were reported in 25 countries. The UK, and Scotland in particular, have a higher incidence than many other European countries, but the reasons for this are not known. In the USA, O157 VTEC is reported separately from other VTEC, but approximately 3500 VTEC cases were reported in 2005, giving an incidence of roughly 2.0 cases per 100 000.

VTEC outbreaks, particularly those caused by *E. coli* O157:H7, have frequently been associated with undercooked minced (ground) beef products such as hamburgers – it has been dubbed "hamburger disease". However, VTEC outbreaks have also been caused by a wide variety of other foods such as cooked meats, raw and recontaminated pasteurised milk, cheese, yoghurt, mayonnaise, unfermented apple cider, unpasteurised apple juice, melon, salad leaves such as lettuce and spinach, parsley, coleslaw, venison jerky, salami and alfalfa sprouts. Contaminated-water sources are also a common source of VTEC outbreaks.

Sources

The main infection reservoir for O157 VTEC is recognised as cattle, which, together with other ruminants such as sheep and camels, are apparently healthy carriers of VTEC. Studies have found that the organism is more likely to be found in cattle faeces during the spring than in the winter. Other animals have also been found to excrete VTEC, including goats, deer, horses, dogs, cats, rats, seagulls, pigeons, and geese. *E. coli* O157 has also been isolated from houseflies. A number of outbreaks have been associated with direct contact with infected animals in petting zoos.

Contamination of water supplies with animal faeces has led to outbreaks linked to drinking water and wells, as well as from recreational waters such as lakes, paddling pools and water parks. Soil manured with animal faeces, or in fields where animals have been grazing, can be contaminated with VTEC and contamination may be transferred to crops.

Person-to-person spread via the faecal–oral route has also occurred causing outbreaks in institutions and child-care settings such as nurseries. Asymptomatic carriers, a state where individuals show no clinical symptom of the disease but are capable of infecting others, have also been reported.

Growth and Survival in Foods

VTEC can grow over the temperature range 7–46 °C (although some sources suggest possibly up to 50 °C) with an optimum of 37 °C. Some isolates of *E. coli* O157:H7 have been reported to grow in raw milk at 8 °C. *E. coli* O157:H7 also grows poorly at 44–45 °C, so that traditional methods to detect *E. coli* in food may not pick up this important pathogen.

VTEC survive well at chilled and frozen temperatures. Low temperature is reported to be the primary trigger for VTEC to enter a "viable non-culturable" state (VNC) in water. A VNC state means that normal methods of detection are unable to recover the organism, but it is still able to cause illness.

VTEC are unusual amongst *E. coli* because they are relatively acid tolerant. The minimum pH for the growth of *E. coli* O157 under otherwise optimum conditions is reported as 4.0–4.4, although the minimum value is affected by the acidulant and acetic and lactic acids are more inhibitory than hydrochloric acid. The organism is able to survive acid conditions (down to 3.6) and has been reported to survive for two months at 4 °C at a pH of 4.5.

The minimum reported water activity for the growth of VTEC is 0.95. Salt (NaCl) at 8.5% inhibits the growth of *E. coli* O157 and growth is retarded at

2.5%. VTEC are very resistant to desiccation and are able to survive many drying and fermentation processes. Outbreaks have been associated with salami and jerky-type meat products.

VTEC are facultative anaerobes (able to grow with or without the presence of oxygen). Modified-atmosphere packaging has little effect on the pathogen although it is reported that it is inhibited on meat packaged under 100% CO_2.

VTEC are not notably resistant to preservatives and sanitisers typically used in the food industry. Organic acids (acetic and lactic acid) are used in the US to decontaminate beef carcasses.

Thermal Resistance

VTECs are not heat-resistant organisms. For *E. coli* O157, $D_{57\,°C}$ values of 5 min, and $D_{63\,°C}$ values of 0.5 min have been reported in meat.

VTEC present on the surface of the product are likely to be inactivated rapidly during cooking, but cells at the centre of ground-meat products and rolled meat joints will only be inactivated if the centre of the product is sufficiently heated. Advice has been given in the US and the UK on the cooking of hamburgers (meat patties, beef burgers) to ensure the complete inactivation of the pathogen. In the US, this advice is that they should reach an internal temperature of 71 °C throughout, and in the UK it is recommended that they be cooked to 70 °C for 2 min, or the equivalent, in all parts of every burger.

Control Options

The control of VTEC starts on the farm with the implementation of good agricultural practices. This can help reduce the shedding of *E. coli* O157 from cattle. Good agricultural practices are extremely important for the production of fresh fruits, salad stuffs and vegetables. It is very important to minimise faecal contamination of all food commodities.

Processing

It is safe to assume that raw products of bovine origin (such as fresh meat and raw milk) are potentially contaminated with VTEC and to treat them accordingly using a HACCP approach. Good hygienic practices should be implemented when handling beef carcasses and the controlled use of chilled temperatures will prevent the growth of VTEC in these products. The possible survival of VTEC should also be considered during the development of products such as bovine-milk cheeses and fermented-meat products. There are published guidelines for producers of such foods, but the use of unpasteurised milk is best avoided.

US regulations require abattoirs and meat-processing establishments to implement a step to eliminate *E. coli* O157:H7 and this can include decontamination. Non-intact raw-beef products (as well as intact raw-beef products intended to be processed into non-intact raw-beef products) found positive for *E. coli* O157:H7 are considered "adulterated" and are recalled.

It is important to ensure that heat processes (where appropriate) are designed to inactivate any VTEC. Cross-contamination between raw and processed product must be avoided.

Product Use

Consumers should be advised of the risks associated with raw-meat products, in particular those made from minced/ground meat, and that all beef products need to be thoroughly cooked. Advice has been given on the required internal cooking temperature for burgers (see *thermal inactivation*). In the US consumers are advised that checking the colour of meat patties or burgers (brown as opposed to pink or red) is not a reliable indication that the product has reached a safe temperature and that they should use a thermometer to check that the required temperature has been reached.

Consumers should be advised to avoid unpasteurised dairy products, juice or cider, and to wash fruit and vegetables well (although washing may not remove all contamination). Vulnerable groups (the young, elderly and the immunocompromised) should be advised not to eat raw or lightly cooked sprouts (such as alfalfa and mung beans).

Legislation

EU regulations have some general requirements for *E. coli* as an indicator of faecal contamination in some products. These requirements giving maximum levels for *E. coli* in some products do not pertain specifically to VTEC, but the presence of VTEC in any product that will not receive a heat treatment prior to consumption is unacceptable. The UK Health Protection Agency has issued guidelines for the microbiological quality of ready-to-eat foods and these state that in these products *E. coli* O157 and other VTEC should be absent in 25 g.

The US Food Code (2005) requires food to be safe and unadulterated and product that will not be heated prior to being consumed would need to be absent from VTEC to conform to this requirement. In addition, *E. coli* O157:H7 is considered an adulterant in non-intact raw-beef products (ground, minced or chopped), as well as intact raw-beef products intended to be processed into non-intact raw-beef products.

Sources of Further Information

Published

Desmarchelier, P.M. and Fegan, N. Enteropathogenic *Escherichia coli*, in Foodborne Microorganisms of Public Health Significance. 6th edn. ed. Australian Institute of Food Science and Technology, Waterloo DC. AIFST, 2003, 267–310.

Bell, C. and Kyriakides, A. *E. coli*: a practical approach to the organism and its control in foods. London. Blackie, 1998.

On the Web

Risk profile: Shiga-like toxin producing *Escherichia coli* in leafy vegetables. Institute of Environmental Science and Research Limited. (February 2006). http://www.nzfsa.govt.nz/science/risk-profiles/FW0456_STEC_in_leafy_veges_February_2006_Final_version_to_NZFSA.pdf

Risk profile: Shiga-like toxin producing *Escherichia coli* in uncooked comminuted fermented-meat products. Institute of Environmental Science and Research Limited. (February 2003). http://www.nzfsa.govt.nz/science/risk-profiles/escherichia-coli-meat.pdf

Risk profile for enterohemorragic *E. coli* including the identification of the commodities of concern, including sprouts, ground beef and pork. Food and Agriculture Organization of the United Nations/World Health Organization. (February 2003). ftp://ftp.fao.org/codex/ccfh35/fh0305de.pdf

Risk profile: Shiga toxin-producing *Escherichia coli* in red meat and meat products. Institute of Environmental Science and Research Limited. (August 2002). http://www.nzfsa.govt.nz/science/risk-profiles/stec-in-red-meat.pdf

Comparative risk assessment for intact (non tenderised) and non intact (tenderised) beef: technical report. US Department of Agriculture's Food Safety and Inspection Service (FSIS). (March 2002). http://www.fsis.usda.gov/PDF/Beef_Risk_Assess_Report_Mar2002.pdf

Draft risk assessment of the public health impact of *Escherichia coli* O157:H7 in ground beef. US Department of Agriculture's Food Safety and Inspection Service (FSIS). (October 2001). http://www.fsis.usda.gov/OPPDE/rdad/FRPubs/00-023N/00-023NReport.pdf

Guidelines for the microbiological quality of some ready-to-eat foods sampled at the point of sale. Health Protection Agency. (September 2000). http://www.hpa.org.uk/cdph/issues/CDPHvol3/No3/guides_micro.pdf

1.1.18 *VIBRIO CHOLERAE*

Hazard Identification

What is Vibrio cholerae*?*

Vibrio cholerae is a gram-negative, non-spore-forming bacterium. It is the causative organism of cholera, a serious human disease responsible for many fatal outbreaks throughout history. Although cholera is usually associated with poor hygiene and faecal contamination, the disease can also be foodborne.

Not all strains of *V. cholerae* cause cholera. Strains (or serotypes) causing classic epidemic cholera are O1 and O139, but there have been rare reports of non-O1/O139 serotypes causing cholera-like disease.

Occurrence in Foods

Vibrio cholerae can be present on food if it is contaminated by polluted water, or by food handlers carrying the pathogen. Contaminated water used to make ice can lead to the contamination of beverages.

In the developed world *V. cholerae* infections are usually associated with the consumption of seafood. Shellfish can become contaminated from environmental sources and most non-O1/O139 cholera infections are associated with the consumption of raw oysters. Other foods implicated in *V. cholerae* infections are fruits and vegetables, grains, meat and legumes.

Hazard Characterisation

Effects on Health

For O1/O139 cholera, symptoms can occur between 5 h and 6 days after infection. The infective dose is thought to be 10^6–10^8 cells. Those most at risk of developing severe cholera are individuals with impaired or undeveloped immunity, such as the immunocompromised and young children, and those suffering from malnutrition. Typically, symptoms start with mild diarrhoea, leading to more severe diarrhoea typified by the production of grey "rice water" stools. Nausea, abdominal pains and low blood pressure can also occur. If untreated, the infection can lead to dehydration, and in severe cases this can result in death. Healthy individuals usually recover in 1–6 days.

For non-O1/O139 *V. cholerae* infections, symptoms usually occur within 48 h of infection and last for around 6–7 days. A much milder form of diarrhoea occurs than with O1/O139 cholera, but it can be bloody and is accompanied by abdominal cramps and fever. Sometimes nausea and vomiting also occur. In rare cases the infection can result in septicemia, and deaths have been reported.

Incidence and Outbreaks

The incidence of infections caused by *V. cholerae* in the developed world is low and is usually caused by serotypes of the organism that cause less severe forms of disease (non-O1/O139 serotypes). However, *V. cholerae* is a major health problem in parts of India, Asia, Latin America and Africa, and in these regions O1/O139 cholera is endemic. In these parts of the world the disease is linked to poverty and poor sanitation, and large waterborne epidemics and foodborne outbreaks occur.

Although most cholera outbreaks are caused by contaminated water, foodborne outbreaks have been reported, but outbreaks are rare in developed regions. Although primarily associated with shellfish, other fish, as well as vegetables, fruit, meat, frozen coconut milk and cooked rice have been implicated as vehicles for the pathogen. A cholera outbreak in Zambia during 2004, in which raw vegetables were implicated as the vehicle, involved an estimated 2500 cases.

Sources

Humans are the main reservoir for *V. cholerae*. Individuals suffering from cholera excrete large numbers of the organism into the environment. In addition, asymptomatic carriers of the organism are known to occur. Contamination of raw or processed food is usually the result of faecal contamination (either directly or indirectly from faecally contaminated water).

V. cholerae O1 survives for short periods in fresh water, but it can survive in seawater for longer periods. Fish and shellfish from contaminated estuarine environments may become colonised by the pathogen and are a particular risk. *V. cholerae* O1 can persist in contaminated shellfish for many weeks without requiring continuous contamination from human faeces.

Non-O1/O139 *V. cholerae* strains are part of the natural marine environment although the existence of a natural aquatic reservoir for O1/O139 strains is uncertain.

Growth and Survival in Foods

V. cholerae can grow over the temperature range 10–43 °C, with an optimum of 37 °C. The organism can increase rapidly in temperature-abused processed foods where there is little competing microflora. It can also survive for extended periods under refrigeration and is reported to survive in moist, low-acid chilled foods for 2 or more weeks. It can also survive for long periods at freezing temperatures.

The pH range for the growth of *V. cholerae* is 5.0–9.6, with an optimum value of 7.6. It is tolerant of high-pH conditions but not acid and is rapidly inactivated at pH values of <4.5 at room temperature.

V. cholerae, unlike other *Vibrio* spp., does not have an absolute requirement for salt to grow, although its growth is enhanced in the presence of low concentrations of salt. The organism is sensitive to desiccation and survives for less then 48 h in dry foods.

V. cholerae is a facultative anaerobe (grows with or without oxygen). It grows best, however, under aerobic conditions.

The organism is not resistant to sanitisers normally used in food-processing environments.

Thermal Resistance

V. cholerae is not heat resistant and is killed by pasteurisation temperatures with $D_{60\,^{\circ}C}$ of 2.65 min and $D_{71\,^{\circ}C}$ of 0.30 min being reported. Cooking to 70 °C is normally adequate to ensure inactivation of *V. cholerae*.

Control Options

Measures to prevent food becoming contaminated with *V. cholerae* should focus on preventing faecal contamination of raw and processed foods and using safe or treated water supplies for irrigation of crops and for food processing. Raw sewage should not be used as a fertiliser for crops.

The World Health Organization (WHO) advises that there need not be an embargo on importing foods from cholera-affected areas. It is suggested that importers agree with food exporters on the good hygienic practices that need to be implemented during food handling and processing to prevent, minimise, or reduce the risk of any potential contamination.

Legislation

EU regulations and the US Food code do not have specific requirements relating to levels of *V. cholerae* in foods.

The presence of *V. cholerae* (toxigenic O1, or non O1) in ready-to-eat fishery products (minimal cooking by consumer) is an action level in the US FDA guidelines for microbiological contaminants in seafoods.

The UK Health Protection Agency (HPA) guideline on the microbiological quality of some ready-to-eat foods at the point of sale states that if *V. cholerae* is detected in 25 g these foods are considered unacceptable/potentially hazardous.

Sources of Further Information

Published

Sakazaki, R., Kaysner, C. and Abeyta, C. *Vibrio* infections in "Foodborne infections and intoxications." ed. Riemann, H.P. and Cliver, D.O. 3rd edn. London. Academic Press, 2005, 185–204.

Nair, G.B., Faruque, S.M. and Sack, D.A. Vibrios in "Emerging foodborne pathogens." ed. Motarjemi, Y. and Adams, M. Cambridge. Woodhead Publishing Ltd, 2006, 332–372.

Rabbani, G.H. and Greenough, W.B. Food as a vehicle of transmission of cholera. *Journal of Diarrhoeal Disease Research*, 1999. 17, 1–9.

On the Web

Risk assessment of choleragenic *Vibrio cholerae* O1 and O139 in warm water shrimp in international trade. Interpretative summary and technical report. The Food and Agriculture Organization of the United Nations/World Health Organization. (2005). ftp://ftp.fao.org/docrep/fao/009/a0253e/a0253e00.pdf

1.1.19 *VIBRIO PARAHAEMOLYTICUS*

Hazard Identification

What is Vibrio parahaemolyticus?

Vibrio parahaemolyticus is a gram-negative, non-spore-forming bacterium normally found in marine environments. It is the most likely *Vibrio* species to be implicated in foodborne disease, although both *V. vulnificus* and *V. cholerae* may also cause foodborne infections and are covered elsewhere in this book. Other *Vibrio* species associated with foodborne disease to a much lesser extent are *V. alginolyticus, V. mimicus, V. damsela, V. hollisae* and *V. fluvialis*.

Not all strains of *V. parahaemolyticus* cause illness and two distinct groups have been defined: pathogenic "Kanagawa-positive" strains, which cause *V. parahaemolyticus* food poisoning, and "Kanagawa-negative" strains, which do not.

Occurrence in Foods

Vibrio parahaemolyticus is found mainly in foods of marine origin, and studies carried out in the USA found that 60–100% of seafood samples were contaminated with the organism. When present, it is usually at levels of around 10 CFU/g, although levels can be around 10^3 CFU/g, or even higher in the warmer summer months. Seafood from warm waters presents a greater risk of *V. parahaemolyticus* food poisoning, with 89% of oysters causing the illness reported as originating from waters where the temperature was above 22 °C.

Cases of illness caused by *V. parahaemolyticus* have also occurred when seafoods have been cross-contaminated by raw fish after cooking and subsequently temperature abused. Implicated seafood in outbreaks include clams, oysters, scallops, shrimp and crab.

Hazard Characterisation

Effects on Health

Kanagawa-positive strains of *V. parahaemolyticus* produce a heat-stable haemolysin, which can be pre-formed in food. This haemolysin is thought to be responsible for the illness although other toxins could also be involved.

Although the minimum infective dose for *V. parahaemolyticus* is unknown, volunteer studies with healthy individuals have shown that high numbers (10^5–10^7) of Kanagawa-positive *V. parahaemolyticus* cells are required to cause illness. The infective dose may be lower when the organism is consumed at the same time as antacids or foods. All individuals are susceptible to infection by *V. parahaemolyticus*.

The incubation time for the infection is 4–96 h (average 15 h). The organism usually causes a mild to moderate form of gastroenteritis with abdominal cramps and watery diarrhoea. Nausea, vomiting, headache and fever can also occur. Some affected individuals can require hospitalisation. Symptoms can last for 1–7 days, although the average is 2.5 days and the illness is usually self-limiting. Deaths rarely occur.

Incidence and Outbreaks

The consumption of raw-seafood products (such as oysters and sashimi/sushi) from "high-risk" waters significantly increases the risk from *V. parahaemolyticus* food poisoning. The pathogen is a major cause of food poisoning in Asian countries, but in the UK illnesses caused by *V. parahaemolyticus* are usually associated with the consumption of imported seafoods, or with foreign travel.

In Japan, *V. parahaemolyticus* reportedly accounts for approximately half of all cases of bacterial foodborne infection. In the USA, *V. parahaemolyticus* illnesses prior to 1997 were infrequently reported; however, during 1997 and 1998 there were 4 multistate outbreaks associated with the consumption of raw or undercooked oysters, affecting over 700 individuals. This dramatic increase in illnesses caused by *V. parahaemolyticus* in the US has been attributed to the emergence of a new pandemic strain (O3:K6), previously this strain had only been associated with illnesses in Asia.

In Europe *V. parahaemolyticus* infections are rarely reported. However, a review of clinical data in Spain published in 2005 has concluded that they are more common than previously thought and a *V. parahaemolyticus* outbreak in Spain in 2004 caused by seafood harvested from European waters has been linked to the pandemic strain O3:K6.

Sources

Vibrio parahaemolyticus is a normal inhabitant of the marine environment and is an obligate halophile (having a minimum requirement for salt to grow). Favourable conditions for its growth are found in tropical and temperate sea-waters. For this reason the organism is usually associated with seafoods from estuarine or coastal marine environments where water temperatures are highest, such as the southern coastal US States and Japan, particularly during the summer months. However, an outbreak of *V. parahaemolyticus* in 2004 was linked to Alaskan oysters and rising seawater temperature is thought to have led to the organism proliferating in shellfish from this Northerly latitude.

Seasonal temperature variations influence the presence of the organism and although levels are highest in shellfish during the warmer months, the organism can over-winter in sediment and can be difficult to detect in water or fish samples during the winter period. However, more than 99% of environmental isolates are not pathogenic (*i.e.* they are Kanagawa-negative).

Human asymptomatic carriers of *V. parahaemolyticus* are known to occur and they can act as a source of environmental contamination.

Growth and Survival in Foods

The temperature range for growth of *V. parahaemolyticus* is 5–43 °C, with an optimum temperature of 37 °C. Under optimal conditions growth can be very rapid. The organism declines (but is not eliminated) in numbers during chilled (0–5 °C) storage.

The organism survives freezing although numbers will initially be reduced.

The pH range for growth is 4.8–11, optimum 7.8–8.6. The organism is not particularly tolerant of low pH environments and the minimum pH for growth decreases as the storage temperature increase towards optimum.

V. parahaemolyticus is unable to grow unless salt (NaCl) is present. The optimum salt concentration for growth is 3% (equating to 0.980 water activity). The organism can grow in salt concentrations from 0.5–10%, representing a water activity range of 0.940–0.996.

The organism is inactivated by desiccation and by exposure to fresh water.

V. parahaemolyticus is a facultative anaerobe (can grow in the presence or absence of oxygen) and can grow in foods that are either vacuum or aerobically packaged. It grows best, however, under aerobic conditions.

Thermal Resistance

V. parahaemolyticus is not heat resistant and is inactivated at temperatures >65 °C. *D*-values of <1 min at 65 °C, and 2.5 min at 55 °C have been reported.

Control Options

Seafood should be considered potentially contaminated with *V. parahaemolyticus*, particularly if it has been harvested from tropical and subtropical waters. However, it should be noted that seafood from what are considered "colder" seawaters may be contaminated, particularly shellfish harvested during the summer months. The risk of *V. parahaemolyticus* food poisoning is increasing with the worldwide growth in the consumption of raw fish.

Processing

Decontamination processes such as depuration or relay technologies are not effective at removing *V. parahaemolyticus* from shellfish, and effective control of the organism should focus on keeping numbers low. Measures to ensure this include, maintenance of the cold chain (<5 °C) from harvest to consumer, minimising delays between harvesting and landing, and avoiding further exposure to untreated seawater and soiled containers. Shellfish-growing areas can also be monitored for the presence of pathogenic strains of *V. parahaemolyticus*, with the closure of waters for harvesting if levels of the pathogens are deemed to be too high.

Seafood should be handled carefully to avoid cross-contamination between raw and cooked product and avoiding temperature abuse is also very important.

Product Use

Consumers should be encouraged to cook seafood thoroughly and not to eat product raw. In the US, raw oysters and restaurants offering raw oysters on their menus are required to carry health warnings about eating raw shellfish.

Legislation

EU regulations and the US Food code do not have specific requirements relating to levels of *V. parahaemolyticus* in foods. A review of published guidelines concluded that in general, levels of 10^2–10^3 CFU/g of *V. parahaemolyticus* are acceptable. The US FDA guidelines for microbiological contaminants in seafoods has an action level of $\geq 10^4$ CFU/g.

The UK Health Protection Agency (HPA) guideline on the microbiological quality of some ready-to-eat foods at the point of sale states that levels of *V. parahaemolyticus* in seafoods of < 100 CFU/g are satisfactory, 100–1000 CFU/g in these products is unsatisfactory, and levels of ≥ 1000/g are unacceptable/potentially hazardous.

Sources of Further Information

Published

Sakazaki, R., Kaysner, C. and Abeyta, C. *Vibrio* infections in "Foodborne infections and intoxications." ed. Riemann, H.P. and Cliver, D.O. 3rd edn. London. Academic Press, 2005, 185–204.

Nair, G.B., Faruque, S.M. and Sack, D.A. Vibrios in "Emerging foodborne pathogens." ed. Motarjemi, Y. and Adams, M. Cambridge. Woodhead Publishing Ltd, 2006, 332–372.

On the Web

Quantitative risk assessment of the public health impact of pathogenic *Vibrio parahaemolyticus* in raw oysters. US Food and Drug Administration's Center for Food Safety and Applied Nutrition. (July 2005). http://www.cfsan.fda.gov/~dms/vpra-toc.html

Risk profile: *Vibrio parahaemolyticus* in seafood. Institute of Environmental Science and Research Limited. (December 2003). http://www.nzfsa.govt.nz/science/risk-profiles/vibrio-parahaemolyticus.pdf

Discussion paper on risk management strategies for *Vibrio* spp. in seafood. Codex Committee on Food Hygiene 35th Session. Food and Agriculture Organization of the United Nations/World Health Organization. (January–February 2003). ftp://ftp.fao.org/codex/ccfh35/fh0305ce.pdf

Opinion of the Scientific Committee on Veterinary Measures relating to public health on *Vibrio vulnificus* and *Vibrio parahaemolyticus* (in raw and undercooked seafood). European Commission. (September 2001). http://ec.europa.eu/food/fs/sc/scv/out45_en.pdf

Draft risk assessment on the public health impact of *Vibrio parahaemolyticus* in raw molluscan shellfish. US Food and Drug Administration's Center for Food Safety and Applied Nutrition. (January 2001). http://www.cfsan.fda.gov/~dms/vprisk.html

1.1.20 *VIBRIO VULNIFICUS*

Hazard Identification

What is Vibrio vulnificus*?*

Vibrio vulnificus is a gram-negative, non-spore-forming bacterium normally found in marine environments. It is an occasional cause of serious infections, which may sometimes be foodborne. *V. vulnificus* is an obligate halophile (having a minimum requirement for salt to grow) and favourable conditions for growth are found in tropical and temperate seawater.

Occurrence in Foods

This pathogen is usually associated with seafoods from estuarine or coastal marine environments where water temperatures are highest, such as the southern coastal US States. Although *V. vulnificus* is most often associated with filter-feeding shellfish, such as oysters, which concentrate the bacteria within the tissues, potentially, the organism could contaminate any fish from the marine environment. It is mostly associated with shellfish and crustacean, but can also be found in the guts of fish feeding on plankton or other fish.

Oysters collected monthly from 14 US states contained *V. vulnificus* levels of 0 to 1 100 000 CFU/g, with water temperature and salinity having a dramatic influence on numbers present. Warm summer temperatures see concentrations of the organism at their highest in oysters. During the summer months it has been estimated that nearly 100% of oysters from the Gulf of Mexico are contaminated with *V. vulnificus*, with levels usually around 10^3–10^4 /g and most infections caused by the organism occur during the summer months when seawater temperatures are between 20–30 °C.

Hazard Characterisation

Effects on Health

V. vulnificus can cause three types of illness. Gastroenteritis (5–10% of cases), primary septicaemia (45% of cases), or wound infections (45% of cases). In healthy individuals the consumption of *V. vulnificus*-contaminated seafood can cause gastroenteritis, but in susceptible individuals (those suffering from some form of chronic disease such as liver disease, or AIDS) it causes primary septicaemia and these infections are very severe (associated with a mortality rate >50%). Around 90% of *V. vulnificus* infections require hospitalisation.

The infective dose for healthy individuals is unknown and the gastroenteritis (diarrhoea, vomiting and abdominal pain) suffered by these individuals usually occurs about 16 h after infection. This form of the disease is considered self-limiting.

The infective dose for at-risk groups could be less than 100 cells and onset of primary septicaemia can occur from 7 h–2 days after exposure. Initial symptoms include chills, fever and malaise, and septicaemia can occur 36 h after symptoms first occur. Secondary lesions may occur, especially in the extremities, which can lead to amputation.

V. vulnificus wound infections occur when an open lesion is infected by contaminated seawater. Seafood handlers are at risk if they cut themselves while cleaning and harvesting oysters and if the lesion is exposed to contaminated seawater.

Incidence and Outbreaks

The consumption of raw-seafood products by susceptible individuals, in particular oysters, from "high-risk" waters significantly increases the risk from *V. vulnificus* food poisoning. Although there are not many reported cases annually (around 90 cases are reported in the USA each year – not all associated with the consumption of contaminated seafood), the high mortality rate associated with *V. vulnificus* infections has made this organism an important public-health issue, particularly in the USA.

No major foodborne outbreaks have been caused by this pathogen and cases tend to be sporadic, the frequency increasing during the summer months. *V. vulnificus* infections are rarely reported during the winter months even though most oysters are eaten during this period. Cases have also occasionally occurred in Europe. Infections due to *V. vulnificus* have also been reported in Korea, Taiwan and other countries.

Sources

V. vulnificus is naturally present in coastal seawater in tropical and temperate regions throughout the world. Numbers of the organism relate to water temperature with higher numbers found during summer months. *V. vulnificus* is thought to enter a viable but non-culturable state (VNC) in cold winter waters and although still present can be difficult to detect. The low numbers of reported illnesses suggests that either many *V. vulnificus* strains are not pathogenic to humans, or that the infective dose is high for healthy individuals.

Growth and Survival in Foods

V. vulnificus can grow over the temperature range 8–43 °C, with an optimum temperature of 37 °C. In live oysters the organism does not grow below 13 °C, indicating the importance of chilling shellfish as soon as possible after harvesting. *V. vulnificus* survives in oysters at chill temperatures (0–4 °C) but can be difficult to culture from chilled environments. This can make the detection and enumeration of the organism from chilled foods unreliable.

Although freezing initially reduces levels of the pathogen in oyster tissue, the surviving *V. vulnificus* population remains stable throughout frozen storage.

The pH range for growth of the pathogen is 5–10, and the optimum is 7.8. The organism is inactivated at pH values <5.0.

V. vulnificus is a halophile and is able to grow at salt levels between 0.5–5%, although the optimum concentration for growth is 2.5%. This equates to a water activity range of 0.96–0.997. The pathogen is sensitive to dehydration.

V. vulnificus is a facultative anaerobe (able to grow in the presence or absence of oxygen). Vacuum packing combined with frozen storage was found to reduce levels of *V. vulnificus* in oysters more effectively than frozen storage alone but cannot be relied upon to completely eliminate the pathogen.

Thermal Resistance

V. vulnificus is not a heat-resistant organism and is easily destroyed during cooking processes. A low-temperature pasteurisation of 10 min at 50 °C for shellstock oysters has been found to ensure inactivation.

Control Options

Processing

Decontamination processes such as depuration or relay technologies are not effective at removing *V. vulnificus* from shellfish, so strategies should focus on keeping levels low and encouraging consumers not to eat raw shellfish. Shellfish should be harvested from approved waters. In California there are restrictions on the sale of oysters from the Gulf of Mexico from April to October unless the oysters are treated with a scientifically validated method to eliminate *V. vulnificus*.

Levels of the pathogen increase in temperature-abused shellfish and the time taken from harvesting to refrigeration is known to be critical. In the USA the time permitted from harvest to refrigeration can depend on whether an area has been associated with *V. vulnificus* infections, as well as the temperature of the seawater, the season and the air temperature. Oysters harvested during the warmer months can be diverted for cooking, pasteurisation or irradiation to avoid the possibility of them being consumed raw.

Product Use

Consumers should be advised of the risks of consuming raw or undercooked shellfish, particularly those with medical conditions that make them more at risk on contracting *V. vulnificus* infections.

Legislation

EU regulations and the US Food Code do not have specific requirements relating to levels of *V. vulnificus* in foods.

The presence of *V. vulnificus* showing mouse lethality in product is an action level in the US FDA guidelines for microbiological contaminants in seafoods.

Sources of Further Information

Published

Sakazaki, R., Kaysner, C. and Abeyta, C. *Vibrio* infections in "Foodborne infections and intoxications." ed. Riemann, H.P. and Cliver, D.O. 3rd edn. London. Academic Press, 2005, 185–204.

Nair, G.B., Faruque, S.M. and Sack, D.A. Vibrios in "Emerging foodborne pathogens." ed. Motarjemi, Y. and Adams, M. Cambridge. Woodhead Publishing Ltd, 2006, 332–372.

On the Web

Risk assessment of *Vibrio vulnificus* in raw oysters. Interpretative summary and technical report. World Health Organization/Food and Agriculture Organization of the United Nations. (2005). http://www.who.int/foodsafety/publications/micro/mra8.pdf

Opinion of the Scientific Committee on Veterinary Measures relating to public health on *Vibrio vulnificus* and *Vibrio parahaemolyticus* (in raw and undercooked seafood). European Commission. (September 2001). http://ec.europa.eu/food/fs/sc/scv/out45_en.pdf

1.1.21 *YERSINIA ENTEROCOLITICA*

Hazard Identification

What is Yersinia enterocolitica*?*

Yersinia species are gram-negative, non-spore-forming, facultatively anaerobic bacteria belonging to the group *Enterobacteriaceae*. Two species of *Yersinia* have been associated with foodborne disease in man, *Yersinia enterocolitica* and *Yersinia pseudotuberculosis* (see Section 1.1.2.2).

Not all strains of *Y. enterocolitica* are pathogenic (*i.e.* cause disease). In fact, only a proportion of isolates can cause disease and these potentially pathogenic isolates carry a piece of genetic material known as a "virulence" plasmid. There are a large number of different serotypes, but the most common cause of disease worldwide is serotype O:3. In Europe and the USA 90% of cases of yersiniosis are caused by this serotype. Other important pathogenic serotypes are O:9, O:8 and O:5,27, although at least another 8 serotypes are recognised as potential causes of yersiniosis.

Occurrence in Foods

Yersinia enterocolitica is most often associated with pork products and milk, because foodborne outbreaks are often linked to these foods. However, the organism has been isolated from other foods such as fruits, vegetables, dairy products, various meats and poultry, oysters, fish, salads, sandwiches, pastries and tofu, although isolates from these sources frequently include non-pathogenic types.

Hazard Characterisation

Effects on Health

The infective dose for *Y. enterocolitica* infection is unknown, but the severity of the symptoms is thought to be related to the number of organisms ingested. Those most at risk of developing the disease and its associated long-term effects are infants, the elderly and the immunocompromised.

The incubation time for *Y. enterocolitica* infections is from 1–11 days (usually 1–2 days). The disease is usually self-limiting and of short duration, and symptoms typically cease after 2–3 days. Occasionally, symptoms can last for 1–3 weeks, or even a few months.

Symptoms vary and in adults can include abdominal pain, fever, vomiting, nausea and diarrhoea. The infection is often confused with appendicitis and unnecessary appendectomies can be carried out as a result of the abdominal pain. *Y. enterocolitica* infections in children usually cause gastroenteritis and inflammation of the lymph glands.

Longer-term effects include reactive arthritis and skin disorders, such as painful red skin lesions. In rare cases, bacteraemia can occur (when the organism enters the blood stream), which may occasionally be fatal. But this tends to affect individuals who have other underlying disease.

Incidence and Outbreaks

Yersiniosis is a relatively common foodborne infection in Northern Europe, Japan and Scandinavia and it is the third most common cause of gastroenteritis in Finland and Norway. Infection is often acquired through the consumption of raw or undercooked pork products, or from contaminated milk and fresh produce.

In the US and Canada, where foodborne outbreaks of yersiniosis are relatively unusual, cases have mainly been linked to the consumption of raw, or recontaminated pasteurised milk. In 1976 an outbreak involving 217 individuals in the US was linked to the consumption of a chocolate milk drink. Chitterlings, a speciality prepared from raw pig intestines, have been associated with outbreaks amongst the African-American community in the US.

Sources

Yersinia enterocolitica is ubiquitous; it can be found in a wide range of animals and in the environment. However, many strains found in soil and water are non-pathogenic. The organism has been isolated from water supplies (drinking and surface) and infections have been caused by contaminated water.

The most common reservoir for the organism amongst food producing animals is the throat and tonsil area of pigs. However, the organism can be carried at a lower rate by sheep, poultry and cattle. Data from the USA suggests that *Y. enterocolitica* in cattle faeces is a potential source of contamination for raw milk.

Low numbers of *Y. enterocolitica*, many of which are non-pathogenic, can be part of the transient intestinal flora of healthy humans. Food handlers have been implicated in cases of foodborne disease, and person-to-person transmission, via the faecal–oral route, has been reported as the cause of yersiniosis infections.

Growth and Survival Characteristics

Yersinia enterocolitica is psychotrophic and is able to grow at chilled temperatures. The organism can grow over the temperature range 0–44 °C, although there have been reports of extremely slow growth at –1.3 °C. The optimum temperature for growth is 28–29 °C. *Y. enterocolitica* survives freezing and there have been reports that it can survive in frozen foods for some time.

The pH range for growth is 4.2–10, although minimum pH values depend on the type of acid present and the storage temperature – the minimum of 4.2 is more likely to occur with inorganic acids. With organic acids, such as acetic or citric acids, the minimum pH for growth is around 5.0. *Y. enterocolitica* is inactivated at lower pH values, but can survive in acid conditions for some days at refrigerated temperatures.

The minimum water activity for growth is 0.945. Levels of salt between 5–7% inhibit growth.

The organism is a facultative anaerobe, and is able to grow with or without oxygen. Vacuum packaging and some modified atmospheres (100% N_2 or CO_2/N_2 mixtures) can slow down or inhibit growth, particularly at chilled temperatures.

Thermal Resistance

Yersinia enterocolitica is sensitive to heat and is easily inactivated at temperatures above 60 °C. *D*-values of around 0.5 min and 2 s at 60 °C, and 65 °C, respectively, have been recorded. Typical pasteurisation treatments should easily ensure that the organism is destroyed.

Control Options

Processing

The level of *Y. enterocolitica* in raw pork can be reduced by using measures to limit the level of faecal contamination on pig carcasses after slaughter. Careful removal of the tongue from the head of pigs soon after slaughter can also help to minimise carcass contamination. Raw pork should always be regarded as a potential source of *Y. enterocolitica* and should be handled as such.

Control of the pathogen on fresh produce should focus on avoiding contamination. Measures include implementing good practices in growing and harvesting that are designed to minimise the risk of faecal contamination. The use of irrigation water from clean, uncontaminated sources is also important.

Cooking and milk-pasteurisation processes are adequate means of destroying the pathogen, and care should be taken to ensure that recontamination of heat-processed foods does not occur after the cooking process. For example, a multistate outbreak in the USA was blamed on the use of dirty, contaminated crates to transport pasteurised milk. The presence of *Y. enterocolitica* in any heat-processed food indicates inadequate heating or post-process contamination, and is unacceptable. The organism may increase during chilled storage and therefore refrigeration is not an effective means of control.

Product Use

The risk of contracting yersiniosis increases with the consumption of raw pork, or pork cooked rare. Consumers should be advised on measures to ensure that pork products are cooked thoroughly and that cross-contamination from raw pork to ready-to-eat products should be avoided.

Consumers should also be advised of the potential health risks from drinking raw milk, and water from untreated sources, particularly in areas where pigs are kept.

Legislation

There are no specific requirements for levels of *Y. enterocolitica* in foods under EC legislation or in the US Food Code.

Sources of Further Information

Published

Nesbakken, T. *Yersinia enterocolitica*, In Emerging Fodborne Pathogens. ed. Motarjemi, Y. and Adams, M. Cambridge. Woodhead Publishing Ltd, 2006, 373–405.

Robins-Browne, R.M. *Yersinia enterocolitica*, In Food Microbiology: fundamentals and frontiers. ed. Doyle, M.P., Beuchat, L.R. and Monteville, T.J. 2nd edn. Washington D.C. ASM Press, 2001, 215–245.

On the Web

Risk Profie: *Yersinia enterocolitica* in pork. Institute of Environmental Science and Research Limited. (March 2004). http://www.nzfsa.govt.nz/science/risk-profiles/yersinia-in-pork.pdf

Bottone, E.J. *Yersinia enterocolitica*: The charisma continues. *Clinical Microbiology Reviews,* 1997, 10, 257–276. http://www.pubmedcentral.nih.gov/picrender.fcgi?artid=172919&blobtype=pdf

1.1.22 *YERSINIA PSEUDOTUBERCULOSIS*

Hazard Identification

What is Yersinia pseudotuberculosis*?*

Yersinia pseudotuberculosis is a gram-negative, non-spore-forming, bacterium belonging to the family *Enterobacteriaceae*. Although another species, *Y. enterocolitica*, is the primary cause of the disease known as yersiniosis in humans, *Y. pseudotuberculosis* has also been associated with causing the condition. There is increasing evidence that disease caused by *Y. pseudotuberculosis* can be foodborne, and in the past decade foodborne outbreaks have been reported in the literature.

Since the beginning of the twentieth century the classification of this species has changed repeatedly and it has been known by a number of names. Initially, it was called *Pasteurella pseudotuberculosis*, and then *Shigella pseudotuberculosis* until the current name *Yersinia pseudotuberculosis* was established in the 1960s. Very old references may still refer to the organism using either of these previous names. Not all strains of *Y. pseudotuberculosis* are pathogenic, but the potential pathogenicity of isolates can only be determined by laboratory testing.

Occurrence in Foods

There is little data on the occurrence of *Y. pseudotuberculosis* in food. A study in Italy, which examined 10 842 food samples for the pathogen failed to recover it from a food source. However, the organism is reported to be difficult to isolate from food and from the environment. Cases of yersiniosis caused by *Y. pseudotuberculosis* have been associated with the ingestion of contaminated drinking water, vegetable juice, pasteurised milk, salad leaves and raw vegetables.

Hazard Characterisation

Effects on Health

The infective dose for *Y. pseudotuberculosis* infections is unknown, but is probably between 10^6–10^9 viable cells. The incubation time for the pathogen to cause illness is uncertain, but the literature suggests that it varies from 3–10 days. The illness manifests itself as fever, a rash, and severe abdominal pain and it is often confused with acute appendicitis. Diarrhoea is uncommon but can also occur. Long-term complications can include reactive arthritis, and in immunocompromised patients with liver disease it can occasionally cause sepsis.

Infections are normally self-limiting, although in patients developing sepsis because of acute liver disease, the mortality rate can be high ($>75\%$). *Y. pseudotuberculosis* infections occur most frequently in children between 5–15 years of age ($>75\%$ of cases). Individuals recovering from *Y. pseudotuberculosis* infections can excrete the pathogen for a number of weeks after the illness.

Incidence and Outbreaks

There is very little published information on the incidence of foodborne *Yersinia pseudotuberculosis* infection, but the organism is mostly a health concern in countries with a temperate climate, such as Japan, Northern Europe and the former Soviet Union, and cases seem to occur more frequently during the winter months.

Outbreaks associated with foods have occurred in Canada, Finland, Japan and the former Soviet Union. In 1998, an outbreak in Canada was associated with the consumption of contaminated homogenised milk. Again in 1998, an outbreak in Finland was linked to the consumption of Iceberg lettuce, and in the same country outbreaks of *Y. pseudotuberculosis* infections in 2003 and 2004 were traced to raw carrots.

Sources

Yersinia pseudotuberculosis is found in the faeces of a wide number of wild and domestic animals in Eurasia and North America, and it is thought that wild mammals and birds are the main reservoir for infection-causing *Y. pseudotuberculosis*. The organism can cause disease in a number of animal species, but is also carried by apparently healthy animals. *Y. pseudotuberculosis* infection is a zoonosis, but not all strains of the organism are pathogenic.

Animals, such as rodents, deer, hares and birds (*e.g.* ducks and geese), can excrete the pathogen leading to the contamination of soil and water sources. However, the organism is isolated from environmental sources far less frequently than *Y. enterocolitica*. In an outbreak of *Y. pseudotuberculosis* infections associated with the consumption of raw carrots it is thought that the vegetables were contaminated via the faeces of rodents, and possibly other wild animals that had access to the barn where the produce was stored in open containers during the winter.

Growth and Survival Characteristics

The physicochemical parameters affecting the growth and survival of *Y. pseudotuberculosis* are probably similar to those relating to *Y. enterocolitica*. The organism is psychrotrophic, and growth may not be prevented by storage of product at chill temperatures. It is thought that cold temperatures during the winter in temperate climates provide an advantage to the organism when present in water and on fresh produce, and may explain why more cases of *Y. pseudotuberculosis* cases occur during these months.

Y. pseudotuberculosis is a facultative anaerobe: it is able to grow with or without oxygen.

Thermal Resistance

Yersinia pseudotuberculosis is not a heat-resistant micro-organism and normal pasteurisation processes used in the food industry should inactivate the cells.

In buffer at pH 7.0, *D*-values of around 23 min and 2.6 min, at 53.9 °C and 57.8 °C, respectively, have been recorded, with a *z*-value of 3.75 °C. These *D*-values are reduced significantly when the organism is heated in fruit (apple or orange) juices.

Control Options

Fresh produce can become contaminated with pathogens at any time during growing, harvesting, packing, shipping and processing. However, the refrigeration temperatures often used during transportation actually favours the survival and growth of *Y. pseudotuberculosis*. Therefore, strategies to reduce the risk of foodborne *Y. pseudotuberculosis* infections need to focus on ensuring that contamination is prevented in the first place, and need to be implemented at all stages of production, including at the farm. These should include preventing the access of wild animals to growing areas and water supplies by the addition of fences, as well as preventing animals accessing fresh-produce storage facilities. Treated water should be used to wash and process fresh produce.

Processing

Equipment used to process produce can spread contamination and processing equipment should be cleaned regularly and thoroughly. It has been recommended that any inadequate cleaning regimes should be identified and corrected by routine inspections of production facilities.

Product Use

To reduce the risk of foodborne disease, including *Y. pseudotuberculosis* infections, consumers should be advised to thoroughly wash fresh produce prior to consumption.

Legislation

There are no specific requirements for levels of *Y. pseudotuberculosis* in foods under EC legislation or in the US Food Code.

Sources of Further Information

Published

Tauxe, R.V. Salad and pseudoappendicitis: *Yersinia pseudotuberculosis* as a foodborne pathogen. *The Journal of Infectious Diseases*, 2004, 189, 761–763.

1.1.23 OTHER ENTEROBACTERIACEAE

Hazard Identification

What are Enterobacteriaceae*?*

The *Enterobacteriaceae* are a family of gram-negative, facultatively anaerobic (able to grow in the presence, or absence of oxygen) non-spore-forming bacteria that includes a number of genera and species (*Salmonella, Escherichia coli, Enterobacter sakazakii, Shigella* spp. and *Yersinia* spp.) that are well-known causes of foodborne disease and are covered in detail elsewhere in this book. However, there are a number of other, less well known species that have also been implicated in foodborne disease. Although more often associated with food spoilage, it is thought that some strains of *Citrobacter* spp. (notably *Citrobacter freundii*), *Klebsiella* spp., *Providencia* spp. *Enterobacter* spp. and *Proteus* spp., may occasionally cause what is often described as opportunistic gastroenteritis.

Occurrence in Foods

Enterobacteriaceae are found as contaminants in a wide variety of raw and processed foods. They are often involved in spoilage of dairy products, meat, poultry, fresh fruits and vegetables, usually as a consequence of temperature abuse. However, the prevalence of potentially pathogenic strains of *Citrobacter freundii, Klebsiella* spp., *Providencia* spp., *Enterobacter* spp. and *Proteus* spp. in foods is unknown.

High numbers of these bacteria in foods may be a cause for concern. For example, fresh sprout products (such as alfalfa) have been recalled in Canada because they have been found to be contaminated with *Klebsiella pneumoniae*.

Hazard Characterisation

Effects on Health

These organisms are considered to be opportunistic pathogens, and healthy adults are not considered to be at high risk of developing infections and illness. Young children, the elderly and the immunocompromised are most at risk in the developed world. People in developing countries with poor sanitation and inadequate nutrition are at higher risk.

The infectious dose of these potentially pathogenic strains is unknown. Typically, onset of illness occurs 12–24 h after the ingestion of the contaminated foodstuff. Symptoms include flu-like symptoms, fever, nausea, stomach cramps, vomiting and watery diarrhoea. The illness can, on occasions be chronic and last for some months. In infants and undernourished children the disease caused by these organisms can result in death.

Incidence and Outbreaks

The incidence of foodborne infection by these bacteria is uncertain, but outbreaks of disease have been reported. Outbreaks associated with *Citrobacter freundii* in the United States have been associated with the consumption of imported semi-soft cheeses (Brie or Camembert). In Germany an outbreak associated with *Citrobacter freundii* caused gastroenteritis amongst nursery children, followed by haemolytic uraemic syndrome with acute renal failure. It was linked to the consumption of green butter sandwiches (butter containing parsley leaves from an organic garden). Contaminated infant formula has also been implicated as the vehicle of infection in an outbreak of *Citrobacter freundii* infection.

Klebsiella pneumoniae infections have been linked to the consumption of a contaminated hamburger and cooked turkey.

In 1996 a large outbreak of gastroenteritis caused by *Providencia alcalifaciens* at three schools in Japan was linked to a lunch cooked at a single catering facility. At least 610 individuals were involved.

Sources

These organisms are found in the environment, in the soil and in freshwater. They have been isolated from fresh vegetables and herbs, such as parsley and alfalfa sprouts. They occur in shellfish-harvesting waters and have been found in raw shellfish. They have also been found in raw milk and dairy products.

These bacteria can be isolated from the stools of healthy individuals and are part of the normal intestinal flora of animals.

Growth and Survival Characteristics

Opportunistically pathogenic *Enterobacteriaceae* are not particularly heat resistant. Normal pasteurisation and cooking processes used by the food industry will inactivate these bacteria.

The organisms survive relatively well in the environment for non-spore-forming bacteria. Some species, including *Citrobacter freundii, Citrobacter koseri* and *Klebsiella pneumoniae* have been isolated from dried infant formula, indicating that they can survive desiccation for some time.

Control Options

Effective control of these bacteria focuses on good hygiene practice and temperature control.

Processing

Fresh produce should be sourced from suppliers implementing good agricultural practices.

The rapid chilling of cooked foods after cooking is extremely important to prevent an increase in numbers of potentially pathogenic micro-organisms.

The implementation of good hygienic practices by food handlers is extremely important to prevent the contamination of foods that will not be further heated prior to consumption.

Product Use

Consumers should be advised to wash fruit and vegetables well prior to consumption. They should also be reminded of the importance of good hygienic practices when preparing and storing foods to reduce the risks associated with foodborne disease.

Legislation

Although there are no specific requirements for each individual micro-organism covered in this section, there may be requirements/standards/guidelines for levels of *Enterobacteriaceae* or coliforms (a group containing some, but not all, genera from the *Enterobacteriaceae*) in some foods and drink as an indication of hygienic status.

Sources of Further Information

Published

Stiles, M.E. Less recognised and suspected foodborne bacterial pathogens, in "The microbiological safety and quality of food, volume 2." ed. Lund, B.M., Baird-Parker, T.C. and Gould, G.W. Gaithersburg. Aspen Publishers, 2000, 1394–1419.

CHAPTER 1.2
Viruses

1.2.1 ASTROVIRUSES

Hazard Identification

What are Astroviruses?

Astroviruses are spherical, positive-sense, single-stranded RNA viruses belonging to the family *Astroviridae*. These viruses are host specific and a number of different astroviruses have been described (*e.g.* bovine astrovirus, feline astrovirus, human astrovirus), many of which cause gastroenteritis in the host. Human astroviruses are classified in the genus *Mamastrovirus* and at least eight human serotypes (human astrovirus 1 through to human astrovirus 8) have been recognised.

Astrovirus infections are mainly spread by person-to-person transmission via the faecal–oral route, however, a very small percentage of infections are estimated to be foodborne (<1%).

Occurrence in Foods

Evidence of astroviruses in naturally grown oysters has been reported in Japan, particularly in product sampled during the winter season.

A food handler infected with astrovirus could potentially contaminate almost any foodstuff. This could present a risk of infection if it is consumed without a further heating step.

Hazard Characterisation

Effects on Health

Astrovirus infections are mostly associated with young children (between 6 months–2 years old), but they can also cause a mild infection in adults.

The Food Safety Hazard Guidebook
By Richard Lawley, Laurie Curtis & Judy Davis

The infective dose is thought to be <100 virus particles and symptoms occur 3–4 days after infection. Astrovirus infections are associated with watery diarrhoea, nausea, fever, abdominal pain and vomiting. The diarrhoea usually lasts for 2–3 days and is self-limiting, but it can sometimes last as long as 14 days. During infection the virus is excreted in high numbers in the faeces of the affected individual.

Incidence and Outbreaks

Although occurring all year round, outbreaks of astrovirus infections peak in temperate climates during the winter and spring. Outbreaks occur mostly in child-care situations, paediatric wards and amongst the institutionalised elderly. In many instances, astroviruses are second only to rotaviruses as a cause of childhood diarrhoea. Based on this fact, immunity to astrovirus infections is thought to be acquired during childhood, be maintained during adult life, and to diminish in old age.

Although astrovirus infections usually occur via person-to-person transmission through the faecal–oral route, foodborne infections and outbreaks associated with these viruses are described in the literature. Infections associated with shellfish and water have occasionally been reported. Probably the largest outbreak reported, involving thousands of children and adults from 14 different schools in Japan in June 1991, was caused by school lunches from a common supplier.

Sources

Humans are the reservoir for human astroviruses and infected individuals can excrete very high numbers of viruses. Infections are usually spread via the faecal–oral route. Faecally contaminated-water sources (both drinking and recreational), shellfish from contaminated water and foods contaminated by infected food handlers can also be sources of human astroviruses.

Survival Characteristics

Viruses, including astroviruses, are unable to multiply outside of the host. Although they cannot grow in food or water, astroviruses can survive for some time in the environment, particularly when protected by organic matter at low temperatures. Astroviruses can survive in unchlorinated water and when dried onto porous and non-porous materials, again particularly at low temperatures. Astroviruses are acid stable, and are resistant to freezing at –20 °C.

Thermal Inactivation

Astroviruses can survive heat treatments of 50 °C for 1 h. A heat process at 60 °C for 15 min is reported to give a 6 $\log_{10}$ reduction in Astrovirus titre.

Control Options

The control of astroviruses should focus on the implementation of strict personal hygiene by food handlers. Ready-to-eat foods that are handled but will receive no further cooking pose the greatest risk.

Processing

Food handlers should be trained in effective hand-washing techniques and should wash hands after visiting the toilet as well as before handling foods. Those suffering from viral gastroenteritis should be excluded from work for at least 48–72 h after symptoms have ceased.

Product Use

Consumers should be educated on the importance of adhering to good personal hygiene during food preparation and should be advised to consume only adequately cooked shellfish, especially oysters, harvested from approved waters.

Legislation

There is no specific legislation in the EC or in the USA regarding levels of enteric viruses, such as astroviruses, in foods.

Sources of Further Information

Published

Greening, G.E. Human and animal viruses in food (including taxonomy of enteric viruses), in Viruses in Foods. ed. Goyal, S. New York. Springer, 2006, 2–42.

1.2.2 HEPATITIS A VIRUS

Hazard Identification

What is the Hepatitis A Virus?

The hepatitis A virus (HAV) is an enteric virus, which causes a liver disease in humans now known as hepatitis A (previously known by other names including infectious jaundice, viral hepatitis and infectious hepatitis). There are a number of different hepatitis viruses but only the HAV, and possibly the hepatitis E virus, can cause foodborne disease. HAV is a single-stranded RNA virus belonging to the *Picornaviridae* family and the genus *Hepatovirus*.

Although HAV is most commonly spread by direct person-to-person contact via the faecal–oral route, there are many documented foodborne outbreaks in the literature. Foodborne outbreaks can often be traced back to an infected food handler or foods that have come into contact with faecally contaminated water.

Occurrence in Foods

The HAV can only be present in foodstuffs as the result of faecal contamination. Although this means that any food that is handled under poor hygienic practices could potentially be contaminated with the pathogen, it is bivalve molluscan shellfish, such as oysters, cockles and mussels, which are most often considered to be a source of foodborne viruses. These shellfish concentrate any virus particles present in their tissues during filter feeding in faecally contaminated water. Depuration techniques used to decontaminate shellfish are more successful in reducing bacterial loading than affecting viral contamination.

In recent years fresh produce, such as salads, fresh fruits and vegetables, has increasingly been implicated in foodborne outbreaks of hepatitis A. These products are likely to be consumed raw or lightly cooked, and can become contaminated with faecal matter at almost any point during growing, harvesting, transport and packing.

Hazard Characterisation

Effects on Health

The infective dose for the HAV is unknown. However, it is thought that as few as 10–100 virus particles could cause disease. The incubation time for symptoms to appear is on average about 4 weeks, but it can vary from 2–6 weeks. This long incubation time before the illness becomes evident can mean that it can be difficult to trace the exact source of the infection, and it can also mean that large numbers of individuals are affected before it is evident that there is viral contamination in the food chain.

Many cases of HAV infection are asymptomatic, particularly in children. When disease is evident, hepatitis A infection is usually a mild illness. Initial symptoms include headache, fatigue, fever, poor appetite, abdominal discomfort, nausea and vomiting. After a week or so, viraemia (where the virus can be detected in the blood stream) and liver disease in the form of jaundice, or liver enzyme elevation, occurs. Hepatitis A is usually a self-limiting disease lasting for up to 2 months, but in a small group of affected people, the HAV can cause long or recurring illness lasting up to 6 months. Infection can be fatal, particularly in people over 50 years old. In the USA, this age group has a mortality rate reported as 1.8%.

During infection individuals can excrete high numbers of virus particles ($>10^6$ particles/g of faeces). The shedding of particles can start in the last 2 weeks of the incubation period and in some individuals can continue for up to 5 months after infection.

Incidence and Outbreaks

In many developing countries the disease is endemic and exposure during early childhood because of poor hygiene is common. Early childhood infections are usually asymptomatic and infer lifelong immunity.

Outbreaks of hepatitis A are more likely to occur in developed nations, or amongst travellers from developed countries to the developing world, because exposure to the virus during early childhood in individuals from developed regions is low. Countries where the adult population has no immunity are at risk of large hepatitis A outbreaks when food or water supplies are contaminated with the virus.

Contaminated water and bivalve shellfish such as oysters, cockles and mussels, are often associated with hepatitis A infections. The largest recorded foodborne outbreak of hepatitis A infections, involving 290 000 cases, was in Shanghai, China in 1988 and was caused by clams harvested from waters polluted by raw sewage.

Fresh fruits, such as strawberries and raspberries harvested by infected pickers, and associated products such as orange juice, have caused outbreaks in the UK and the USA, respectively. Imported lettuce and, more recently, imported raw/undercooked green onions have also caused large outbreaks in the USA.

Other foods linked to outbreaks include bakery products, sandwiches, iced beverages, milk and milk products, beer and soft drinks.

Sources

The human intestine is the main reservoir for the HAV and asymptomatic infected individuals, especially children, are an important source of the virus.

Transmission can occur by the faecal–oral route by direct person-to-person contact, or from the ingestion of faecally contaminated food or water. It has been reported that transmission of the virus can occur as the result of using contaminated drinking glasses. Infected food handlers with poor hygiene are a potential source of the virus in food. The virus could potentially be present in any water source or soil that is faecally contaminated.

Survival Characteristics

Viruses, including the HAV, are unable to multiply outside of the host. Although the HAV cannot grow in food or water, it can survive in many environments for some time. When excreted in human faeces the HAV can survive in the environment in water or soil for at least 12 weeks at 25 °C. The HAV has a high resistance to many chemicals and solvents and it is more resistant to heat and drying than other enteroviruses. It can survive refrigeration and freezing for up to two years and it is resistant to acid (pH 1 for 2 h at room temperature).

The HAV is resistant to low levels of free chlorine (0.5–1 mg free chlorine/l for 30 min). It is also resistant to perchloroacetic acid (300 mg/l) and chloramines (1 g/l) for 15 min at 20 °C. The virus can be inactivated on surfaces with a 1:100 solution of sodium hypochlorite, or household bleach in tap water.

Thermal Inactivation

The HAV is relatively heat resistant, although thorough cooking at higher temperatures will usually inactivate the virus. It is resistant at 70 °C for up to 10 min but is inactivated at temperatures of 85 °C for 1 min. In the UK it has been recommended that cockles are heated to an internal temperature of 85–90 °C for 1.5 min to inactivate HAV and data from the World Health Organization suggests that shellfish from HAV-contaminated areas should be heated to 90 °C for 4 min or steamed for 90 s.

Control Options

Strategies to reduce the risk of foodborne outbreaks of hepatitis A should focus on preventing foods from becoming contaminated. In developing countries young children should be kept away from areas where fresh produce is grown and harvested, and clean water should be used for the irrigation, washing and processing of foods. Shellfish-harvesting areas should be monitored for sewage contamination.

Processing

Food handlers should implement frequent hand washing and the wearing of gloves particularly at points in the food chain where foodstuffs that will receive no further cooking are handled. In addition, those suffering from symptoms of hepatitis A should be removed from the food-production area until they have a medical release. In some parts of the USA food handlers are immunised against hepatitis A, but the effectiveness of such a policy is uncertain.

Product Use

If food could be contaminated with the HAV, consumers should be advised only to eat thoroughly cooked foods from known sources and not to eat uncooked fruits or vegetables that they have not peeled or prepared themselves.

Legislation

There is no specific legislation in the EC or in the USA regarding levels of enteric viruses such as the HAV in foods.

Sources of Further Information

Published

Cook, N. and Rzezutka, A. Hepatitis viruses in "Emerging foodborne pathogens." ed. Motarjemi, Y. and Adams, M. Cambridge. Woodhead Publishing Ltd, 2006, 282–308.

Fiore, A.E. Hepatitis A transmitted by food. *Clinical Infectious Diseases*, 2004. 38, 705–715.

Koopmans, M. and Duizer, E. Foodborne viruses: an emerging problem. *International Journal of Food Microbiology*, 2004. 90, 23–41.

Koopmans, M., von Bonsdorff, C-H., Vinjé, J., de Medici, D. and Monroe, S. Foodborne viruses. *FEMS Microbiology Reviews*, 2002. 26, 187–205.

On the Web

Hepatitis A. World Health Organization (2000). http://www.who.int/csr/disease/hepatitis/HepatitisA_whocdscsredc2000_7.pdf

1.2.3 HEPATITIS E VIRUS

Hazard Identification

What is the Hepatitis E Virus?

The hepatitis E virus (HEV) is an enteric virus, which causes a liver disease in humans now known as hepatitis E (other names for the disease include, enterically transmitted non-A non-B hepatitis and faecal-oral non-A non-B hepatitis). The HEV is distinctly different from the hepatitis A virus and is a single-stranded RNA virus, which has recently been classified in the family *Hepeviridae* and the genus *Hepevirus*.

Studies have found that there are distinct similarities between HEV strains affecting humans and HEV strains found in pigs in developed countries. This has led to the conclusion that HEV is a zoonosis and, potentially, a foodborne pathogen.

Occurrence in Foods

The virus is most often associated with pigs, and surveys to determine the frequency of the HEV in pig populations and in pork livers have been conducted in a number of countries. Pigs carrying the HEV do not show any signs of disease and the virus is now known to be present in most pig populations throughout the world. It is reported to have been present in pigs in the UK since at least 1986 and it is estimated that it is present in 75% of pigs in the country.

Studies in Japan and the Netherlands to determine the incidence of HEV in raw retail pig livers found detectable levels of the virus in three of 197 (1.9%), and four of 62 (6.5%) of samples, respectively. However, in the UK a recent study of samples of retail pig livers from 80 outlets in Cornwall found none positive for the virus.

Hazard Characterisation

Effects on Health

Although all individuals are susceptible to contracting hepatitis E, the disease is most frequently seen in the 15–40-year-old age group. The infective dose for the HEV is unknown and the incubation time for the disease can vary from 2 to 9 weeks. Many HEV infections are asymptomatic, cause no sign of disease, and where hepatitis E does occur it is usually a mild illness lasting 3–4 weeks.

The symptoms for this mild form of the disease include general fatigue, jaundice, production of darker urine and pale stools, abdominal pain, vomiting and nausea. However, the virus can occasionally cause a severe disease with complete liver failure and even death, especially amongst individuals who are pregnant or immunocompromised, suffering from chronic liver disease, or from older age groups. In pregnant women the disease may also cause a miscarriage.

In the general population the mortality rate associated with hepatitis E is 0.5–2.0%, but amongst groups susceptible to the more severe form of the disease the fatality rate can be as high as 30%.

Incidence and Outbreaks

In developing countries with poor sanitation hepatitis E is common. In these regions most cases of the infection are sporadic, although large outbreaks associated with contaminated water are not infrequent.

In industrialised countries, cases of hepatitis E have traditionally been associated with foreign travel to the developing world and large outbreaks of the disease have not occurred. However, there is an increasing body of evidence to suggest that a significant number of hepatitis E infections in developed countries are acquired "at home" (in the UK up to 50% of cases may be domestically acquired).

In recent years there have been reports in the literature of cases of foodborne transmission of hepatitis E. These have been associated with the consumption of raw or undercooked meat (pork liver, deer and wild boar) and unpasteurised milk. There is also some evidence suggesting that the infection may also be acquired from the consumption of raw, or poorly cooked, shellfish.

Sources

In developing regions the main source of the virus is drinking water contaminated with human faecal material.

In developed countries the main source of the virus is from direct or indirect contact with animals. In these regions the main reservoir for the HEV is pigs and pig faeces. Pork and associated products may also be contaminated. Transmission of the virus between pigs is thought to be via the faecal–oral route.

Other animals have also been reported to have antibodies to the HEV, and these include deer, wild boar, cattle, goats, chickens and sheep, domestic animals such as dogs and cats, and rodents such as rats and mice.

Survival Characteristics

Viruses, including the HEV, are unable to multiply outside of the host. Although the HEV cannot grow in food or water, it can survive and still remain infective. There is, however, very limited data on factors affecting the survival of the HEV in the environment and in food.

The virus is known to survive frozen storage for extended periods and is also able to survive in the gastrointestinal tract, indicating that it is relatively resistant to acid conditions. It does appear, however, to be very sensitive to high salt concentrations and is inactivated in chlorinated water.

Thermal Inactivation

Only a few studies to determine the thermal inactivation of the HEV have been conducted. It has been concluded that although the HEV is less heat resistant than the hepatitis A virus, some HEV is likely to survive the internal temperatures reached in rare-cooked meat.

Control Options

Processing

The risk of acquiring hepatitis E through the ingestion of contaminated food is considered low. However, the risk can be reduced further by ensuring that all pork and pork products (including liver) are cooked thoroughly during processing.

Product Use

The risk of acquiring travel-associated hepatitis E can be reduced by avoiding drinking water or drinks containing ice made from water of an unknown purity in areas where the disease is endemic. In addition, travellers should be advised not to eat uncooked shellfish, or uncooked fruits or vegetables that they have not peeled or prepared themselves.

In industrialised countries, where sanitary conditions are good, it has been recommended that consumers should be advised that pork products should not be consumed rare.

Legislation

There is no specific legislation in the EC or in the USA regarding levels of enteric viruses such as the HEV in foods.

In the UK the Advisory Committee on the Microbiological Safety of Food (ACMSF) has concluded that the risk of acquiring hepatitis E through the food chain in the UK is likely to be low. However, the expert committee concluded that searing the outside of meat joints would be insufficient to destroy viruses, such as hepatitis E, that may be present in meat muscle, and it has been recommended therefore that pork and pig products (including liver) should be cooked all the way through prior to consumption.

Sources of Further Information

Published

Cook, N. and Rzezutka, A. Hepatitis viruses in "Emerging foodborne pathogens." ed. Motarjemi, Y. and Adams, M. Cambridge. Woodhead Publishing Ltd, 2006, 282–308.

On the Web

Advisory committee on the microbiological safety of food. Risk assessment of the role of foodborne transmission of hepatitis E in the UK. UK Food Standards Agency. (September 2005). http://www.food.gov.uk/multimedia/pdfs/acm766.pdf

Hepatitis E. World Health Organization. (2001) http://www.who.int/csr/disease/hepatitis/whocdscsredc200112/en/index.html

1.2.4 HIGHLY PATHOGENIC AVIAN INFLUENZA VIRUSES

Hazard Identification

What are Highly Pathogenic Avian Influenza Viruses?

Highly pathogenic avian influenza (HPAI) viruses belong to the family *Orthomyxoviridae*, and within this family these viruses are in the group known as influenza type-A viruses. Influenza type-A viruses are classified into subtypes, and are named according to two main surface proteins, haemagglutinin ("HA") and neuraminidase ("NA"). For example, the subtype H5N1 has an HA 5 protein and a NA 1 protein. To date, 16 HA subtypes, and 9 NA subtypes have been described and many different combinations of HA and NA proteins are known to exist (*e.g.* H5N1, H1N1, H7N3 and H7N7).

Although influenza A viruses can infect many animals including birds, humans, pigs, dogs, cats and horses, wild birds are the natural hosts for these viruses. Avian influenza A virus strains are grouped based on genetic and pathogenic criteria as either low pathogenic avian influenza (LPAI), causing mild disease in birds, or HPAI, having enhanced virulence and causing the rapid onset of severe disease with high mortality rates in birds.

Some avian influenza viruses can be transmitted to humans and cause illness. LPAI viruses cause mild symptoms in humans, whereas HPAI can cause severe disease with high mortality rates. The type of HPAI virus that causes the most severe form of avian influenza (AI) in humans is the H5N1 virus. In recent years this virus has crossed the species barrier between birds and humans on a number of occasions, and an outbreak that began in South-East Asia during 2003 has become widespread, even reaching a number of European countries. The presence of HPAI H5N1 virus in birds is of concern for a number of reasons: it can cause severe disease in domestic poultry flocks resulting in up to 100% mortality; it can be spread to humans from infected birds; and it could potentially develop the ability to spread easily from human to human, resulting in a severe influenza pandemic.

There have been concerns that humans may become infected with the H5N1 virus by the handling and consumption of contaminated poultry and poultry products, and this has led to research into the virus and its potential as a foodborne pathogen. However, it is important to note that, although there is a theoretical potential for foodborne transmission of the virus, this has not yet been conclusively demonstrated. Most public health authorities, including the World Health Organization (WHO), do not currently consider HPAI H5N1 to be a food safety hazard.

Occurrence in Foods

Poultry, such as chickens and turkeys are particularly susceptible to HPAI viruses such as H5N1. All parts of the infected bird, including blood, meat and bones, are potentially contaminated with virus. The virus is also present in the saliva, nasal excretions and faeces of infected birds resulting in the contamination of feathers. Evidence suggests that the risk of exposure to the H5N1 virus is high during the slaughtering and handling of affected birds, or meat prior to cooking. There have also been reports of two cases of H5N1 infections in humans possibly linked to the consumption of uncooked poultry products (raw blood-based dishes), and cats are thought to have contracted the H5N1 virus through eating uncooked infected chicken carcasses, or possibly infected wild birds.

The HPAI H5N1 virus is also present on the inside and on the surface of eggs laid by infected birds. To date, there is no evidence to suggest that humans have contracted the H5N1 virus through the consumption of eggs or egg products.

Hazard Characterisation

Effects on Health

There are many strains of avian influenza A viruses, however, only four subtypes (H5N1, N7N3, H7N7 and H9N2) are known to cause illness in humans. Usually these viruses cause mild influenza-like symptoms such as fever, muscle aches, a cough and a sore throat or sometimes conjunctivitis.

However, in many individuals infected with the H5N1 virus the course of the disease is different. Most reported cases of H5N1 infections have occurred in previously healthy children and young adults and the infectious dose is unknown. It is thought that the incubation period for the H5N1 virus in humans is between 2–8 days but may be as long as 17 days (the WHO advises that an incubation time of 7 days be used to monitor patient contacts for the disease).

Initial effects may include influenza-like symptoms, a temperature of greater than 38 °C, or acute encephalitis. Sometimes, watery diarrhoea without blood, vomiting, chest pain, abdominal pain, and bleeding from the nose and gums have been described. Typically initial symptoms are followed around 5 days later by lower respiratory tract illness such as breathing difficulties, respiratory distress, a hoarse voice, a crackling sound when inhaling and sometimes the production of sputum, which may contain blood. Deterioration is rapid with the development of acute respiratory distress and possibly multiorgan failure. The disease has an associated mortality rate of 55%.

The majority of reported cases occur as the result of close contact with H5N1-infected poultry or H5N1-contaminated surfaces. There have been a few reports of person-to-person transmission occurring between family members, suggesting that very close contact for prolonged periods is needed to contract H5N1 AI from this source.

Incidence and Outbreaks

The outbreak of H5N1 in poultry that began in Asia during 2003 is the largest and most severe on record. It is known to have spread to birds in more than

50 countries in Africa, Europe, Asia and the Near East, and has resulted in at least 277 reported human cases with 168 deaths across 12 countries. Two cases may possibly have been caused by the consumption of infected raw duck-blood products. However, contact with infected live birds or carcasses cannot be ruled out, so the infected product may not have been the only source of infection.

Sources

Wild water birds are thought to be the main reservoir for the H5N1 virus, and some species, particularly ducks, are thought to act as asymptomatic carriers. Pigeons may also play a role in the spread of the virus. Mammals such as cats have also been infected with H5N1 virus and have died from the disease. Other mammals, such as dogs, have also tested serologically positive for the virus in outbreak areas, indicating that they too can become infected.

Contaminated bird faeces can lead to the contamination of the environment, where the virus can survive for some time, particularly at low temperatures. The virus can also cause infection by airborne transmission if birds are close together. However, there is no evidence to confirm that waterborne transmission of the virus occurs between birds, and it is thought that the risk of waterborne transmission of the virus to humans is small.

There is evidence to suggest that the HPAI H5N1 virus is excreted in the faeces of infected humans. However, data is limited on the extent of H5N1 virus excretion in urine and faeces in all infected mammals, including humans. It is not yet known whether this is another possible source of the virus.

Survival Characteristics

Viruses, including influenza viruses, are unable to multiply outside the host. However, the H5N1 virus is able to survive, sometimes for extended periods, in the environment.

The survival of AI viruses in water is dependent on the temperature, pH and salinity. Specific data on the survival of the H5N1 virus in water is limited, but generally for AI viruses, the survival in natural water (fresh, brackish and seawater) decreases with increasing salinity and increasing pH values above neutral. Different strains of avian influenza have been shown to survive in water at 17 °C, and at 28 °C, for up to 207 days and 102 days, respectively.

The WHO suggests that the avian influenza virus cannot generally be detected in birds 4 weeks after infection. However, the survival of the highly pathogenic H5N1 virus in bird faeces is dependent on initial concentration, temperature and pH. Studies using H5N1 viruses circulating during 2004 found that in faeces held at 4 °C and 37 °C, live viruses survived for 35 days and 6 days, respectively. On surfaces such as that found in poultry house environments, avian influenza viruses are reported to survive for a few weeks.

If the H5N1 virus is present in poultry meat, it can survive in this environment under chilling and freezing conditions with little affect on levels or the viability of the virus. In general, low temperatures actually prolong the survival of the virus in poultry tissue.

HPAI viruses are reported to be inactivated at extremes of pH, and are sensitive to desiccation. All avian influenza viruses are reported to be relatively susceptible to most disinfectants, including chlorine.

Thermal Inactivation

HPAI viruses are inactivated when held at 121 °C for 15 min, 60 °C for 30 min, or at 56 °C for 3 h. In foods, the H5N1 virus is inactivated when all parts of the item reach 70 °C or above. Therefore, properly cooked poultry products are safe to eat. The WHO advises that the virus is inactivated during conventional cooking practices used to cook poultry products where temperatures reach 70 °C or above at the centre of the product.

It has been reported that most standard pasteurisation temperatures for eggs used by industry will inactivate HPAI viruses (*e.g.* whole egg, 60 °C, 210 s; liquid egg white, 55.6 °C, 372 s; 10% salted yolk, 63.3 °C, 210s). However, the industry standard of treating dried egg white of 54.4 °C for 7–10 days would not be sufficient to inactivate HPAI viruses.

Control Options

Control of HPAI viruses currently focuses on containing outbreaks in poultry by culling infected birds, implementing strict biosecurity measures and limiting movement of poultry within designated areas. However, there are also sensible preventative measures that may be relevant to the food industry.

Processing

Although there is no evidence to suggest that there is a risk of acquiring infection of the HPAI H5N1 virus through the consumption of properly cooked poultry and egg products, there are risks associated with the slaughtering, de-feathering and eviscerating of infected birds, or the handling of raw or partially cooked contaminated eggs. In outbreak areas, diseased birds or those found dead should never be used for human consumption. In addition, good hygiene practices are essential during slaughter and the post-slaughter handling of poultry carcasses to prevent any possible exposure via raw poultry meat, or cross-contamination from poultry to other foods, food-preparation surfaces, or equipment. Good hygiene is also essential when handling prepared poultry meat and eggs from outbreak areas and thorough cooking of all egg and poultry-meat products should be ensured.

It should be noted that the likelihood of the HPAI virus being present in poultry in non-outbreak areas is negligible, and the possibility of infected meat being sold and handled by a consumer in most regions is extremely low.

Product Use

Poultry meat should be thoroughly cooked (heated to 70 °C in all parts) to ensure the inactivation of foodborne pathogens in general. Similar comments also apply to eggs and egg products.

Legislation

There is no specific legislation in the EC or in the USA regarding avian influenza viruses in foods. However, it is highly likely that there will be import and animal movement restrictions applying to areas affected by avian influenza outbreaks.

Sources of Further Information

Published

Swayne, D.E. Microassay for measuring thermal inactivation of H5N1 high pathogenicity avian influenza virus in naturally infected chicken meat. *International Journal of Food Microbiology*, 2006. 108, 268–71.

Swayne, D.E. and Beck, J.R. Heat inactivation of avian influenza and Newcastle disease viruses in egg products. *Avian pathology*, 2004. 33, 512–18.

On the Web

Scientific Report of the Scientific Panel on Biological Hazards on "Food as a possible source of infection with highly pathogenic avian influenza viruses for humans and other mammals." European Food Safety Authority. (June 2006). http://www.efsa.europa.eu/en/science/biohaz/biohaz_documents.html

Highly pathogenic H5N1 avian influenza outbreaks in poultry and in humans: Food safety implications. World Health Organization, International Food Safety Authorities Network (INFOSAN). (November 2005). http://www.who.int/foodsafety/fs_management/No_07_AI_Nov05_en.pdf

1.2.5 NOROVIRUSES

Hazard Identification

What are Noroviruses?

Noroviruses is the name given to a group of related non-enveloped, single-stranded RNA viruses that have recently been classified in the family *Caliciviridae*, genus *Norovirus*. These highly infectious enteric viruses are a major cause of acute gastroenteritis in humans (the infection is often called viral gastroenteritis). Although many cases are caused by person-to-person spread, the ingestion of contaminated food or water also plays a significant part in their transmission.

Noroviruses were first described following an outbreak of gastroenteritis in a school in Norwalk, Ohio in 1968. For many years they were known as the Norwalk group, as Norwalk-like viruses (NLV), or as "small round structured viruses" (SRSVs), because of their morphological characteristics. However, the name Norovirus (NoV) has recently been recognised as the official genus for this group of human caliciviruses. NoV strains are named after the location from which they were first associated, *e.g.* Norwalk virus, Southampton virus, Snow Mountain virus and Mexico virus.

Occurrence in Foods

Noroviruses are non-culturable in the diagnostic laboratory and there is no known animal model. Until relatively recently they could only be detected when present in high numbers using electron microscopy. Recent technological advances have enabled noroviruses to be detected and characterised by molecular methods, but the detection of these viruses in foods is extremely difficult and has only been successful in shellfish.

Food vehicles for noroviruses are thought to include sewage-contaminated bivalve shellfish, foodstuffs that are contaminated by an infected handler, fruits and vegetables contaminated during irrigation or washing, and water (including drinking water and ice).

Infected food handlers can contaminate any foodstuff, and outbreaks of NoV infections can be associated with any food that is handled and will be eaten without a further cooking step. Contamination can occur during the preparation of foods as well as during the harvesting of fresh produce such as soft fruits.

Hazard Characterisation

Effects on Health

Noroviruses can cause illness in any age group, although the elderly and the immunocompromised are particularly susceptible. Recent evidence suggests that susceptibility to NoV infection could be genetically determined, and people with blood group O seem more likely to develop a severe infection. Illness can occur at any time of year but in temperate climates is more common during the winter months. Noroviruses are very contagious, however, the illness is usually mild and self-limiting.

The infective dose is low, and as few as 10 virus particles may be sufficient to infect an individual. Signs of infection first appear from between 10–50 h, typically 24–28 h, after ingestion of the virus. The onset of illness is abrupt and typical symptoms are vomiting (often projectile), diarrhoea, abdominal pains, nausea, headache, stomach cramps and occasionally low-grade fever. The illness is typically relatively short, lasting from 12–60 h, although there are reports that symptoms in some individuals last for more than 2 weeks. Recovery is usually complete with no long-lasting effects.

During the illness high numbers of the virus are generated in the vomit of affected individuals as well as being shed in their faeces. Virus shedding appears to occur before symptoms start and continue for up to two weeks after symptoms have ceased. Outbreaks associated with an infected food handler have been associated with foods prepared before the onset of symptoms.

Incidence and Outbreaks

Norovirus outbreaks are very common, but there is little published information on the incidence of foodborne infection. In the USA 382 confirmed outbreaks (not necessarily foodborne) were recorded in the period from October to December 2006 alone and rising incidence is thought to be linked to the appearance of new strains of the virus. In the UK, the incidence of norovirus infections has also been rising steadily since the 1980s, and in 2006, nearly 4500 confirmed cases were recorded, although there is no indication of the proportion that were foodborne.

Contaminated water is the most common source of a NoV outbreak and has caused very large outbreaks of viral gastroenteritis. Outbreaks have been linked to water from wells, municipal water supplies, swimming pools, lakes and water stored on cruise ships. In the USA, commercially prepared ice from a production facility that was contaminated during flooding was associated with a widespread outbreak.

Foodborne outbreaks of NoV infections are frequently caused by infected food handlers. Foods associated with this source of contamination are cold, ready-to-eat foods such as prepared salads, fresh cut fruits, sandwiches and bakery products. Large outbreaks have been caused when liquid foods such as icings or salad dressings have become infected during preparation and then mixed leading to widespread distribution of the virus.

Shellfish, in particular oysters, from sewage-contaminated water, when eaten raw, or lightly cooked, have also caused large outbreaks of NoV illness.

Contaminated fresh produce, in particular salads and raspberries, has been associated with large foodborne outbreaks of NoV infections. These foods may be contaminated either from irrigation water, during washing or spraying, or during harvesting by infected handlers. In recent years frozen raspberries have caused extensive foodborne outbreaks in Canada and in Europe. The viruses are able to survive the freezing process and frozen fruits are often exported to other countries resulting in the wide distribution of the virus.

Sources

Humans are the only known reservoir for noroviruses. It has been hypothesised that there may also be an animal reservoir, but, although related caliciviruses have been found in many animal species, there have not been any documented cases of cross-species transmission.

Faeces or vomit from infected individuals can lead to the environmental contamination of soil, water and surfaces. Airborne droplets produced during vomiting are a particularly effective method of distribution for viruses.

Noroviruses can accumulate and concentrate in the guts of bivalve molluscs, such as oysters and mussels, growing in sewage-contaminated waters. Depuration processes designed to reduce the bacterial contamination of these shellfish are ineffective for removing viruses. Faecal contamination of water supplies can be a potential source of noroviruses. Live viruses have even been detected in commercially available bottled mineral water, although cases of illness have not yet been traced to this possible source of infection.

Survival Characteristics

Viruses, including noroviruses, are unable to multiply outside of the host. Although noroviruses cannot grow in food or water, they can survive in many environments for significant periods. The virus can remain infective when held at ambient, chilled and freezing temperatures. In chilled and frozen environments survival can be measured in months or even years. Noroviruses are resistant to acid and can survive gastric acid at pH 3–4. They have also been shown to still be infective when exposed to a pH of 2.7 for 3 h at ambient temperature. The virus can survive in water environments and in shellfish for extended periods (possibly months). It is resistant to drying, and is reported to persist on environmental surfaces, such as carpets, for up to 12 days.

Noroviruses can survive exposure to up to 10 ppm free chlorine, and can therefore survive the usual chlorination processes used to treat public water supplies.

Thermal Inactivation

Noroviruses have been shown to remain infective when held at 60 °C for 30 min. The virus is able to survive some pasteurisation processes and has also caused illness after it was steamed in shellfish. It is inactivated by boiling.

Control Options

To reduce the risk of foodborne transmission of noroviruses, controls should focus on ensuring the use of potable water for food processing, strict hygiene control, and using shellfish from approved waters.

Processing

Food handlers or fruit pickers suffering from viral gastroenteritis should not return to work for at least 48–72 h after symptoms have ceased. Effective training in adequate personal hygiene practices is essential. Thorough cleaning with an effective sanitiser should follow any episode of vomiting in a food-processing environment.

Shellfish should be gathered from approved harvesting waters and should be thoroughly cooked prior to consumption.

Product Use

Consumers should be advised not to eat raw shellfish and to ensure these products are thoroughly cooked prior to consumption. In addition, consumers should be advised to thoroughly wash all fruits and vegetables that will be eaten raw or lightly cooked in potable water.

Legislation

There is no specific legislation in the EC or in the USA regarding levels of enteric viruses, such as noroviruses, in foods.

Sources of Further Information

Published

Koopmans, M. and Duizer, E. Foodborne viruses: an emerging problem. *International Journal of Food Microbiology*, 2004. 90, 23–41.

Koopmans, M., von Bonsdorff, C-H., Vinjé, J., de Medici, D. and Monroe, S. Foodborne viruses. *FEMS Microbiology Reviews*, 2002. 26, 187–205.

Lopmann, B.A., Brown, D.W. and Koopmans, M. Human caliciviruses in Europe. *Journal of Clinical Virology*, 2002. 24, 137–160.

On the Web

Risk profile: Norwalk-like viruses in mollusca (raw). Institute of Environmental Science and Research Limited. (January 2003). http://www.nzfsa.govt.nz/science/risk-profiles/norwalk-like-virus-in-raw-mollusca.pdf

1.2.6 PARVOVIRUSES

Hazard Identification

What are Parvoviruses?

The parvoviruses are very small, single-stranded DNA viruses belonging to the family *Parvoviridae*. These viruses have a smooth surface with no discernable features and were previously included in the group of viruses known as small round viruses (SRVs) or featureless viruses.

Data on these viruses as a cause of human gastroenteritis is limited, but it is known that parvoviruses may cause gastroenteritis in other animal species (*e.g.* canine parvovirus).

Occurrence in Foods

Data is very limited, although parvovirus or parvovirus-like particles have been linked to a number of outbreaks associated with the consumption of shellfish. Parvovirus-like particles similar to those found in patients have been detected in shellfish.

Hazard Characterisation

Effects on Health

Gastroenteritis caused by parvovirus has been described as "winter vomiting virus", suggesting similarities with norovirus infections.

During some outbreaks it has been found that large numbers of virus particles are excreted in the faeces of many patients. It is also known that the shedding of virus particles can continue for a number of weeks after symptoms subside. Low numbers of parvovirus-like particles can also be found in the faeces of healthy individuals.

Incidence and Outbreaks

A parvovirus serotype, known as the "cockle agent parvovirus" has been linked to a large outbreak (>800 cases) of gastroenteritis in the UK associated with the consumption of cockles.

Other parvovirus-like particles, the Parramatta agent and the Wollan/Ditchling group, have been linked to outbreaks of gastroenteritis in schools.

Sources

Parvoviruses causing gastroenteritis in humans are likely to be found in environments that are faecally contaminated. The cockle agent parvovirus was linked to cockles harvested during the winter, much closer to sewage outlets than was usual.

Survival Characteristics

Due to the infrequency with which parvoviruses are associated with gastrointestinal disease in humans there is very little data of the survival characteristics of these agents.

Control Options

To reduce the risk from viral gastroenteritis associated with the consumption of shellfish it is important to ensure that shellfish are harvested from approved waters and that these products are properly cooked prior to consumption.

Legislation

There is no specific legislation in the EC or in the USA regarding levels of enteric viruses, such as parvoviruses, in foods.

Sources of Further Information

Published

Appleton, H. Norwalk virus and the small round viruses causing foodborne gastroenteritis, in Foodborne Disease Handbook, volume 2: viruses, parasites, pathogens and HACCP. ed. Hui, Y.H., Sattar, S.A., Murrell, K.D., Nip, W.K., and Stanfield, P.S. 2nd edn. New York, Marcel Dekker, 2000, 77–97.

1.2.7 ROTAVIRUSES

Hazard Identification

What are Rotaviruses?

Rotaviruses are non-enveloped, double-stranded RNA viruses, which are classified as belonging to the family *Reoviridae*, genus *Rotavirus*. There are seven described species or "serotypes" of rotavirus (known by the letters A–G). The name rotavirus is derived from the characteristic wheel like appearance of the viruses when viewed under an electron microscope. Groups A, B and C rotaviruses are known to infect humans, and of these, group A rotaviruses are the most significant. Group A rotaviruses are the leading cause of severe diarrhoea in infants and young children worldwide.

Although group A rotaviruses are a major cause of acute diarrhoea it is thought that only a small percentage (around 1%) of cases are actually food-borne, the main route of transmission is person-to-person through the faecal–oral route.

Occurrence in Foods

Potentially, an infected food handler could contaminate any food prepared and consumed without a subsequent heating step. Salads, cold foods (such as sandwiches and hors d'oeuvres), fruits and contaminated water (including ice cubes) have all been implicated in cases of foodborne rotavirus infections. Rotaviruses have also been detected in shellfish.

Hazard Characterisation

Effects on Health

In countries with a temperate climate, such as the UK and the USA, rotavirus infections usually occur in the winter and spring months, whereas in tropical regions infections occur throughout the year.

Rotaviruses are highly infectious and as few as 10 rotavirus A particles (possibly a single virus particle) can cause illness in a child. Although individuals of all ages are susceptible to rotavirus A infections, the disease usually occurs in infants and young children and the most severe symptoms are seen in the very young, the immunocompromised and the elderly. Infection usually confers limited immunity to further rotavirus infections. When infections do occur in adults the disease is often very mild, or even asymptomatic.

The incubation time is 1–3 days and initial symptoms include vomiting and watery diarrhoea for about 2–3 days, often leading to dehydration. The diarrhoea can sometimes persist for 5–8 days. Without electrolyte replacement and adequate fluids, severe, potentially fatal, dehydration can result.

Other symptoms include abdominal discomfort, headaches, chills and low-grade fever. In most cases, the infection is self-limiting, and in developed countries most children make a full recovery.

During infection, affected individuals shed high numbers of virus particles in their stools (up to 10^{11}/g) and asymptomatic carriers of the virus also occur.

Incidence and Outbreaks

In developing countries rotaviruses cause an estimated 125 million cases annually in infants and young children. Some 18 million of these are severe cases resulting in almost 900 000 deaths each year. In industrialised countries deaths from rotavirus infections are extremely rare. However, in the US it is estimated that as many as 70 000 children, and in England and Wales 18 000 children, require hospitalisation annually as a result of the illness. Most of these cases are not caused by foodborne infection, but in the USA it has been estimated that approximately 39 000 cases of viral diarrhoea annually are actually caused by foodborne rotaviruses.

Foodborne outbreaks of rotavirus infections have occasionally been documented in the literature. Suspected vehicles include sandwiches, lettuce, salads, cold foods, strawberry shortcake and shepherd's pie. Contaminated water has been associated with outbreaks in many countries.

Sources

Infected individuals act as a reservoir for human rotaviruses. Individuals suffering from the disease, as well as asymptomatic cases, excrete high numbers of the virus into the environment in their faeces. Most infections occur as the result of person-to-person transmission through the faecal–oral route. However, the virus can contaminate environmental surfaces and objects and these can act as reservoirs for the disease, particularly in institutions such as hospitals and nursing homes.

Foods can be contaminated by infected food handlers, by the use of faecal matter to fertilise crops, or through the use of contaminated water for irrigation of fresh produce.

Water contaminated with infected faeces can also act as a source of the virus. Shellfish cultivated in contaminated waters can accumulate rotavirus particles.

Survival Characteristics

Viruses, including rotaviruses, are unable to multiply outside of the host. However, rotaviruses can persist in the environment, and they are known to survive in river water at 20 °C and at 4 °C for several weeks. Rotaviruses can survive for some time on hard surfaces and can remain infective in anaerobically stored animal waste for up to 6 months. Bovine rotaviruses have been shown to survive processes used to produce soft cheese.

Rotaviruses are reported to be sensitive to drying and to extremes of pH.

Rotaviruses are relatively resistant to many disinfectants, but they are susceptible to 95% ethanol, 2% sodium hypochlorite (with a long contact time), and to 5% Lysol.

Thermal Inactivation

Rotaviruses are reported to be relatively heat sensitive. Although there is little data on the heat inactivation of these viruses, it is thought that normal cooking processes should inactivate them. A study found that rotavirus infectivity is reduced by 99% when heated at 50 °C for at least 30 min.

Control Options

Strategies to reduce the risk of foodborne outbreaks of rotavirus infections should focus on preventing foods from becoming contaminated by the use of clean water for the irrigation, washing and processing of foods, and preventing shellfish-harvesting areas from becoming contaminated with sewage.

Processing

Food handlers should implement frequent hand washing (rotaviruses are most effectively controlled using alcohol-based hand-cleaning agents) and the wearing of gloves, particularly at points in the food chain where foodstuffs that will receive no further cooking are handled. Food handlers suffering from viral gastroenteritis should be excluded from work and advised not return for at least 48–72 h after symptoms have ceased.

Product Use

Consumers should be advised not to eat raw or inadequately cooked shellfish.

Legislation

There is no specific legislation in the EC or in the USA regarding levels of enteric viruses, such as rotaviruses, in foods.

Sources of Further Information

Published

Cliver, D.O., Matsui, S.M. and Casteel, M. Infections with viruses and prions, in Foodborne infections and intoxications. 3rd edn. ed. Reimann H.P., Cliver, D.O. London. Academic Press, 2005, 367–448.

Sattar, S.A., Springthorpe, V.S. and Tetro, J.A. Rotavirus, in Foodborne disease handbook, volume 2: viruses, parasites, pathogens and HACCP. 2nd edn. ed. Hui, Y.H., Sattar, S.A., Murrell, K.D., Nip, W.K. and Stanfield, P.S. New York, Marcel Dekker, 2000, 99–125.

1.2.8 SAPOVIRUSES

Hazard Identification

What are Sapoviruses?

The sapoviruses are a group of single-stranded, positive-sense, RNA viruses recently classified in the family *Caliciviridae*, genus *Sapovirus*. Previously, these human caliciviruses were known as Sapporo-like viruses (SLV), or referred to as classic, or typical caliciviruses. Sapoviruses can be distinguished from the other group of human caliciviruses, the noroviruses, by their six pointed "Star of David" morphological appearance when viewed with an electron microscope.

Sapoviruses are commonly associated with causing mild viral gastroenteritis in infants and children worldwide.

Occurrence in Foods

Recent studies in Japan have isolated sapoviruses from clams collected from supermarkets and fish markets, as well as from environmental fresh waters during both summer and winter months. However, sapovirus infections are not generally associated with the consumption of seafood.

Human caliciviruses, including sapoviruses, could potentially be present in any food or water supply where faecal contamination is present. Contaminated water supplies could result in the contamination of foods grown, irrigated, or washed with the water, such as shellfish, fruits and vegetables.

Hazard Characterisation

Effects on Health

The infective dose for caliciviruses, including sapoviruses, in low (estimated to be between 10–100 virus particles). Sapoviruses usually cause infections in infants and young children, although in neonates infections are often subclinical. It is thought that sapovirus infection in children may confer long-lived immunity against further infection. Occasionally, infections and outbreaks are reported amongst adults and the elderly and it is thought that these illnesses are associated with weakened immunity. Although illness caused by the viruses can occur throughout the year, sapovirus infections peak in the winter months.

The incubation time for sapovirus infections is 1–3 days, and symptoms persist for about 4 days. Typically, the illness is characterised by watery stools, mild or acute diarrhoea, vomiting, nausea, stomach cramps and sometimes a low fever. Sapovirus infections are not well understood, but it is known that the infection is self-limiting, and individuals in developed countries usually make a full recovery. Deaths are very rare and occur mainly in those vulnerable to dehydration.

During infection individuals excrete very high numbers of the virus in their stools. In addition, asymptomatic carriers of these viruses can occur.

Incidence and Outbreaks

Transmission of sapoviruses generally occurs via the faecal–oral route. Secondary infections between close contacts (person-to-person transmission) such as in schools and child-care settings are also common. Most sapovirus infections occur as sporadic infections in young children and definite food vehicles have yet to be determined.

Foodborne outbreaks have occasionally been associated with sapoviruses, but they occur far less frequently than foodborne outbreaks associated with noroviruses. The data on foodborne sapovirus outbreaks is limited, but an outbreak in Maryland, USA, was thought to have been caused by food prepared by infected food handlers.

Sources

Humans are the reservoir for sapoviruses and infected individuals can excrete very high numbers of virus particles. Contaminated environmental sources such as sewage and water (both drinking and recreational) could also be potential sources of sapoviruses, as could foods contaminated by infected food handlers.

Survival Characteristics

Sapoviruses have not been as intensively studied as the noroviruses, and little is known about their survival characteristics. Like other viruses, they are unable to multiply outside of the host, but they are thought to survive for some time in the environment.

High levels of chlorination are required to inactivate human caliciviruses in drinking water. Levels of around 10 ppm, or 10 mg/L of chlorine for more than 30 min have been reported as being required for adequate disinfection.

Thermal Inactivation

Human caliciviruses are thought to be inactivated by "adequate cooking processes" (*e.g.* >1 min at 90 °C).

Control Options

The control of sapoviruses should focus on the implementation of strict personal hygiene by food handlers. Ready-to-eat foods that are handled but will receive no further cooking, such as sandwiches and salads, pose the greatest risk.

Legislation

There is no specific legislation in the EC or in the USA regarding levels of enteric viruses, such as sapoviruses, in foods.

Sources of Further Information

Published

Lopmann, B.A., Brown, D.W. and Koopmans, M. Human caliciviruses in Europe. Journal of Clinical Virology, 2002. 24, 137–160.

CHAPTER 1.3
Parasites

1.3.1 PROTOZOA

1.3.1.1 *Cryptosporidium*

Hazard Identification

What is Cryptosporidium*?*

Cryptosporidium is a single-celled protozoan parasite belonging to the subclass *Coccidia*. Until recently, the only species thought to be important in human illness was classified as *Cryptosporidium parvum*. However, recent taxonomic studies have shown that several species can infect humans, including *C. hominis*, which is specific to humans, and *C. parvum*, which infects both humans and ruminants. Other species that have been reported to infect humans include *C. felis*, *C. canis*, *C. meleagridis* and *C. suis*.

Cryptosporidium is an obligate parasite and requires a host in order to multiply. It was first discovered almost 100 years ago, but was not associated with human illness until 1976. It is a cause of gastrointestinal infection in humans and some other animals, especially calves and lambs, and is found worldwide.

Cryptosporidium has a complex life cycle, most of which takes place within the gastrointestinal tract (mainly in the small intestine) of a single host. The transmissible stage in the cycle is a highly resistant, thick-walled spore, known as an oocyst.

Occurrence in Foods

Cryptosporidium is mainly associated with water that has been polluted by human or animal faeces, but oocysts have also been found in a number of unprocessed foods, notably raw milk, meat and shellfish and fresh fruit and vegetables. *Cryptosporidium* cannot grow in foods or in water and does not multiply in the environment outside of a suitable host.

The Food Safety Hazard Guidebook
By Richard Lawley, Laurie Curtis & Judy Davis

Oocysts are easily destroyed by heat and *Cryptosporidium* is not normally associated with cooked and processed foods. Any food that may come into contact with contaminated water during production, and where there is no subsequent process that will destroy oocysts, is at risk from *Cryptosporidium* contamination. However, food is not considered to be a major vehicle for the transmission of the parasite. The person-to-person and animal-to-human (zoonotic) transmission routes are likely to be much more common.

Hazard Characterisation

Effects on Health

Cryptosporidium can cause an acute gastrointestinal infection in humans. It invades the epithelial cells lining the gut causing inflammation and loss of fluid. The incubation time for the infection is usually between 5–7 days, but it may vary from 2–14 days, possibly depending on the number of oocysts ingested. The main symptom is profuse watery diarrhoea, often accompanied by abdominal pain. Vomiting, fever and weight loss may also occur. Symptoms are most severe in the very young, the elderly and in immunocompromised adults, such as AIDS patients. In healthy adults, symptoms typically last for 2–4 days, but may last for up to 2–3 weeks in some cases. The infection is usually self-limiting and is resolved without medical treatment. However, in vulnerable individuals, infection can be more serious and long lasting, requiring hospital treatment, and deaths have been recorded. *Cryptosporidium* is also capable of invading other organs, such as the respiratory system, in some cases.

The infective dose is uncertain, but may be as low as 10 oocysts, or even less. A single oocyst is thought to be capable of causing disease in young lambs, and possibly also in very young children and immunocompromised adults. Infected individuals shed very large numbers of infectious oocysts in their faeces, and this may continue at a low level for several weeks after symptoms have subsided. This shedding of oocysts is the main reason why person-to-person and zoonotic transmission of the parasite are so common. Asymptomatic cases of infection have also been reported.

Incidence and Outbreaks

Cases of *Cryptosporidium* infection are not particularly common. For example, in England and Wales between 1986 and 2006, the number of reported cases each year generally ranges from 3000 to 6000, with a peak of nearly 8000 cases in 1989. The most recent data for the EU refers to 2005 and shows a total of 7960 reported cases of cryptosporidiosis from 16 countries. However, 70% of these were from the UK, suggesting significant under-reporting in many other countries. The European country with the highest reported incidence was Ireland with 13.7 cases per 100 000 people. The results also show that peaks of infection commonly occur in the autumn, or occasionally in spring. Cryptosporidiosis is a notifiable disease in the EU and in the USA.

There were 8269 reported cases of cryptosporidiosis in the USA in 2005. This represented an increase of more than 100% over the previous year, but this is considered to be mainly due to a single waterborne outbreak. Most cases were reported from the northern states and there was a peak in the summer and early autumn.

The incidence of cryptosporidiosis in New Zealand is reported to be relatively high (21.2 cases per 100 000 in 2000), with peaks of cases in spring and autumn.

There is little or no information about the proportion of reported cases that are foodborne, but it is thought likely that the majority are caused by contact with infected animals, people, or contaminated water.

Most recorded outbreaks are associated with contaminated drinking water, or recreational waters. For example, in 1993 a waterborne outbreak occurred in Milwaukee in the USA, which affected more than 400 000 people and caused an estimated 69 deaths. Foodborne outbreaks have also been recorded, usually caused by an infected food handler, or by faecal contamination, either direct or through polluted water. Outbreaks have been linked to raw produce, chicken salad, green onions and raw milk. In the USA, there have been several outbreaks linked to unpasteurised apple cider. For example, in 2003, cider made from contaminated apples caused illness in 144 people. The cider had reportedly been treated with ozone, but this had clearly not been effective.

Sources

Cryptosporidium spp are all obligate parasites and thus originate from the host animal. *C. hominis* is thought to primarily infect humans, while *C. parvum* infects humans and ruminants. The primary source of *Cryptosporidium* is therefore the faeces of infected humans and animals, which may contain up to 10^9 oocysts in a single bowel movement. Infected cattle are a particularly important reservoir of *C. parvum*. The oocysts are extremely infectious and may be transferred to food via an infected food handler, or through polluted water used for crop irrigation or processing.

Cryptosporidium oocysts are quite difficult to remove from water, even by modern water-treatment methods. Their small size (4–6 μm diameter) and resistance to chlorine enable them to pass through some water-treatment plants, especially if they are present in high numbers. This can happen when heavy rains cause run-off from agricultural land used for grazing. Under these circumstances it may not be possible to guarantee that public water supplies are free from *Cryptosporidium* oocysts.

Stability in Foods

Cryptosporidium oocysts are very resistant to most environmental factors, with the exception of heat and desiccation. Oocysts can persist for months in water and in soil and have been shown to survive for hours on wet surfaces, including stainless steel. However, they are not resistant to drying and die rapidly on dry surfaces.

The oocysts are also remarkably resistant to many sanitisers and disinfectants, notably chlorine. One study reported survival for two hours on exposure to chlorine at 50 000 ppm. 18-hour exposure to 4% iodophore and 10% benzalkonium chloride solutions has also been demonstrated to be ineffective in inactivating oocysts.

Cryptosporidium oocysts are not especially heat resistant and are destroyed by conventional milk pasteurisation. A temperature of greater than 73 °C will cause instantaneous inactivation of oocysts. Therefore, most controlled cooking processes used in food production should destroy any viable oocysts in the product.

Oocysts can survive for short periods at temperatures below 0 °C, especially in water, but the commercial ice cream freezing process has been shown to cause inactivation and eventual die-off occurs at temperatures below −15 °C.

There is little information on the effect of pH, but some loss of viability has been shown in acid conditions below pH 4.0. It has been reported that oocysts lost 85% of viability in 24 h when contaminated water was used to brew beer and produce a carbonated beverage.

Control Options

Processing

Control measures for *Cryptosporidium* in food processing focus largely on the control of contamination in the water supply. Food processors using potable water from the public supply network should carry out a risk assessment on the consequences of mains water contamination and a "Boil Water Notice" issued by the water supplier. Where there is a high risk, as in the production of raw food products, such as fresh-cut produce and salads, it may be worthwhile considering the introduction of additional on-site water-treatment measures, such as charcoal or membrane filtration. Treatment with biocides such as hydrogen peroxide and chlorine dioxide may be effective, but only at concentrations well above those usually used in water treatment.

Heat processing is an effective control against *Cryptosporidium* oocysts in food. Normal milk-pasteurisation processes are effective, as are recommended "Listeria cook" processes for meat products (70 °C for at least 2 min). Reheating cooked foods to at least 74 °C will destroy oocysts immediately.

Freezing foods for at least 7 days is an effective control, as is drying. Oocysts were reported to lose infectivity in 7 days when stored at a water activity of 0.85 at 7 °C.

Hygiene

Infected food handlers are also a major *Cryptosporidium* contamination risk for foods that do not undergo any further processing, such as sandwiches and salads. Good personal hygiene practice, especially hand washing, is an essential control and any staff suffering from gastroenteritis should be excluded from processing areas.

Legislation

Cryptosporidium is generally considered to be a waterborne pathogen rather than foodborne. It may therefore be covered in drinking-water regulations, as is the case in the UK, but is not usually mentioned specifically in food safety and hygiene law.

Sources of Further Information

Published

Foodborne Parasites. ed. Y.R. Ortega. New York, Springer Science and Business, 2006.

Dawson, D. Foodborne protozoan parasites. *International Journal of Food Microbiology*, 2005, 103(2), 207–27.

Erickson, M.C. and Ortega, Y.R. Inactivation of protozoan parasites in food, water, and environmental systems. *Journal of Food Protection*, 2006, 69(11), 2786–808.

Water quality for the food industry: management and microbiological issues CCFRA Guideline No. 27 (2000).

On the Web

CDC parasitic disease information – cryptosporidiosis. http://www.cdc.gov/NCIDOD/DPD/parasites/cryptosporidiosis/default.htm

IFST Information Statement – Cryptosporidium (2001). http://www.ifst.org/uploadedfiles/cms/store/ATTACHMENTS/cryptosporidium.pdf

NZFSA fact sheet – Cryptosporidium parvum. http://www.nzfsa.govt.nz/science/data-sheets/cryptosporidium-parvum.pdf

Food Research Institute Briefing – Foodborne Parasites. http://www.wisc.edu/fri/briefs/parasites.pdf

Society of Food Hygiene and Technology – Hygiene Review 1997, Cryptosporidium. http://www.sofht.co.uk/isfht/irish_97_cryptosporidium.htm

1.3.1.2 *Cyclospora*

Hazard Identification

What is Cyclospora*?*

Cyclospora is a single-celled protozoan parasite belonging to the subclass *Coccidia*. The only species known to cause human illness is *Cyclospora cayetanensis*. This species has also been reported in chimpanzees and other non-human primates, rodents and a few other animals, but it is possible that humans are the primary host.

Cyclospora is an obligate parasite and requires a host in order to multiply. It was first discovered in 1881, but was not associated with human illness until the late 1970s. It is a cause of gastrointestinal infection (cyclosporiasis) in humans, and is endemic in some developing countries, notably in Central and South America and some parts of Asia.

Cyclospora has a complex life cycle, most of which takes place within the gastrointestinal tract (mainly in the small intestine) of a single host. The transmissible stage in the cycle is a highly resistant, thick-walled spore, known as an oocyst.

Occurrence in Foods

Cyclospora was not considered to be a foodborne pathogen until 1996 when a large *C. cayetanensis* outbreak occurred in the USA. This was linked to imported raspberries from Guatemala. Until then, most reported cases in the USA were associated with foreign travel. Where *Cyclospora* is endemic, it is mainly associated with water that has been polluted by human or animal faeces. There has been very little attempted surveillance of *Cyclospora* oocysts in foods and effective test methods have been developed only recently. However, oocysts have been isolated from fresh basil implicated in a foodborne outbreak and epidemiological evidence from other outbreaks suggests that it may have been present in other fresh fruits and vegetables.

Cyclospora cannot grow in foods or in water and does not multiply in the environment outside of a suitable host. The parasite has not been reported to be associated with cooked and processed foods.

Contaminated water and food are thought to be the main routes for transmission of infection. Direct person-to-person transmission of *Cyclospora* is thought unlikely.

Hazard Characterisation

Effects on Health

Cyclospora cayenatensis can cause an acute gastrointestinal infection in humans. It invades the epithelial cells lining the gut, especially in the jejunum, causing inflammation and loss of fluid. The incubation time for the infection is typically between 5–7 days from ingesting sporulated oocysts, but it may vary from 1–14 days. The main symptom is watery diarrhoea, which may alternate with periods of constipation and persist for long periods (1–2 months in some cases). Other reported symptoms include abdominal pain, vomiting, fatigue, fever and weight loss. Diarrhoea is usually self-limiting in healthy adults, but may be more prolonged and debilitating in young children and the immunocompromised. Asymptomatic and mild cases of infection are reported to be common and immunity may be developed in areas where the disease is endemic.

The infective dose is uncertain, but is probably low. Infected individuals shed moderate numbers of oocysts in their faeces, but at this stage the oocysts are unsporulated and are not infectious. This is the main reason that person-to-person transmission is considered unlikely. Sporulation only takes place outside the body at higher concentrations of oxygen than those found in the gut and requires a period of 7–10 days at 30 °C. However, the process takes much longer at lower ambient temperatures. This may be why cyclosporiasis is not endemic in temperate regions.

Incidence and Outbreaks

Cases of cyclosporiasis are rare in developed countries, and until recently were generally associated with travel to countries where the disease is endemic, such as Peru, Haiti and Nepal. It is likely that *Cyclospora* is prevalent worldwide, but the incidence of disease is not known in most countries.

In England and Wales, approximately 60 cases of cyclosporiasis a year have been reported since the mid 1990s, but many of these are known to have been acquired abroad. There is little published information on the incidence of the disease elsewhere in Europe and few documented reports of cases of foodborne infection. The lack of awareness of *C. cayetanensis* and the absence of surveillance suggests that the disease is likely to be substantially under-reported.

Surveillance for *Cyclospora* in the USA is more developed following several large outbreaks in the 1990s. The overall incidence of cyclosporiasis in the USA in 2006 was estimated to be approximately 0.1 cases per 100 000 people. This equates to around 300 cases per year, but it is not known how many of these result from contaminated foods. There is usually a peak in reported cases in summer when high temperatures help the oocysts to sporulate. Cyclosporiasis is a notifiable disease in the USA.

Most recorded foodborne cyclosporiasis outbreaks have occurred in North America, including the first recorded outbreak in 1996, which affected almost 1500 people in the USA and Canada and was linked to imported raspberries from Guatemala. Since then, there have been a further 10 or more outbreaks in the USA, almost all linked to contaminated produce, such as mesclun lettuce, fresh basil, and imported berries. In 2000 an outbreak affecting 34 people was reported in Germany associated with consumption of contaminated salad.

These outbreaks are generally thought to be caused by the use of contaminated water for irrigation rather than by infected food handlers.

Sources

Cyclospora cayetanensis is an obligate parasite and thus originates from the host animal. Humans may well be the primary host for the parasite and human faeces are therefore the main source of *C. cayetanensis* oocysts. The oocysts may be transferred to food crops via polluted surface water used for irrigation or to dilute pesticides for application by spraying. Once sporulation has taken place the oocysts become infectious if ingested.

Stability in Foods

Like the closely related *Cryptosporidium* oocysts, *Cyclospora* oocysts are probably resistant to most environmental factors, with the likely exception of heat and desiccation. However, there is little published information to confirm this.

The oocysts are probably quite resistant to chlorine and cases of cyclosporiasis have been associated with chlorinated water supplies in Nepal. It is likely that the normal chlorination levels used in water treatment would be insufficient to inactivate oocysts.

There is no evidence that *Cyclospora* oocysts are any more heat resistant than those of *Cryptosporidium* and it seems probable that they too are inactivated by milk pasteurisation and other cooking processes.

Cyclospora oocysts are larger than those of *Crytosporidium* (9–10 μm diameter) and are therefore more easily removed from water supplies by conventional treatment. However, their apparent resistance to chlorination means that there is a risk that they may pass into public water supplies if treatment, especially filtration systems, is not well controlled.

Control Options

Control measures for *Cyclospora* in food focus largely on good agricultural practice in fruit and vegetable production in countries where the parasite is endemic and on ensuring that contaminated surface water is not used in irrigation or the application of pesticides and fertilisers. For example, the US Food and Drug Administration has worked with Guatemalan raspberry growers since the 1996 outbreak to improve standards and has developed a code of practice that includes filtration of all water used in cleaning and sanitation. The expansion of supply chains for fresh fruit and vegetables into countries where *Cyclospora* is prevalent means that this approach is likely to become more important in the future to prevent foodborne outbreaks of cyclosporiasis.

Processing

Heat processing is probably an effective control against *Cyclospora* oocysts in food and normal milk-pasteurisation processes are likely to inactivate them, as are cooking processes that raise the product temperature to 70 °C or more.

Freezing and drying of foods may also be effective controls, as is the case for *Cryptosporidium*.

Legislation

Cyclospora is not mentioned specifically in food safety and hygiene law in most countries. The US government has adopted import restrictions for high-risk foods such as raspberries grown in Guatemala. Only growers approved by the FDA may export to the USA.

Sources of Further Information

Published

Foodborne Parasites. ed. Y.R. Ortega. New York, Springer Science and Business, 2006.

Dawson, D. Foodborne protozoan parasites. *International Journal of Food Microbiology*, 2005, 103(2), 207–27.

Erickson, M.C. and Ortega, Y.R. Inactivation of protozoan parasites in food, water, and environmental systems. *Journal of Food Protection*, 2006, 69(11), 2786–808.

Water quality for the food industry: management and microbiological issues. CCFRA Guideline No. 27 (2000).

On the Web

CDC parasitic disease information – cyclosporiasis. http://www.cdc.gov/NCIDOD/DPD/parasites/cyclospora/default.htm

IFST Information Statement – Cyclospora (2003). http://www.ifst.org/uploadedfiles/cms/store/ATTACHMENTS/cyclospora.pdf

Food Research Institute Briefing – Foodborne Parasites. http://www.wisc.edu/fri/briefs/parasites.pdf

1.3.1.3 *Entamoeba*

Hazard Identification

What is Entamoeba*?*

Entamoeba is a single-celled protozoan parasite belonging to the subphylum *Sarcodina*. The species important in human illness is *Entamoeba histolytica*, but at least five other species are also found in humans, notably *Entamoeba dispar*, which is morphologically indistinguishable from *E. histolytica*, but much more common and non-pathogenic. *E. histolytica* is also found in non-human primates and other mammals, including cats and dogs.

E. histolytica is an obligate parasite and requires a host in order to multiply. It has been recognised as a cause of gastrointestinal disease (amoebiasis) in humans for many years and is found worldwide, but is particularly prevalent in developing countries.

E. histolytica has a two-stage life cycle, and exists in two forms. The active trophozoite stage exists and multiplies within the gastrointestinal tract of the host. Some of these form spore-like resistant cysts within the small intestine. Both forms may be excreted in the host's faeces, but the trophozoites die quickly and the transmissible stage in the cycle is the cyst.

Occurrence in Foods

E. histolytica is mainly associated with surface water that has been polluted by human faeces, but cysts may also be present in a number of unprocessed foods, including fruit and vegetables, if polluted water has been used for irrigation or processing. *E. histolytica* does not grow in foods or in water and does not multiply in the environment outside of a suitable host.

Cysts are destroyed by heat and *E. histolytica* is not normally associated with cooked and processed foods, unless recontamination from an infected food handler has occurred. Any food that may come into contact with contaminated water or infected food handlers during production, and where there is no subsequent process that will destroy cysts, may be at risk from *E. histolytica* contamination. However, food is not considered to be a major vehicle for the transmission of the parasite. The waterborne and person-to-person transmission routes are thought to be much more common.

Hazard Characterisation

Effects on Health

E. histolytica can cause an acute gastrointestinal infection (amoebiasis) in humans, and may become invasive in a few cases. The trophozoites multiply in the gastrointestinal tract, particularly in the colon, and occasionally invade the cells of the intestinal mucosa by producing proteases. The trophozoites have also been reported to produce toxins. The incubation time for the infection is very variable, but is usually between 1 to 4 weeks from ingestion of cysts. The majority of cases are asymptomatic, but about 10% of those infected suffer mild gastroenteritis symptoms of slight diarrhoea and abdominal discomfort. In some cases, more severe symptoms of acute colitis develop, characterised by bloody diarrhoea, high temperature, fever and severe lower abdominal pain. This condition is generally referred to as amoebic dysentery. Symptoms can be long lasting and may persist for several weeks, or even months. Very rarely, other tissues, notably the liver, may be invaded and abscesses can be formed. Chronic invasive amoebiasis is a serious disease and can be fatal. Immunocompromised individuals are particularly vulnerable to severe infections.

The infective dose is thought to be very low and, in theory, ingestion of a single cyst may be enough to cause amoebiasis. Infected individuals shed large numbers of infectious cysts in their faeces, and this may continue long after symptoms have subsided. Asymptomatic carriers have also been reported to shed cysts in their faeces over long periods, possibly several years in some cases.

Incidence and Outbreaks

E. histolytica is probably the most commonly reported intestinal parasite worldwide. It was previously estimated that approximately 500 million people worldwide were infected with *E. histolytica*, but it is now accepted that the majority of those people are carriers of non-pathogenic *E. dispar*. The true figure for the number of cases of infection with *E. histolytica* is now estimated to be about 50 million worldwide. The infection is also estimated to cause between 50 000 and 100 000 deaths each year, mostly in developing countries.

In England and Wales between 1990 and 2006, there has been a downward trend in the number of confirmed cases of *E. histolytica* infection from a peak of 1017 cases in 1991 to just 89 in 2006. Most of these cases are thought to be associated with foreign travel.

Between 1990 and 1994 (the most recent national figures) approximately 3000 cases of amoebiasis were reported each year in the USA. The majority of these are associated with foreign travel or occurred in recent immigrants. The incidence is reported to be higher in states along the southern border with Mexico.

There have been few documented outbreaks of amoebiasis in developed countries and none that were definitely foodborne, despite the high incidence of the disease in many developing countries. A large outbreak associated with contaminated drinking water occurred in Chicago in 1933. This affected at least 1000 people with 58 deaths. Infected food handlers have been suspected of causing isolated cases of amoebiasis, but the incubation period for the infection is often too long to identify the source with much certainty. Foodborne outbreaks are probably quite common in developing countries where there is a high incidence of the disease, but waterborne and person-to-person transmission are thought to be more important.

Sources

E. histolytica is an obligate parasite and thus originates from the host. The primary source of *E. histolytica* cysts is therefore the faeces of infected humans, many of whom do not display symptoms. Carriers may shed up to 15 million cysts each day in faeces. The cysts are infectious and may be transferred to food via an infected food handler, or through polluted water used for crop irrigation or processing.

E. histolytica cysts are larger than those of *Cryptosporidium* (10–15 μm diameter) and are not so difficult to remove from water using modern water-treatment methods, such as filtration. Amoebiasis is most often associated with conditions of poor sanitation and inadequate treatment of drinking water.

Stability in Foods

E. histolytica are relatively resistant to environmental factors, other than heat and desiccation. Cysts can remain infectious for some time in cool, moist conditions. However, there is relatively little published information on their survival and inactivation in foods.

E. histolytica cysts are not especially heat resistant and are reported to be destroyed by heating at temperatures above 50 °C and by conventional milk pasteurisation. Therefore, most controlled cooking processes used in food production should destroy any viable cysts in the product.

The cysts are relatively resistant to chlorine at the levels used in conventional water treatment, but are reported to be destroyed by 1% solutions of sodium hypochlorite.

Control Options

Control measures for *E. histolytica* in food processing focus largely on the control of contamination in water and the management of infected food handlers.

Processing

Care should be taken to ensure that raw-food ingredients and products that do not undergo further processing do not come into contact with contaminated surface water. In high-risk areas, fresh produce should be obtained from suppliers practicing good agricultural practice. Fresh produce and other raw foods should only be washed/processed using potable quality water.

Heat processing is an effective control against *E. histolytica* cysts in food. Normal milk-pasteurisation processes are effective, as are recommended "Listeria cook" processes for meat products (70 °C for at least 2 min). Reheating cooked foods to at least 74 °C will destroy cysts immediately.

Hygiene

Infected food handlers are also a major *E. histolytica* contamination risk for foods that do not undergo any further processing, such as sandwiches and salads, and for the recontamination of cooked foods. Good personal hygiene practice, especially hand washing, is an essential control and any staff suffering from gastroenteritis, especially following foreign travel, should be excluded from processing areas.

Legislation

E. histolytica is not usually mentioned specifically in food safety and hygiene law.

Sources of Further Information

Published

Leber, A.L. Intestinal amebae. *Clinics in Laboratory Medicine*, 1999, 19(3), 601–19.

Foodborne Parasites. ed. Y.R. Ortega. New York, Springer Science and Business, 2006.

Erickson, M.C. and Ortega, Y.R. Inactivation of protozoan parasites in food, water, and environmental systems. *Journal of Food Protection*, 2006, 69(11), 2786–808.

On the Web

The Entamoeba homepage. http://homepages.lshtm.ac.uk/entamoeba/

CDC parasitic disease information – amebiasis. http://www.cdc.gov/NCIDOD/DPD/parasites/amebiasis/default.htm

Food Research Institute Briefing – Foodborne Parasites. http://www.wisc.edu/fri/briefs/parasites.pdf

1.3.1.4 *Giardia*

Hazard Identification

What is Giardia?

Giardia is a single-celled flagellate protozoan parasite belonging to the order *Diplomonadida*. The cells are unusual in having two nuclei. The species important in human illness is *Giardia intestinalis* (previously referred to as *G. lamblia*, or *G. duodenalis*). *G. intestinalis* is also found in a number of domestic and wild animals, including cattle, cats and dogs.

G. intestinalis is an obligate parasite and requires a host in order to multiply. It was first discovered in 1859, but was not confirmed as a human pathogen until the late 1970s. It is a cause of gastrointestinal infection (giardiasis) in humans and some other animals, and is found worldwide.

G. intestinalis has a two-stage life cycle, and exists in two forms. Pear-shaped flagellated trophozoites exist and multiply within the gastrointestinal tract of the host. Some of these form spore-like resistant cysts within the small intestine. Both forms may be excreted in the host's faeces, but the trophozoites die quickly and the transmissible stage in the cycle is the resistant, thick-walled cyst.

Occurrence in Foods

G. intestinalis is mainly associated with surface water that has been polluted by human or animal faeces, but cysts have also been found in a number of unprocessed foods, including root crops, lettuce, herbs and strawberries. *G. intestinalis* cannot grow in foods or in water and does not multiply in the environment outside of a suitable host.

Cysts are destroyed by heat and *G. intestinalis* is not normally associated with cooked and processed foods. Any food that may come into contact with contaminated water during production, and where there is no subsequent process that will destroy cysts, may be at risk from *G. intestinalis* contamination. However, food is not considered to be a major vehicle for the transmission of the parasite. The waterborne and person-to-person transmission routes are thought to be much more common. Animal-to-human (zoonotic) transmission may also occur, but the significance to human health of *G. intestinalis* in livestock and domestic animals is not clear.

Hazard Characterisation

Effects on Health

G. intestinalis can cause an acute gastrointestinal infection in humans, and children are especially vulnerable to infection. The mechanism by which it causes disease is unclear. The trophozoites attach to the epithelial cells lining the gut, but do not seem to invade the cells. They may produce a toxin in the small intestine, but this has not been confirmed. The incubation time for the infection is usually between 1–3 weeks from ingestion of cysts. The main symptom is diarrhoea, often accompanied by abdominal pain. Flatulence, fever and loss of appetite may also occur. In healthy adults, symptoms typically last for 1–2 weeks, but may last for up to 6 weeks in some cases. The infection is generally self-limiting in most cases, but drug treatment is sometimes required. However, in immunocompromised individuals, infection can be more serious and long lasting, requiring hospital treatment, and occasional deaths have been recorded. Complications of chronic giardiasis may include, severe weight loss, the development of lactose intolerance and possibly reactive arthritis.

The infective dose is thought to be very low and ingestion of as few as 10 cysts (trophozoites are virtually non-infective) may be enough to cause giardiasis. Infected individuals shed very large numbers of infectious cysts in their faeces, and this may continue for months after symptoms have subsided. Asymptomatic cases of infection are quite common and asymptomatic carriers have been reported to continue shedding cysts for years.

Incidence and Outbreaks

G. intestinalis is probably the most commonly reported intestinal parasite in the developed world. In England and Wales between 1986 and 1996, the number of reported cases each year generally ranged from 5000 to 7000, but from 1996 to 2006 the number of confirmed cases fell and now averages around 3000 cases each year.

The most recent data for the EU refers to 2005 and shows a total of 15 103 reported cases of giardiasis from 18 countries. However, there are large differences between surveillance systems in different European countries and there is likely to be significant under-reporting. The European countries with the highest reported incidence were Estonia (24.28 cases per 100 000 people) and Iceland (14.65 cases per 100 000 people). The results also show that children aged 0–4 years were most commonly infected and that there are seasonal peaks of infection in spring and autumn. Giardiasis is a notifiable disease in much of the EU and in the USA.

There were 20 075 reported cases of giardiasis in the USA in 2005. This figure has been relatively stable in recent years. Most cases were reported from the northern states and there was a peak in the summer and early autumn.

The incidence of giardiasis in New Zealand is reported to be relatively high (46.5 cases per 100 000 in 2000), with a peak of infection in the autumn.

There is little or no information about the proportion of reported cases that are foodborne, but it is thought likely that the majority are caused by contact with contaminated water, infected people, and occasionally animals.

Most reported outbreaks of giardiasis are associated with contaminated surface water, or person-to-person transmission. Most of the documented outbreaks have been recorded in the USA, and outbreaks in Europe appear to be rare. Foodborne outbreaks have also been recorded in the USA, usually caused by an infected food handler, or by faecal contamination, either direct or through polluted water. Outbreaks have been linked to salad, lettuce and tomatoes, noodle salad, canned salmon, cheese dip, sandwiches, fruit salad and ice.

Sources

G. intestinalis is an obligate parasite and thus originates from the host. The primary source of *G. intestinalis* is therefore the faeces of infected humans and animals, which may contain up to 10^9 cysts in a single day. The cysts are extremely infectious and may be transferred to food via an infected food handler, or through polluted water used for crop irrigation or processing.

G. intestinalis cysts are larger than those of *Cryptosporidium* (9–12 μm diameter) and are not so difficult to remove from water using modern water-treatment methods. They are also less resistant to chlorine, but are not inactivated by the concentrations normally used to treat water. They are much less likely to pass through water-treatment plants into the public water supply system.

Stability in Foods

G. intestinalis cysts are generally resistant to environmental factors. Cysts can persist for months in cool, moist conditions and have been shown to survive for 8 days on the leaves of herbs. However, there is little information on their survival and inactivation in foods.

The cysts are relatively resistant to some sanitisers and disinfectants, notably chlorine and ozone, but are reported to be inactivated by phenolic disinfectants.

G. intestinalis cysts are not especially heat resistant and are destroyed by conventional milk pasteurisation. A temperature of 60–70 °C for 10 min is reported to inactivate cysts completely. Therefore, most controlled cooking processes used in food production should destroy any viable cysts in the product.

Oocysts can survive for significant periods at temperatures below 0 °C, especially in water, but frozen storage is reported to cause inactivation.

There is little information on the effect of pH, but it has been reported that cysts are resistant to low pH values down to about 3.0.

Control Options

Control measures for *G. intestinalis* in food processing focus largely on the control of contamination in water and the management of infected food handlers.

Processing

Care should be taken to ensure that raw-food ingredients and products that do not undergo further processing do not come into contact with contaminated surface water. Fresh produce should be obtained from suppliers practicing good agricultural practice. Fresh produce and other raw foods should only be washed/processed using potable quality water.

Heat processing is an effective control against *G. intestinalis* cysts in food. Normal milk-pasteurisation processes are effective, as are recommended "Listeria cook" processes for meat products (70 °C for at least 2 min). Reheating cooked foods to at least 74 °C will destroy cysts immediately.

Freezing foods for at least 7 days is also an effective control.

Hygiene

Infected food handlers are also a major *G. intestinalis* contamination risk for foods that do not undergo any further processing, such as sandwiches and salads. Good personal hygiene practice, especially hand washing, is an essential control and any staff suffering from gastroenteritis should be excluded from processing areas.

Legislation

G. intestinalis is generally considered to be a waterborne pathogen rather than foodborne. It may therefore be covered in drinking-water regulations, as is the case in the UK, but is not usually mentioned specifically in food safety and hygiene law.

Sources of Further Information

Published

Dawson, D. Foodborne protozoan parasites. *International Journal of Food Microbiology*, 2005, 103(2), 207–27.

Foodborne Parasites. ed. Y.R. Ortega. New York, Springer Science and Business, 2006.

Erickson, M.C. and Ortega, Y.R. Inactivation of protozoan parasites in food, water, and environmental systems. *Journal of Food Protection*, 2006, 69(11), 2786–808.

Water quality for the food industry: management and microbiological issues CCFRA Guideline No. 27 (2000).

On the Web

CDC parasitic disease information – giardiasis. http://www.cdc.gov/NCIDOD/DPD/parasites/giardiasis/default.htm

NZFSA information sheet – Giardia intestinalis. http://www.nzfsa.govt.nz/science/data-sheets/giardia-intestinalis.pdf

Food Research Institute Briefing – Foodborne Parasites. http://www.wisc.edu/fri/briefs/parasites.pdf

1.3.1.5 *Toxoplasma*

Hazard Identification

What is Toxoplasma*?*

Toxoplasma is a single-celled protozoan parasite belonging to the subclass *Coccidia*. The species of significance to human health and food safety is *Toxoplasma gondii*.

Toxoplasma is an obligate parasite and requires a host in order to multiply. It has been known as the cause of a disease (toxoplasmosis) in humans for many years. *Toxoplasma* is able to infect humans, most other mammals and also birds, and has a worldwide distribution. However, the definitive hosts for *Toxoplasma gondii* are members of the cat family, including domestic cats.

Toxoplasma has a very complex life cycle, consisting of several stages and forms, and a wide range of intermediate host species, including humans. There are two transmissible stages in the cycle. One is a resistant, thick-walled spore, known as an oocyst, which is only present in the faeces of cats and becomes infective following sporulation in the environment. The second transmissible stage is microscopic infective tissue cysts, which are found in the muscles of a number of intermediate hosts.

Occurrence in Foods

Toxoplasma oocysts may be present on raw foods, such as home-grown fresh produce, that have been contaminated by cat faeces. Contaminated water has also been implicated as a source of infection and it has been suggested that shellfish may retain oocysts when growing in contaminated seawater.

However, the presence of tissue cysts in meat is probably of more significance from a food safety point of view. Infective tissue cysts have been found in a wide range of domestic and wild species, but infected pork is considered to be particularly important in the transmission of toxoplasmosis to humans. Tissue cysts have also been found in sheep and goat meat, rabbit, horse and deer meat and in poultry, but have rarely been observed in meat from cattle. Beef and veal are considered to be much less significant than pork as a source of infection, but there is some uncertainty about their true importance. The number of tissue cysts in the meat of infected animals is generally low and has been estimated as approximately one cyst per 100 g of meat.

Unpasteurised goat's milk has been implicated as a source of toxoplasmosis, but there are no reports of cow's milk causing infection.

Oocysts are destroyed by heat and *Toxoplasma* is not normally associated with cooked and processed foods, although raw and undercooked meats containing tissue cysts carry a high risk of infection. Cured pork has also been identified as a risk factor in epidemiological studies. The main routes of transmission are from animal to human (zoonotic), either by ingestion of oocysts through direct contact with cat faeces, contaminated water, or food, or by ingestion of tissue cysts in raw or undercooked meat from an infected animal. Infection can also occur by handling infected meat and subsequent ingestion of tissue cysts. Direct person-to-person transmission has not been reported. *Toxoplasma* cannot grow in foods or in water and does not multiply in the environment outside of a suitable host.

Hazard Characterisation

Effects on Health

Toxoplasma gondii infection in humans is thought to be very common, but is usually asymptomatic. On ingestion of sporulated oocysts, or viable tissue cysts, an invasive stage of the parasite, known as tachyzoites, are eventually released in the gut and enter the body through the wall of the intestine. They then migrate through the body and invade various tissues, subsequently multiplying and forming cysts. This process is not usually noticed by the host and no clinical symptoms are reported, but in about 15% of cases, invasion of the tissues is accompanied by self-limiting mild flu-like symptoms and swelling of the lymph nodes. In a very few cases, more serious symptoms may develop, including visual impairment and brain damage, sometimes proving fatal. Where symptoms do occur, the incubation time is generally from 3–20 days.

Certain specific groups of the population are at risk of serious disease from infection by *Toxoplasma*. Infection in pregnant women may result in the tachyzoites crossing the placenta and invading the developing foetus. This infection can cause the death of the foetus in 3–4% of cases and often leads to long-term disease (congenital toxoplasmosis) in the rest. This may take various forms, most commonly visual impairment or blindness, but also including mental retardation, convulsions, and in a few cases, hydrocephalus. In some countries, including France and Austria, pregnant women are routinely screened for *Toxoplasma gondii* infection.

Immunocompromised individuals are also at serious risk from toxoplasmosis, particularly those suffering from AIDS. In these cases the brain and central nervous system are often affected and symptoms may include encephalitis, but other organs may also be affected. Between 10 and 30% of AIDS patients with toxoplasmosis are estimated to die from the infection.

There is also some epidemiological evidence that infection with *Toxoplasma gondii* may be involved in behavioural changes in humans and may have a role in the development of some psychotic illnesses, particularly schizophrenia.

The infective dose is uncertain, but is probably quite low.

Incidence and Outbreaks

Toxoplasma gondii is one of the commonest parasitic infections in humans, and it has been estimated that at least a third of the world's population has been exposed to the parasite. Approximately 30% of adults in the UK are estimated to carry antibodies to *Toxoplasma gondii*, with the estimates for other European countries ranging from 50–80%. The figure for the USA is thought to be around 23%. The vast majority of these cases are asymptomatic. Recent studies using data from the USA and Europe have estimated that 50–60% of cases may be associated with the food chain. The number of foodborne cases is probably higher in countries where raw or rare cooked meat is a regular part of the diet.

The incidence of clinical toxoplasmosis is much lower. For example, in the UK, only 79 cases were recorded in the last year for which figures are available (2004). In the EU there were 1519 cases reported in 16 countries in 2005, with the highest incidences being reported in Lithuania (6.86 cases per 100 000 people) and Slovakia (4.85 per 100 000) the overall incidence for Europe is estimated to be 0.84 per 100 000. However, it is likely that there is considerable under-reporting of the disease.

Estimates for the incidence of acute toxoplasmosis in the USA suggest that as many as 1.5 million people each year suffer symptoms. It is also estimated that there are between 400 and 4000 cases of congenital toxoplasmosis each year.

It is difficult to identify foodborne outbreaks of toxoplasmosis because of the relatively long incubation time and the high proportion of asymptomatic cases. However, outbreaks have been reported in a number of countries, including the UK, the USA, Australia, Korea and Brazil, usually associated with raw, or undercooked meat. A large waterborne outbreak, in which more than 100 people suffered acute toxoplasmosis, was reported in Canada in 1994–5. The outbreak was caused by a contaminated water supply that was chlorinated but not filtered.

Sources

Toxoplasma gondii is an obligate parasite and thus originates from the host animal. The only source of infectious oocysts is the faeces of members of the cat family, with domestic cats being the commonest source in most parts of the world. Infected cats shed very large numbers of oocysts in their faeces, but usually only for short periods (1–2 weeks). The oocysts persist in the environment for long periods and may be present in surface water and on fruit and vegetables grown in contaminated soil. Insect activity may also help to distribute the oocysts from contaminated soil.

The tissue cysts can be present in the flesh of any infected mammal and also in poultry. The most important source of tissue cysts for human infection is considered to be pig meat, but all other food animals are also potential sources of infection, although beef and veal are considered to present a much lower risk.

Stability in Foods

Toxoplasma oocysts are relatively resistant to most environmental factors. Oocysts have been reported to remain infectious for up to 400 days in water and also persist for long periods in soil. Sporulated oocysts are inactivated by freezing at –21 °C for 1–7 days and unsporulated oocysts are also inactivated to some extent at this temperature. Sporulated oocysts are reported to be gradually inactivated by drying.

The oocysts are relatively resistant to some sanitisers and disinfectants, and may not be inactivated by levels of chlorine normally used in drinking-water treatment. *Toxoplasma* oocysts are not reported to be especially heat resistant and are likely to be destroyed by conventional milk pasteurisation.

Tissue cysts in meat are able to survive at refrigeration temperature (4 °C) for several weeks, but are not heat resistant and will be destroyed by proper cooking processes. Cysts in pork are reported to be killed in 44 s at 55 °C and in 6 s at 61 °C. However, rare cooked meats may not achieve an internal temperature sufficient to kill all cysts.

Tissue cysts are also inactivated by freezing at temperatures of less than 10 °C and are destroyed by irradiation at a dose of 1 kGy. The cysts are thought to have some susceptibility to curing agents used in meat, but raw cured pork has been identified as a risk factor for human infection.

Control Options

Control of *Toxoplasma* in food is achieved principally by implementing good practice in meat production and by proper cooking of high-risk meats, such as pork.

Primary Production

Infection of pigs and other food animals by *Toxoplasma gondii* can be controlled to some extent by minimising potential exposure to cat faeces using best practice biosecurity measures. However, this is difficult to achieve for animals kept outdoors.

Fruit and vegetable growers should also adopt measures to exclude cats from fields where produce for human consumption is grown.

Processing

Good hygiene practice at slaughter and in meat processing is important to prevent cross-contamination between infected carcasses and *Toxoplasma*-free animals, since the cysts can be carried on the skin or on soiled equipment and utensils.

Tissue cysts in meat, especially in pork and mutton, are destroyed by heat and ideally all meat should be cooked to an internal temperature of at least 70 °C to ensure inactivation of cysts. Inactivation of cysts can also be achieved by freezing meat at –12 °C or less.

Fruit and vegetables should be washed thoroughly before consumption to remove occysts.

Hygiene

Good personal hygiene practice, especially hand washing, is an essential cross-contamination control when handling raw meat and is also important when preparing fruit and vegetables.

Vulnerable consumers, such as pregnant women and the immunocompromised should avoid direct contact with raw meat, especially pork.

Legislation

Toxoplasmosis is a notifiable disease in some developed countries, but *Toxoplasma gondii* is not usually mentioned specifically in food safety and hygiene law.

Sources of Further Information

Published

Montoya, J.G. and Liesenfeld, O. Toxoplasmosis. *The Lancet*, 2004, 363(9425), 1965–76.

Foodborne Parasites. ed. Y.R. Ortega. New York, Springer Science and Business, 2006.

Dawson, D. Foodborne protozoan parasites. *International Journal of Food Microbiology*, 2005, 103(2), 207–27.

On the Web

CDC parasitic disease information – toxoplasmosis. http://www.cdc.gov/NCIDOD/DPD/parasites/toxoplasmosis/default.htm

USDA Agricultural Research Service – toxoplasmosis. http://www.ars.usda.gov/Main/docs.htm?docid=11013

NZFSA information sheet – Toxoplasma gondii. http://www.nzfsa.govt.nz/science/data-sheets/toxoplasma-gondii.pdf

NZFSA risk profile: Toxoplasma gondii in red meat and meat products. http://www.nzfsa.govt.nz/science/risk-profiles/toxoplasma-gondii-in-red-meat.pdf

Food Research Institute Briefing – Foodborne Parasites. http://www.wisc.edu/fri/briefs/parasites.pdf

1.3.2 NEMATODES

1.3.2.1 Anisakids

Hazard Identification

What are Anisakids?

The anisakids are a family of parasitic marine nematode worms that can cause a potentially serious infection (anisakiasis or anisakidosis) in humans following consumption of infected seafood. The principal species identified in human infection is *Anisakis simplex* (whale worm or herring worm), but the closely related species *Pseudoterranova decipiens* (seal worm or cod worm) may also be found in humans. Other related marine nematodes, such as *Contracaecum* spp. and *Hysterothylacium* spp., have been implicated in human infections, but these have only very rarely been reported in developed countries.

Anisakids are found in the marine environment worldwide and have a very complex life cycle involving a number of hosts. Humans are only an incidental host to the infective third stage (L3) larvae, which may occur in the viscera and muscle tissue of infected fish. The larvae rarely reach the adult stage in humans and are eventually expelled from the gut, or die in the tissues.

Occurrence in Foods

The infective L3 larvae of *Anisakis simplex* and other species occur in the viscera and muscle tissue of infected fish as small, but visible cysts containing the coiled, 2–3 cm long larva. There is evidence that the larvae migrate from the viscera into the muscle tissue when the intermediate host dies. A number of food-fish species are known to act as intermediate hosts, including whitefish such as cod, whiting and haddock, herring, monkfish, mackerel and salmon. Some species of squid may also contain live larvae. Where infection is heavy, it may be obvious on examination of the fish flesh, especially in whitefish, but for fish with pigmented flesh the presence of the larvae may be much less obvious.

Fresh fish is the principal vehicle for *A. simplex* infection in humans, especially if it is eaten raw or undercooked. The larvae die quite quickly in fish that is frozen and do not survive effective cooking, and so processed fish and seafood products present a negligible risk of infection. However, the larvae may survive in some fermented, lightly salted, or cold-smoked and marinated fish products, such as pickled herrings and gravadlax. The growing trend for consumption of raw and lightly cooked fish, such as sushi and sashimi, in the West is thought to be increasing the likelihood of human infection with anisakid worms.

Wild fish are considered to carry a much higher risk of infection than farmed fish. Surveys of fish on sale in markets around the world generally show that a significant proportion (approximately 10–30%) is infected with live L3 *A. simplex* larvae. However, a survey of Norwegian farmed salmon found no infected fish, even though the parasite is quite common in wild salmon. This may be because farmed fish are not able to feed on infected intermediate hosts, such as copepods and other small invertebrates.

Hazard Characterisation

Effects on Health

The gastrointestinal tract of humans resembles that of marine mammals sufficiently for ingested live *A. simplex* and other anisakid larvae to survive for a short time, but most ingested larvae die in the gastrointestinal tract. However, in some cases they may cause a potentially serious acute infection known as anisakiasis, or anisakidosis. This occurs when the L3 larvae burrow into the wall of the digestive tract in the stomach or intestine and occasionally penetrate the gut wall completely, entering the body cavity. This process is often accompanied by severe abdominal pain, nausea and vomiting and the larvae may sometimes be coughed up. Symptoms usually occur within a few h of ingestion. An inflammatory response is also produced, which occasionally leads to the formation of an abscess (eosinophilic granuloma) surrounding the worm. When this occurs in the intestine, symptoms similar to those of Crohn's disease (abdominal pain, diarrhoea and bleeding) may develop after 7–14 days. Abdominal pain can persist for several weeks until the larvae in the gut are expelled, or those that have penetrated the tissues die. In severe cases, the pain is extreme and may require surgical removal of the larvae.

There is also evidence that ingestion of the L3 larva of *A. simplex* can cause a hyperimmune allergic reaction in some individuals. This may be associated with symptoms such as skin rashes (urticaria), asthma and even anaphylactic shock in a few cases.

Incidence and Outbreaks

It is estimated that approximately 2000 people worldwide suffer from the symptoms of anisakiasis each year and the incidence is thought to be increasing as consumption of raw fish becomes more popular. The highest incidence is in Japan (1000 cases per year), where fresh fish makes up a high proportion of the diet. The annual number of cases in Europe is estimated to be about 70, with the highest incidences being recorded in the Netherlands, Germany, France and Spain. According to the FDA about 10 cases of anisakiasis are reported in the USA each year. Outbreaks have been reported in Japan, the Netherlands and Spain.

Although the global incidence of anisakiasis is quite low, there is evidence that exposure to the parasite is much higher in some countries. Many individuals who ingest *A. simplex* larvae do not develop acute symptoms, but may develop specific antibodies to the larvae. A survey of over 34 000 people with skin rashes, or symptoms of seafood allergy, in Japan found that almost 30% had antibodies specific to *A. simplex* in their blood. Similar findings have been reported from Spain. This appears to indicate a more widespread exposure in the population and suggests that allergy caused by *A. simplex* L3 larvae may be more common than expected. However, *A. simplex* allergy is highly cross-reactive with other allergies and is difficult to diagnose.

Sources

The definitive hosts for the adult worms are marine mammals, including whales and dolphins (*Anisakis*) and seals (*Pseudoterranova*), but the various larval stages infect intermediate hosts, including copepods and other small invertebrates, fish and squid. The adult worms live in the gut of marine mammals and eggs are expelled in the faeces. Free-swimming larvae hatch from the eggs once in the marine environment and may be eaten by small crustaceans. The larvae then develop into L3 third stage larvae, which are infective to fish and squid that feed on the infected crustaceans. The larvae penetrate the gut of the infected fish and grow in the viscera, but appear to migrate to muscle tissue when the host dies. The life cycle is then completed when the infected fish are consumed by marine mammals.

Anisakids are found in seawater worldwide, but are less common in fish populations in areas where marine mammals are rare. The rate of infection may also be seasonal and may be affected by water temperature. They do not occur in fresh water.

Stability in Foods

Infective L3 larvae are able to survive in the flesh of dead fish for some time, but are killed by freezing. They are not heat resistant and are killed by temperatures above 60 °C. However, they may survive "cold smoking", marinating and fermentation processes applied to fish.

Control Options

The principal control for anisakid infections in wild fish is visual inspection. The larvae can be seen by "candling" or inspection on a light table, but this is less effective for fish such as salmon that have pigmented flesh. It is possible to physically remove the larvae, but obviously infected fish should not be consumed. Inspection cannot be guaranteed to detect all larvae in infected fish.

Processing

Since the larvae may migrate from the viscera of infected fish into the muscular tissue after death, it is important to ensure that fish are gutted as soon as possible after capture to minimise this migration.

Fish that will be eaten raw or lightly cooked should be frozen at –20 °C or less for at least 24 h to kill the larvae. This should also apply to fish intended to be cold-smoked, fermented, or marinated before consumption.

Hot smoking processes where an internal temperature of at least 60 °C is attained will destroy the larvae, as will cooking to a temperature of 70 °C for at least 2 min. However, cooked and frozen fish may still cause an allergic reaction, as the allergen appears to be quite heat stable.

Legislation

In EU legislation measures to protect consumers against anaskiasis are contained in a directive (91/493/EEC) on sanitary measures for the production and sale of seafood. This legislation requires inspection of fish for parasites, and the removal of obviously infected fish from sale. Fish to be eaten raw must be frozen at –20 °C or less for at least 24 h, as must certain species intended for cold smoking, marinating or salting.

In the USA, the FDA Food Code recommends rapid freezing followed by storage at –20 °C or less for at least 24 h for fish intended for consumption without cooking.

Sources of Further Information

Published

Valls, A., Pascual, C.Y. and Martin Esteban, M. Anisakis and anisakiosis (in Spanish). *Allergologia et Immunopathologia*, 2003, 31(6), 348–55.

Foodborne Parasites. ed. Y.R. Ortega. New York, Springer Science and Business, 2006.

Chai, J.Y., Darwin Murrell, K. and Lymbery, A.J. Fish-borne parasitic zoonoses: status and issues. *International Journal of Parasitology*, 2005 35(11–12), 1233–54.

On the Web

CDC parasitic disease information – anisakiasis. http://www.cdc.gov/NCIDOD/DPD/parasites/anisakis/default.htm

European Commission Scientific Committee opinion on allergic reactions to ingested Anisakis simplex (1998). http://ec.europa.eu/food/fs/sc/scv/out05_en.html

FAO manual on assessment and management of seafood safety and quality – parasites. http://www.fao.org/docrep/006/y4743e/y4743e0c.htm

Food Research Institute Briefing – Foodborne Parasites. http://www.wisc.edu/fri/briefs/parasites.pdf

1.3.2.2 *Trichinella*

Hazard Identification

What is Trichinella*?*

Trichinella is a genus of parasitic nematode worms that can cause a potentially serious infection (trichinellosis or trichinosis) in humans following consumption of infected meat. *Trichinella* was first described as a cause of disease in man as early as 1865. Up to ten species (or genotypes) have been described, at least seven of which can infect man, but the principal species identified in human infection, and the species of most concern to the food industry is *Trichinella spiralis*. The other recognised species identified in human cases are *T. britovi, T. pseudospiralis, T. nativa, T. murrelli, T. papuae* and *T. nelsoni*, but these are less commonly found than *T. spiralis* and are usually associated with wild animals.

Trichinella species are found worldwide and infect a wide variety of animal hosts, mostly carnivorous and omnivorous wild mammals, especially those that scavenge, such as foxes, bears, pigs and wild boar. Rodents, such as rats and mice are also thought to play an important role as hosts in areas where the infection is endemic. The entire life cycle normally occurs within a single host species and consists of an adult worm and two larval stages. Humans are not definitive hosts, but may become infected by ingesting the infective second-stage larvae, which may occur in cysts in the striated muscle tissue of infected animals.

Occurrence in Foods

The infective second stage larvae of *Trichinella* occur in the muscle tissue of infected animals as very small, but detectable, cysts containing the larva. *T. spiralis* cysts are found in highest numbers in the diaphragm and tongue of the infected animal but can also occur in the skeletal muscles. Historically, infected pork from pigs fed with feed containing animal waste was the principal source of *Trichinella* infection in Europe and North America, but successful controls in pork production have greatly reduced the prevalence of infection in commercial herds. The prevalence in commercial pig herds in the EU has recently been estimated at less than 1 in 100 000 animals, with some variation between countries. In the USA, the prevalence of infection in commercial production has been reduced from an estimated 1.41% in 1900, down to 0.013% (13 in 100 000 animals) in 1995.

However, there is still some risk from home-raised pigs and from pigs that are allowed to forage for food in the natural environment, which may include organically produced pigs. There is also a significant risk of infection from wild animals, especially wild boar in parts of Europe and bears in the USA. Imported horsemeat is also now a very significant source of infection in parts of Europe, especially France and Italy.

Raw, or undercooked meat is the principal vehicle for *Trichinella* infection in humans. The larvae do not survive effective cooking, and properly cooked pork and other meats present a negligible risk of infection. However, the larvae may survive in raw cured meats and some *Trichinella* species larvae are not killed by freezing. Therefore, lightly processed and frozen pork or wild game products may still carry the risk of infection.

Hazard Characterisation

Effects on Health

The severity of trichinellosis infection in humans is highly variable. It may be asymptomatic in some cases, while in others complications may prove fatal. The severity of infection seems to be correlated with a number of factors, including the *Trichinella* species involved, the number of encysted larvae ingested and the strength of the immune response in the patient. The minimum infective dose is uncertain but has been estimated at between 100 and 300 live larvae.

After ingestion the larvae are released from the cysts by stomach acid and digestive enzymes and invade the lining of the small intestine, where they develop into adults. This process may be accompanied by gastrointestinal symptoms, including abdominal pain, vomiting and diarrhoea. Onset of symptoms typically occurs 24–48 h after ingestion, but may take longer. After about seven days the adult females release live larvae that migrate through the tissues to the striated muscles where they form cysts. This stage usually takes 4–8 weeks to complete and produces a different range of symptoms, including swelling of the face and around the eyes, fever, muscle pain, conjunctivitis and rashes. The production of the cysts usually causes muscle pain and weakness, but once it has been completed, the symptoms largely disappear.

However, in some cases potentially serious neurological and/or cardiovascular complications may occur, producing a variety of symptoms, such as headache, apathy, dizziness, chest pains and an irregular heartbeat. Rarely, complications may be fatal, especially in elderly people.

Incidence and Outbreaks

It has been estimated that as many as 11 million people worldwide could be affected by trichinellosis. However, the incidence of the disease in most European and North American countries has been decreasing for many years. For example, in the USA between 1947 and 1951, the average number of reported cases each year was 393 and 57 people died from the disease. But from 1997–2001 the annual average was only 12 cases, with no deaths. In the EU, there has been a general downward trend in the incidence of trichinellosis over the last 12 years, and the number of reported cases has been stable since 2000. However, incidence varies considerably between different countries. In 2005, 153 cases were reported in 25 countries, with the highest incidences being recorded in Latvia, Lithuania, Poland and France. Many countries reported no cases, including the UK and

Germany. Elsewhere, relatively high incidences have been reported in Argentina (600 cases per year) Bulgaria (892 cases) and Romania (1744 cases).

(Note: These last two countries are now in the EU, but their incidence figures are not yet included in the annual total.)

Many outbreaks of trichinellosis have been reported all around the world. In the EU there have been significant outbreaks in the last 20 years. Most of these have occurred in Spain, France, Italy and Germany and were caused either by horsemeat imported from third countries, wild boar, or non-intensively raised pigs. An outbreak affecting 124 people in Poland in 2003 was also believed to have been caused by wild-boar meat. However, 52 cases reported in Germany in 1998–99 were linked to commercially produced raw sausages and minced meat.

Outbreaks in the USA have also been reported. In 1990, 105 people were affected in two outbreaks associated with raw sausages made from commercially produced pork. However, since that time, most outbreaks have involved foods prepared from wild game meat, including wild boar and bear.

Sources

Two distinct cycles for *Trichinella* are recognised by epidemiologists. The natural, or sylvatic cycle occurs in wild animals, especially carnivores that scavenge or exhibit cannibalistic behaviour. In this cycle, a number of the recognised *Trichinella* species are involved. The parasites develop in one host and infective encysted larvae are passed to another when infected tissues are consumed. In the domestic cycle, *Trichinella* (most commonly *T. spiralis*) circulate in farm-raised pigs that are fed with feed containing infected animal tissue, or are allowed to come into contact with other infected animals.

The domestic cycle is now much less important in developed countries than was once the case, following improvements in pig husbandry and in statutory controls. For example, in the USA between 1997 and 2001, 72 cases of trichinellosis were reported and only 12 of these were associated with commercial pork products. The remaining cases were caused by consuming wild game, or pork raised under unregulated conditions. In the EU, the most important sources of trichinellosis are now wild-boar meat, and horse meat imported from Eastern Europe. Some EU countries, including the UK, Ireland and Sweden, have not reported cases of human trichinellosis caused by locally produced meat products for at least 20 years.

Stability in Foods

The encysted larvae of *Trichinella* species are extremely persistent in the live host and may survive for many years in striated muscle tissue. Encysted larvae of *T. spiralis* are not resistant to freezing and are killed by rapid freezing and storage at –20 °C or below for at least 48 h. However, this may not be the case for other species of *Trichinella*. Infective *Trichinella* spp. larvae have been found in frozen bear meat after storage for more than two years. The larvae may also be able to survive some curing processes used for pork products. They are not heat resistant and are killed by temperatures above 60 °C for 2 min.

Control Options

The principal control for *Trichinella* in commercial meat products is inspection by a recognised direct detection method, usually tissue digestion followed by microscopic examination of the remaining sediment. This is mandatory for pork, horsemeat and game in the EU and in other developed countries. Infected meat is designated unfit for human consumption.

Primary Production

Improved animal husbandry has been very effective in reducing *Trichinella* infection in commercial pig herds. Measures include ensuring that all pig feed is adequately heat processed to destroy infective larvae, effective separation of pigs from rodents and other potentially infected animals and good on-farm hygiene practices.

Processing

The larvae of *T. spiralis* can be destroyed by freezing, cooking and by some curing procedures. The USDA has produced specific freezing and cooking times and temperatures for pork products and has also specified curing methods. Freezing times and temperatures are dependent on the size of the pieces of meat involved, but for cooking processes, fresh pork should reach a minimum internal temperature of 71 °C. The EU has also specified several freezing treatments that can be used to kill *T. spiralis* larvae in meat. These are detailed in the relevant legislation (see below).

Freezing cannot be relied upon to destroy the larvae of other *Trichinella* species that may be found in game meat and horses.

Legislation

In EU legislation measures to protect consumers against trichinellosis are contained in a Commission Regulation (EC) No 2075/2005. This covers inspection of meat at slaughter, detection methods and freezing procedures.

The US Code of Federal Regulations contains similar requirements and includes recommendations for freezing, cooking and curing of pork products.

Many countries have introduced legislation regulating aspects of animal husbandry, meat inspection and pork processing designed to protect consumers form trichinellosis.

Sources of Further Information

Published

Murrell, K.D. and Pozio, E. Trichinellosis: the zoonosis that won't go quietly. *International Journal of Parasitology*, 2000, 30(12–13), 1339–49.

Foodborne Parasites. ed. Y.R. Ortega. New York, Springer Science and Business, 2006.

On the Web

CDC parasitic disease information – trichinellosis. http://www.cdc.gov/NCIDOD/DPD/parasites/trichinosis/default.htm

New Zealand Ministry of Health – Trichinosis fact sheet. http://www.moh.govt.nz/moh.nsf/pagesmh/1232?Open

Trichinellosis surveillance – United States 1997–2001. http://www.cdc.gov/mmwr/preview/mmwrhtml/ss5206a1.htm

Food Research Institute Briefing – Foodborne Parasites. http://www.wisc.edu/fri/briefs/parasites.pdf

1.3.3 OTHER PARASITES

There are a large number of parasites that can cause human infection and many have the potential to be foodborne. However, most of these are now found only in tropical and subtropical regions, or in areas where standards of sanitation are poor. They are rarely found in developed countries, where infection is only likely to occur in people who have travelled to areas where these parasites are endemic. The preceding sections have dealt with those parasites that are known to present a foodborne risk to public health in developed countries, but there are certain other species that may present a food safety risk as a consequence of the growing globalisation of food supply chains.

Brief details are given below of some parasites that may have food safety significance. All are known to infect humans and may occur as contaminants in certain food commodities.

Protozoa

Balantidium Coli

Balantidium coli is a large (70 μm diameter) ciliate protozoan parasite that is normally associated with pigs, although other mammals, including rodents and non-human primates, may also act as reservoirs of infection. It occurs worldwide, but is most commonly reported in areas where pigs are raised in unsanitary conditions. Balantidiasis is endemic in some countries, such as Bolivia and the Philippines.

The infective stage in the life cycle is a cyst, which is passed in the host's faeces and may be present as a contaminant in polluted water or on food that has been contaminated by human or animal faeces. After ingestion, the cysts rupture to release trophozoites that colonise the large intestine and may invade the wall of the colon. Most cases of human infection are asymptomatic, but where symptoms occur, they generally include persistent diarrhoea, abdominal pain and weight loss. The illness resembles amoebic dysentery and can be severe, or even fatal in some cases.

Control of *Balantidium coli* infection can be achieved by effective water-sanitation measures and good food-hygiene practices.

Sarcocystis

Sarcocystis species are coccidian parasites that have a complex life cycle requiring two hosts, one a definitive predatory host and the other an intermediate prey species host. A number of species associated with specific hosts have been described and several of these can infect humans, including *S. hominis* and *S. suihominis*. *Sarcocystis* species have a widespread distribution and are common parasites in commercially raised cattle and pigs.

Humans can become infected by ingestion of infective oocysts excreted in the faeces of the definitive host, or by consumption of the meat of an intermediate host containing encysted larvae (sarcocysts). Consumption of raw or undercooked pork or beef containing sarcocysts may result in gastrointestinal illness with symptoms including nausea, diarrhoea and abdominal pain lasting for 24–48 h. In rare cases the parasites may invade the body causing a variety of more serious symptoms, including inflammation of muscular and vascular tissue, abortion and congenital disorders. Human outbreaks have been recorded in Europe and North America.

Controls include inspection of meat for the presence of sarcocysts, and effective cooking of beef and pork before consumption.

Nematodes

Ascaris

Ascaris lumbricoides is a very common nematode parasite, for which humans are the host. Infection is endemic in many developing countries, and it is estimated that 25% of the world's population may carry the infection.

The adult worms live in the intestine of the host and produce eggs that are passed out of the body in faeces. The eggs may be ingested in polluted water, or on foods contaminated with human faeces by irrigation or washing with polluted water. The ingested eggs hatch in the intestine and the larval stages may migrate to other tissues, including lungs and liver before they return to the intestine and mature. Many infections are asymptomatic, but the intestinal mucosa may be irritated, causing diarrhoea and affecting protein uptake. Very young children may suffer from diarrhoea and stunted growth if infected soon after birth. When tissue invasion occurs, infection of the liver or lungs can produce a severe acute illness.

Control of *Ascaris* infection can be achieved by proper water sanitation and good hygiene practice in food preparation.

Trematodes (Flukes)

Fasciola Hepatica

Fasciola hepatica is a parasitic liver fluke that commonly infects cattle and sheep in many developing countries. This parasite has a complex life cycle involving a larval stage in water snails. It also causes human infection in areas where water sanitation is inadequate, especially in parts of South America and North Africa. Large outbreaks have also occurred in the Middle East. Cases may sometimes occur in developed countries following consumption of contaminated fresh produce, especially watercress and other green vegetables grown in or near contaminated water. There may be some risk from imported salad greens.

Humans become infected when they ingest cysts in contaminated water or food. The cysts hatch and develop into adult flukes that inhabit the liver. Symptoms of infection include fever, abdominal pain and weight loss and there is some evidence for a link with liver tumours.

Control of *Fasciola hepatica* can be achieved largely by adequate water sanitation.

Paragonimus (Lung Fluke)

There are at least nine species of *Paragonimus* lung flukes that can infect the lungs of humans and other animals, including pigs, dogs and cats. They have a widespread distribution and a complex life cycle with at least two intermediate hosts, including freshwater snails and crabs, or crayfish. They may also infect other animals that feed on crustaceans.

Humans usually become infected by eating raw, or undercooked, crustaceans, but wild-boar meat has also been implicated in human infection in Japan. Infection is usually followed by gastrointestinal symptoms of diarrhoea, fever and abdominal pain. Later, coughing and chest pains may occur as the immature worms pass through the diaphragm and into the lungs. If large numbers of worms are ingested, they can cause chronic lung disease and may enter the central nervous system. In rare cases, infection can be fatal.

Paragonimus is quite resistant and is not destroyed by salting or pickling, but control can be achieved by adequate cooking of crabs and crayfish.

Cestodes (Tapeworms)

Taenia

Taenia species are tapeworms that parasitise a number of animals. Humans are the definitive hosts for two species, *Taenia solium* (the pork tapeworm) and *Taenia saginata* (the beef tapeworm), and are commonly infected by both. Other species have been reported to infect man on rare occasions. The intermediate hosts for *T. saginata* are cattle and pigs act as the intermediate host for *T. solium*, although some other species may be infected. Both species have a widespread distribution and human infection is common in areas where sanitation is inadequate. *T. solium* is rare in countries where pork is not eaten for religious reasons. It has been estimated that as many as 50 million people worldwide could be infected by both species each year.

Intermediate hosts of *Taenia* species become infected by the ingestion of eggs in human faeces. These hatch in the gut, producing larvae that migrate to the muscles and other tissues and form persistent cysts (cysticerci). Humans become infected by eating raw, or undercooked meat from an infected animal and ingesting viable cysticerci. Once in the human gut, these develop into the long-lived adult, which grows to a length of several metres and produces a continuous supply of eggs in the faeces. Infection may be asymptomatic, or may be accompanied by a range of symptoms, such as abdominal pain, constipation, or diarrhoea.

Humans can also serve as the intermediate host for *T. solium* if eggs are ingested. This can have serious, or even fatal, consequences as the larvae encyst in the tissues. Cysticercosis can affect the eyes and the brain and may cause various neurological symptoms, including severe pain, convulsions and paralysis. It has been estimated that cysticercosis may cause as many as 50 000 deaths each year worldwide.

Control of *Taenia* species in most developed countries has been achieved by improved sanitation and animal husbandry practices, together with effective meat inspection and adequate cooking processes, especially for pork.

Diphyllobothrium

Diphyllobothrium species are usually associated with freshwater fish and are often referred to as the fish tapeworms. *Diphyllobothrium latum* is the species most commonly associated with humans, who are one of the definitive hosts for the parasite, along with other fish-eating mammals such as bears. It has a complex life cycle, often involving several intermediate hosts, including copepods, small freshwater fish and larger predatory fish, such as pike and perch. It is common in some temperate regions of the northern hemisphere, such as the Great Lakes of North America, the Baltic and Russia. Infection is most common in countries where raw freshwater fish is eaten, such as Finland and Japan.

Humans become infected by eating raw, or undercooked, fish infected with *D. latum* larvae (plerocercoids). The plerocercoids develop into adult worms in the human gut and can grow very large (up to 10 metres in length). Infection is often asymptomatic, but may be accompanied by various symptoms, such as weight loss, abdominal pain and a type of anaemia. In some individuals, multiple infections with many worms can occur. Symptoms are more likely in these cases.

Control of *Diphyllobothrium* can be achieved by proper cooking of freshwater fish to kill the plerocercoids before consumption.

Echinococcus

There are four recognised species of *Echinococcus*, small tapeworms that normally parasitise members of the dog family. Two of these are of importance in human health in developed countries. *E. granulosus* is a tapeworm of dogs that can cause potentially serious disease (echinococcosis, or hydatid disease) in humans. Intermediate hosts are usually cattle, sheep and other grazing animals. The definitive host of *E. multilocularis* is the fox and the intermediate hosts are usually rodents. This species causes rare, but highly pathogenic alveolar echinococcosis in humans. *E. granulosus* is prevalent in many parts of the world, especially areas where animals are grazed, but *E. multilocularis* is largely restricted to the northern hemisphere. Human echinococcosis is regularly reported in Southern Europe, notably in Spain, Greece, Italy and Portugal and is also quite common in some parts of Eastern Europe. Cases are also occasionally reported in North America.

Humans become infected when they ingest the eggs of the tapeworm in contaminated water, or on unprocessed vegetables. The eggs hatch in the gut, releasing a larval stage called an oncosphere that penetrates the gut wall and migrates to other tissues, especially the liver and lungs. Once in place the oncospheres form cysts that gradually enlarge and produce daughter cysts. Symptoms are slight at first, but as the cysts grow, their size may eventually cause pain and other symptoms. Hydatid cysts caused by *E. granulosus* may finally rupture, causing hypersensitivity reactions, including anaphylactic shock, and the dissemination of new cysts. In alveolar echinococcosis caused by *E. multilocularis*, the cysts invade the tissues, usually the liver, in the same way as a slow-growing destructive tumour. Alveolar echinococcosis is normally fatal if not treated.

Control of *Echinococcus* species can be achieved by the proper destruction of the viscera of infected intermediate host species and by effective hygiene measures when washing and preparing vegetables. There is some concern that growing red fox numbers in Europe may cause an increase in cases of alveolar echinococcosis.

Sources of Further Information

Published

Foodborne Parasites. ed. Y.R. Ortega. New York, Springer Science and Business, 2006.

Chai, J.Y., Darwin Murrell, K. and Lymbery, A.J. Fish-borne parasitic zoonoses: status and issues. *International Journal of Parasitology*, 2005 35(11–12), 1233–54.

On the Web

FAO manual on assessment and management of seafood safety and quality – parasites. http://www.fao.org/docrep/006/y4743e/y4743e0c.htm

CDC parasitic disease information – index. http://www.cdc.gov/NCIDOD/DPD/parasites/index.htm

Cambridge University Schistosomiasis Research Group pages on cestodes (tapeworms). http://www.path.cam.ac.uk/~schisto/Tapes/Tapes_Gen/Tape.html

Food Research Institute Briefing – Foodborne Parasites. http://www.wisc.edu/fri/briefs/parasites.pdf

CHAPTER 1.4

Prions

Hazard Identification

What are Prions?

The term prion (pronounced "pree-on") is now used as a generic term for a group of small glycosylated proteins found mainly in the brain-cell membranes of humans and other mammals. The name was first used by Stanley Prusiner in 1982 to describe the infective agents for a group of invariably fatal diseases known as transmissible spongiform encephalopathies (TSEs), so called because of the "sponge-like" appearance of the brain in the later stages of the disease. The word prion was derived from the term "proteinaceous infectious particle." The role of prions in human disease is still the subject of some controversy, but the consensus of scientific opinion is that abnormal forms of these proteins can act as unconventional infective agents that can replicate without associated DNA or RNA, and are therefore not a form of life in the accepted sense.

Normal non-infective prions are benign, and like other proteins in that they have a three dimensional α-helical structure. Infective prions differ in that their structure is flattened into a form referred to as a β-sheet. These abnormal proteins are much less soluble than the normal version and much more resistant to enzymes. The hypothesis for prion infectivity proposes that when these abnormal proteins reach the brain, they are able to cause the normal prions to change their shape, so that they too assume a β-sheet structure. These altered prions then also become infective, resulting in a progressive change in conformation of the normal prion proteins in brain-cell membranes. This leads to the changes in brain structure and function that are characteristic of TSEs. The evidence for this hypothesis is strong and is growing steadily.

A number of spongiform encephalopathies have been described, affecting a wide range of animals, including humans, cattle, sheep and goats (scrapie), deer (chronic wasting disease), elk, cats and mink. Most of these conditions

The Food Safety Hazard Guidebook
By Richard Lawley, Laurie Curtis & Judy Davis

occurring in humans, such as classic Creutzfeldt–Jakob disease (CJD) are considered to be inherited genetic diseases, or caused by sporadic mutations. However, a few are thought to be transmissible by ingestion of an infective agent (probably a prion), and it is these TSEs that are of concern in food safety terms.

By far the best known and most significant of these is variant Creutzfeldt–Jakob disease (vCJD), a condition first described in 1996, which is now widely considered to be a human form of bovine spongiform encephalopathy (BSE, or "mad cow disease"), a TSE found in cattle. The hypothesis is that human cases of vCJD may be caused by the ingestion of infective prions in meat from BSE-infected cattle in the food chain. The possibility that some prions are able to cross the species barrier is a major concern.

This section will focus on BSE and vCJD as there is no evidence to suggest that the causative agents of other TSEs have caused disease in humans.

Occurrence in Foods

The infective prion thought to be the causative agent for foodborne vCJD in humans is present in certain tissues of cattle suffering from BSE. High levels of BSE prions are known to occur in the central nervous system, particularly in the spinal cord and the brain. Lower levels are also considered to be present in other tissues, such as the tonsils, eyes, large and small intestines, mesentery, skull and vertebral column. These tissues are now known collectively as specified risk material (SRM) and are not allowed to enter the food supply in most countries. Before the introduction of BSE controls, some of these materials were present in meat products, such as mechanically recovered meat (MRM), used in some low-grade beef products, including pies and burgers. BSE prions have not been detected in bovine milk.

Hazard Characterisation

Effects on Health

The disease vCJD differs markedly from classic CJD in terms of the age of those affected and the length of the illness. vCJD affects younger individuals, with the average age being 29 years (classic CJD is 65 years). For vCJD, the usual duration of the illness until death is on average 14 months, whereas for classic CJD it is much shorter, usually 4.5 months on average.

The minimum infectious dose of BSE prions needed to cause vCJD in humans is unknown. However, it is known that the infectious oral dose of the BSE agent for cattle is ≤ 1 g homogenised infected brain tissue, but it is difficult to establish the effect that the species barrier has on the infectivity of the agent from cattle to humans. These experiments cannot be conducted for obvious reasons and the infectivity of various bovine materials is still the subject of investigation. The incubation period for vCJD is also unknown but has been suggested to vary from a few years to more than 25 years. It is thought that some individuals have a genetic factor that may make them more susceptible to infection and to rapid onset of the disease. All of those who have so far died of vCJD in the UK were found to have this factor.

Early symptoms of the disease include psychiatric symptoms such as behavioural changes, depression or schizophrenia-like psychosis. About 50% of affected individuals experience unusual sensory symptoms, *e.g.* stickiness of the skin. As the disease progresses, patients experience unsteadiness and difficulty walking, as well as involuntary movements. Eventually the patient is totally immobile and mute. There is no cure for vCJD and the prognosis for all patients displaying clinical symptoms of this progressive disease is eventual death.

Incidence and Outbreaks

Although the disease was first described in 1996, the first patient to develop symptoms of what is now known as vCJD became ill in 1994. The majority of cases of vCJD worldwide are in individuals who live, or have lived, in the UK, reflecting the fact that the UK is the country where the population has had the highest exposure to BSE prions.

By mid-2007 the UK had reported 165 cases of vCJD, 158 of whom have died. Elsewhere in the world, a further 37 cases of vCJD have been reported, of whom 32 have died. Some of these cases had a history of visiting or living in the UK. France has the second highest number of reported cases of vCJD in the world (22 as of mid- 2007) and imported relatively large quantities of cattle products from the UK before the introduction of import restrictions.

Sources

The first outbreak of BSE in cattle was recognised in the UK in 1986, but the first cases probably occurred at least a year earlier. The original source of the disease has been suggested as being scrapie-infected meat-and-bone meal (MBM) used as a protein supplement in cattle feed, but this has not been confirmed. However, it is thought that the practice of feeding MBM made from infected cattle to young calves may have amplified the outbreak and accelerated its spread. It is estimated that a total of more than two million cattle in the UK have been infected with BSE, and that at least 750 000 of these were slaughtered and potentially consumed by the UK population between 1980 and 1996, when BSE controls were introduced.

Since the first identification of BSE in the UK in the 1980s, other countries have also reported BSE in cattle, in many cases probably caused by the importation of contaminated feed or infected animals. An additional 23 countries have reported BSE in cattle to date, including many European countries, Japan, Israel, Canada and the USA. However, the current incidence of BSE in these countries is far lower than that reported in the UK.

Growth and Survival Characteristics

Infective prions are only capable of "replicating" in the tissues of the host. However, they have been found to be extremely resistant to a wide range of environmental factors, including heat, chemical sterilants, extremes of pH and radiation. For example, the long-term infectivity of prions in rendered MBM made from diseased animals is a demonstration of their stability.

Thermal Resistance

The heat resistance of infective prions is considerable. At high temperatures, the survival of infectivity is greater in dry conditions (dry heat at <300°C cannot be guaranteed to inactivate infective prions), but experiments have shown that, if large amounts of infective material are present, a heat treatment of 133°C for 20 min under 3 bar pressure may still be inadequate even at high moisture contents. A heat treatment of 140°C for 30 min at 3.6 bar pressure has been suggested as an alternative.

Control Options

The control of vCJD in humans is inextricably linked to the control of BSE in cattle. Attempts to eradicate the disease in cattle in affected countries have focused on banning the use of protein derived from ruminants in all farmed animal feed. This was introduced in 1988 and enhanced in 1994 and 1996, and has been very successful in restricting the spread of the disease in cattle in the UK, with the result that only 114 BSE infected cattle were reported in the whole of 2006. At the peak of the BSE epidemic, more than 850 cows were diagnosed with BSE every week.

The other main thrust of vCJD/BSE control is to protect consumers from exposure to BSE-infected materials. To this end, since 1989, a wide range of statutory controls have been introduced in the UK and other affected countries designed to prevent SRM from entering the food chain. Between 1996 and 2005, cattle of more than 30 months of age were also been banned from the food chain in the UK, following the discovery that older animals are more likely to develop the disease. A comprehensive programme of BSE testing at slaughter has now been introduced in the EU, and only animals that test negative can be allowed to enter the food chain.

A very wide range of BSE controls (usually mandatory) have been implemented in affected countries. These are beyond the scope of this book and readers are referred to the web links below and to their national food safety and animal health authorities for specific details of controls that apply at each stage in the meat supply chain.

Legislation

A substantial and ever-changing raft of legislation designed to control BSE in cattle and to protect the public from exposure to BSE-infected materials has been introduced in the EU, North America and other affected countries. The specifics of BSE legislation are beyond the scope of this book and readers are referred to some of the web links below and to their national food safety and animal-health authorities for information on current legislation.

Sources of Further Information

Published

Cliver D.O., Matsui S.M., Casteel M. Infections with viruses and prions, in Foodborne infections and intoxications. ed. Riemann H.P., Cliver D.O. 3rd edn. London. Academic Press, 2005, 367–448.

Hueston, W., Bryant, C.M. Transmissible spongiform encephalopathies. *Journal of Food Science*, 2005, 70 (5), R77–R87.

Brown, P., Will, R.G., Bradley, R., Asher, D.M., Detwiler, L. Bovine Spongiform Encephalopathy and Variant Creutzfeldt–Jakob Disease: Background, Evolution, and Current Concerns. *Emerging Infectious Diseases*, 2001, 7 (1), 6–16.

On the Web

"BSE control explained: Main controls on beef production." UK Food Standards Agency. http://www.food.gov.uk/bse/what/beef/controls

"BSE – Scientific advice." Europa (European Commission web site). http://ec.europa.eu/food/food/biosafety/bse/sci_advice_en.htm

"Quantitative risk assessment on the residual BSE risk in sheep meat and meat products" European Food Safety Authority's Scientific Panel on Biological Hazards. (January 2007). http://www.efsa.europa.eu/etc/medialib/efsa/science/biohaz/biohaz_opinions/ej442_bse_sheep.Par.0001.File.dat/biohaz_op_ ej442_qra_sheep_en.pdf

"Harvard risk assessment on Bovine Spongiform Encephalopathy Update." Harvard Center for risk assessment. (October 2005). http://www.fsis.usda.gov/PDF/BSE_Risk_Assess_Report_2005.pdf

"Information Statement – Bovine Spongiform Encephalopathy (BSE) and Variant Creutzfeldt–Jakob Disease (vCJD) (November 2004)." Institute of Food Science & Technology. http://www.ifst.org/uploadedfiles/cms/store/ATTACHMENTS/BSE.pdf

Section 2: Chemical Hazards

CHAPTER 2.1

Biological Toxins

2.1.1 FUNGAL TOXINS

2.1.1.1 Aflatoxins

Hazard Identification

What are Aflatoxins?

The aflatoxins are a group of chemically similar toxic fungal metabolites (mycotoxins) produced by certain moulds of the genus *Aspergillus* growing on a number of raw-food commodities. Aflatoxins are highly toxic compounds and can cause both acute and chronic toxicity in humans and many other animals. Their importance was first established in 1960 when 100 000 turkeys and other poultry in the UK died in a single event. The cause of this was eventually traced to a toxic contaminant in groundnut meal used in the bird's feed. The contaminant was later named aflatoxin.

The aflatoxins consist of about 20 similar compounds belonging to a group called the difuranocoumarins, but only four are naturally found in foods. These are aflatoxins B_1, B_2, G_1 and G_2. Aflatoxin B_1 is the most commonly found in food and also the most toxic. When lactating cattle and other animals ingest aflatoxins in contaminated feed, toxic metabolites can be formed and may be present in milk. These hydroxylated metabolites are termed aflatoxin M_1 and M_2 and they are potentially important contaminants in dairy products.

Occurrence in Foods

Aflatoxins may be present in a wide range of food commodities, particularly cereals, oilseeds, spices and tree nuts. Maize, groundnuts (peanuts), pistachios, brazils, chillies, black pepper, dried fruit and figs are all known to be high-risk

The Food Safety Hazard Guidebook
By Richard Lawley, Laurie Curtis & Judy Davis

foods for aflatoxin contamination, but the toxin has also been detected in many other commodities. Milk, cheese and other dairy products are also known to be at risk of contamination by aflatoxin M. The highest levels are usually found in commodities from warmer regions of the world where there is a great deal of climatic variation.

It is important to recognise that, although it is primary food commodities that usually become contaminated with aflatoxins by mould growth, these toxins are very stable and may pass through quite severe processes. For this reason they can be a problem in processed foods, such as peanut butter.

Hazard Characterisation

Effects on Health

At high enough exposure levels, aflatoxins can cause acute toxicity, and potentially death, in mammals, birds and fish, as well as in humans. The liver is the principal organ affected, but high levels of aflatoxin have also been found in the lungs, kidneys, brains and hearts of individuals dying of acute aflatoxicosis. Acute necrosis and cirrhosis of the liver is typical, along with haemorrhaging and oedema. LD_{50} (lethal dose) values for animals vary between 0.5 and 10 mg/kg bodyweight.

Chronic toxicity is probably more important from a food safety point of view, certainly in more developed regions of the world. Aflatoxin B_1 is a very potent carcinogen and a mutagen in many animals, and therefore potentially in humans, and the liver is again the main target organ. Ingestion of low levels over a long period has been implicated in primary liver cancer, chronic hepatitis, jaundice, cirrhosis and impaired nutrient conversion. Aflatoxins may also play a role in other conditions, such as Reye's syndrome and kwashiorkor (a childhood condition linked to malnutrition). Less is known about the chronic toxicity of aflatoxin G_1 and M_1, but these are also thought to be carcinogens, though probably a little less potent than B_1.

Little is known about the level of dietary exposure to aflatoxins necessary to affect health, especially in humans, and diagnosis of chronic toxicity is very difficult. It is generally agreed that the best approach is to minimise the levels in all foods as far as is technically possible and to assume that any dietary exposure is undesirable.

Incidence and Outbreaks

The incidence of chronic aflatoxicosis in humans is unknown and is almost impossible to estimate because the symptoms are so difficult to recognise. However, human liver cancer is quite common in parts of the world where aflatoxin contamination of food is likely and there may be a link, although this remains unproven.

Acute human aflatoxicosis is rare, especially in developed countries, where contamination levels in food and monitored and controlled. However, there have been outbreaks in some developing countries, notably in sub-Saharan Africa, where maize and groundnuts can be an important part of the diet and where the climate is suitable for rapid mould growth on crops in the field and in storage.

A notable outbreak occurred in India in 1974 when almost 400 people became ill with fever and jaundice after eating maize contaminated with between 0.25 and 15 mg/kg aflatoxin and more than 100 died. At least two major outbreaks have also occurred in Kenya, most recently in 2004 when 317 people were affected and 125 died, probably as a result of eating contaminated maize.

Sources

Aflatoxins are produced by at least three *Aspergillus* species. These are *A. flavus*, *A. parasiticus* and the much more rare *A. nomius*. These moulds are able to colonise a wide range of crops both in the field as non-destructive pathogens and in storage and can grow and produce aflatoxins at quite low moisture levels (approximate minimum a_w 0.82) and over a broad temperature range (13–37 °C).

Their growth is strongly influenced by climate and, although they are found all over the world, they are more common in tropical regions with extreme variations in temperature, rainfall and humidity. *A. flavus* invasion of groundnut crops in the field is known to be favoured by drought stress and maize crops are vulnerable if damaged by insect pests.

Mould growth and aflatoxin production during storage of crops is also important, especially if drying is inadequate, or storage conditions allow access for insect or animal pests.

Stability in Foods

Aflatoxins are quite stable compounds and survive relatively high temperatures with little degradation. Their heat stability is influenced by other factors, such as moisture level and pH, but heating or cooking processes cannot be relied upon to destroy aflatoxins. For example, roasting green coffee at 180 °C for 10 min gave only a 50% reduction in aflatoxin B_1 level.

The stability of aflatoxin M_1 in milk fermentation processes has also been studied and although appreciable losses do occur, significant quantities of the toxin were found to remain in both cheese and yoghurt.

Aflatoxins can be destroyed by alkaline and acid hydrolysis and by the action of oxidising agents. However, in many cases, the resulting by-products also carry a risk of toxicity, or have not been identified.

Control Options

The ability of aflatoxin-producing fungi to grow on a wide range of food commodities and the stability of aflatoxins in foods mean that control is best achieved by measures designed to prevent the contamination of crops in the field and during storage, or detection and removal of contaminated material from the food supply chain.

Pre-harvest

Pre-Harvest control of aflatoxins is best achieved through general good agricultural practice (GAP) to include such measures as:

- land preparation, crop waste removal, fertiliser application and crop rotation;
- use of fungus- and pest-resistant crop varieties;
- control of insect pests;
- control of fungal infection;
- prevention of drought stress by irrigation;
- harvesting at the correct moisture level and stage of maturity.

Post-harvest Handling and Storage

The most important and effective control measure in post-harvest handling and storage is the control of moisture content and hence, the water activity of the crop. Ensuring that susceptible crops are harvested at a safe moisture level, or are dried to a safe level immediately after harvest is vital to prevent mould growth and aflatoxin production during storage. The safe moisture level varies between crops – for maize it is approximately 14% at 20 °C, but for groundnuts it is much lower, about 7%. These moisture levels must be maintained during storage and transport.

It is also important to ensure that the moisture content does not vary too much in a bulk-stored crop. Small localised "wet spots" can develop mould growth and these can extend to neighbouring areas as the fungus produces metabolic water during respiration. Insect and animal pest damage can also act as focal points for fungal growth.

Decontamination

Physical separation of contaminated material can be an effective means of reducing aflatoxin levels in contaminated commodities. For example, colour sorting is often used to remove mouldy peanuts from bulk shipments. Density segregation, mechanical separation and the removal of fines and screenings from grain and nut shipments can also be effective measures.

Chemical decontamination methods have been investigated, especially for material used in animal feed, but most of the methods investigated are impractical, or produce toxic by-products. So far, an ammoniation process has shown the most promise and has been successfully used to remove aflatoxins from feed in the USA.

Biological decontamination has also been considered, and a single bacterial species, *Flavobacterium aurantiacum*, has been shown to remove aflatoxin B_1 from peanuts and corn.

Although decontamination methods for aflatoxin M_1 in milk and dairy products have also been investigated, most of these are not practical for the dairy industry. The only really effective control is to minimise the contamination of materials used in animal feed for dairy cows.

Testing

Many countries monitor imported commodities that are susceptible to aflatoxin contamination, such as pistachios and Brazil nuts, by sampling and analysis. A number of analytical methods have been developed based on TLC (thin layer chromatography), HPLC (high-performance liquid chromatography) and ELISA (enzyme-linked immunosorbent assay) and there are also rapid-screening kits available. However, moulds and aflatoxins in bulk food shipments tend to be highly heterogeneous in their distribution and it is essential to ensure that an adequate sampling plan is used to monitor imported materials.

In some commodities, such as figs, aflatoxins fluoresce strongly under UV light and this can be used as a rapid screening test for high concentrations.

Legislation

Around 100 countries around the world have regulations governing aflatoxins in food and most include maximum permitted, or recommended levels for specific commodities.

European Union

The EU sets limits for aflatoxin B_1 and for total aflatoxins (B_1, B_2, G_1 and G_2) in nuts, dried fruits, cereals and spices. Limits vary according to the commodity, but range from 2–8 μg/kg for B_1 and from 4–15 μg/kg for total aflatoxins. There is also a limit of 0.050 μg/kg for aflatoxin M_1 in milk and milk products. Sampling and analytical methods are also specified.

More recently limits of 0.10 μg/kg for B_1 and 0.025 μg/kg for M_1 have been set for infant foods.

USA

US food safety regulations include a limit of 20 μg/kg for total aflatoxins (B_1, B_2, G_1 and G_2) in all foods except milk and a limit of 0.5 μg/kg for M_1 in milk. Higher limits apply in animal feeds.

Others

Both Australia and Canada set limits of 15 μg/kg for total aflatoxins (B_1, B_2, G_1 and G_2) in nuts. This is the same as the international limit recommended for raw peanuts by the Codex Alimentarius Commission.

More information can be found at the FAO web link below.

Sources of Further Information

Published

The mycotoxin factbook: Food & feed topics. ed. Barug, D. *et al.* Wageningen, Wageningen Academic Publishers, 2006.

Mycotoxins: Bennett, J.W. and Klich, M. *Clinical Microbiology Reviews*, 2003, 16(3), 497–516.

On the Web

Food-Info.net: Overview of foodborne toxins – mycotoxins (aflatoxins). http://www.food-info.net/uk/tox/afla.htm

European Mycotoxin Awareness Network (EMAN). http://www.mycotoxins.org/

WHO Food Additives Series 40 – JECFA monograph on aflatoxins. http://www.inchem.org/documents/jecfa/jecmono/v040je16.htm

FAO Food and Nutrition Paper 81 – Worldwide regulations for mycotoxins in food and feed 2003. http://www.fao.org/docrep/007/y5499e/y5499e00.htm

Aflatoxin.info. http://www.aflatoxin.info/aflatoxin.asp

2.1.1.2 Citrinin

Hazard Identification

What is Citrinin?

Citrinin is a toxic fungal metabolite (mycotoxin) produced by some moulds of the genera *Penicillium, Aspergillus* and *Monascus* growing on certain food commodities, especially cereals and fruit. It was first isolated from a culture of *Penicillium citrinum* in 1931. Citrinin exhibits a number of toxic effects in animals and its presence in food is undesirable.

Citrinin is a relatively small molecule ($C_{13}H_{14}O_5$) and is slightly soluble in water.

Occurrence in Foods

Citrinin has been found in a range of cereals, including rice, wheat, barley, maize, rye and oats. Co-occurrence with ochratoxin A in cereals has been reported. It has also been found in wheat flour and there is some evidence that it may survive to some extent in processed cereal products. Citrinin has also been found in peanuts and in mouldy fruit.

Citrinin also occurs in some fermented foods that are susceptible to surface mould growth, such as cheeses and fermented sausages. There is evidence that it may penetrate two or more centimetres into cheese showing surface mould growth. Recently, citrinin has also been found in certain vegetarian foods that have been coloured with pigments derived from *Monascus* species fungi.

It is likely that the occurrence of citrinin in foods is under-reported, since it is not often looked for and has a tendency to partially degrade during analysis.

Hazard Characterisation

Effects on Health

Most of the information on the toxicity of citrinin is derived from animal studies and there is little or no experimental, or epidemiological, data on acute or chronic toxicity in humans.

At relatively high doses, citrinin is acutely nephrotoxic in mice and rats, rabbits, pigs and poultry causing swelling and eventual necrosis of the kidneys and affecting liver function to a lesser extent. LD_{50} values (lethal dose) are variable, but values of about 50 mg/kg bodyweight have been reported for oral administration in rats.

The International Agency for Research on Cancer (IARC) has reviewed the available data and concluded that there is limited evidence for carcinogenicity in animals.

It has been proposed that citrinin may be implicated in human disease, such as "yellow rice" disease in Japan and Balkan Endemic Nephropathy, when present with other mycotoxins, especially ochratoxin A.

Sources

Citrinin is produced by at least 12 species of *Penicillium*, including *P. citrinum*, some strains of *P. camembertii* (used in cheese production) and *P. verrucosum*, which also produces ochratoxin A. Some *Aspergillus* species, such as *A. terreus* and *A. niveus* are also reported to produce citrinin and the toxin has also been detected in cultures of *Monascus ruber* and *Monascus purpureus*, used to make red pigments.

P. citrinum has been isolated from a very wide range of food commodities worldwide. It is able to grow in a temperature range of 5–37 °C and at water activity values as low as 0.80.

Stability in Foods

Citrinin is not particularly stable and is degraded by heat and by alkaline conditions. There is little published information on the fate of citrinin during food processing, but it seems unlikely that it persists in significant amounts in bakery products and other processed cereal foods. However, there is some evidence that toxic breakdown products may be formed when citrinin degrades in wet environments.

Citrinin is unlikely to survive the brewing process and more than 90% is reported to be destroyed during barley germination, with the remainder being degraded during mashing.

Citrinin produced by mould growth on cheese appears to be quite stable with more than 50% still being present after storage for eight days.

Control Options

There are few specific documented control measures for citrinin, but its co-occurrence with ochratoxin in cereals means that the pre- and post-harvest control measures recommended for ochratoxin may also provide indirect control of citrinin.

Processing

Control of citrinin in fermented foods, such as cheese and sausage can be achieved by good hygienic practice to prevent surface contamination and growth of toxin-producing mould species. Where potentially citrinin-producing species of *Penicillium* or *Aspergillus* (*e.g. P. camembertii*) are used in the production of fermented foods, it is important to select non-toxin-producing strains as starter cultures.

Cheese that has undergone surface mould spoilage is often trimmed to remove mould growth before sale, but it is important to remember that some citrinin may still be present in the surface layers of trimmed cheese.

Testing

Quantitative analysis of citrinin in agricultural products down to levels of about 10 ppb can be achieved using HPLC methods, but it is important to ensure that degradation does not occur during analysis. There are also screening methods based on ELISA techniques.

Legislation

There are no current specific regulations setting mandatory or recommended maximum limits for citrinin in food or feed.

Sources of Further Information

Published

The mycotoxin factbook: Food & feed topics, ed. Barug, D. *et al.* Wageningen, Wageningen Academic Publishers, 2006.

Mycotoxins. Bennett, J.W. and Klich, M. *Clinical Microbiology Reviews*, 2003, 16(3), 497–516.

On the Web

European Mycotoxin Awareness Network (EMAN). http://www.mycotoxins.org/

2.1.1.3 Cyclopiazonic Acid

Hazard Identification

What is Cyclopiazonic Acid?

Cyclopiazonic acid (CPA) is a toxic fungal metabolite (mycotoxin) produced by some moulds of the genera *Penicillium* and *Aspergillus* growing on a wide range of food commodities. As it can be produced by *Aspergillus flavus*, it has the potential to co-occur with aflatoxins, but there is comparatively little data about its occurrence in foods. At high concentrations it exhibits a number of toxic effects in animals and its presence in food is undesirable.

CPA is an indole tetramic acid with a molecular weight of 336.4. Cyclopiazonic acid imine occurs as a related metabolite in culture, but this is considered to be much less toxic than the parent compound.

Occurrence in Foods

CPA acid has been detected at levels of up to 10 mg/kg in a wide variety of food and feed commodities, including maize and other cereals, pulses, peanuts, cheese, ham and sausages, tomatoes, milk, hay and mixed animal feeds. It has also been found to co-occur with aflatoxins in some samples of peanuts.

Natural occurrence in foods and the potential for human exposure from the diet appear to be quite low.

Hazard Characterisation

Effects on Health

Most of the information on the toxicity of CPA is derived from a limited number of animal studies and there is little or no experimental, or epidemiological, data on acute or chronic toxicity in humans. Its significance for human health is therefore still unclear.

The toxicity of CPA is probably based on its ability to interfere with the uptake of calcium ions. It is reported to be neurotoxic when injected intraperitoneally into rats and an LD_{50} of 2.3 mg/kg bodyweight has been observed. However, higher oral doses appear to be necessary to cause significant toxic effects, and an LD_{50} of 36–63 mg/kg bodyweight has been reported for rats when CPA was administered by feeding. A number of toxic effects have been observed, notably lesions in the liver, kidneys and spleen, with varied symptoms, including diarrhoea, dehydration, hypokinesis, convulsion and death. It may also be toxic to poultry, but interpretation of published studies is complicated by the possible presence of other mycotoxins.

CPA displays some mutagenic activity and it may also contribute to overall toxicity when it co-occurs with aflatoxins.

It has been proposed that CPA is implicated in "Kodua" poisoning in India, a neurological condition associated with eating mouldy millet. Symptoms include somnolence, tremors and giddiness.

Sources

CPA is produced by several species of *Penicillium*, including *P. cyclopium*, *P. commune* and *P. camembertii*. Some strains of *Aspergillus flavus* and *A. versicolor* have also been demonstrated to produce the toxin.

The species known to be capable of producing CPA have a widespread distribution, are able to colonise a very wide range of food commodities and can grow over a wide range of temperatures and water activities. There is therefore a potential for CPA to be produced in a number of foods intended for human consumption. Furthermore, one CPA-producing species, *P. camembertii*, is used in the production of some types of cheese as a surface-ripening agent.

Stability in Foods

Relatively little is known about the stability of CPA during food processing. It has been found to be quite stable on dry-cured ham and in milk stored at chill temperatures. It also survives spray-drying processes used in milk powder production. Approximately 40% of CPA was lost during the manufacture of condensed milk using temperatures of 100 °C.

Control Options

There are few specific documented control measures for CPA, but its co-occurrence with other aflatoxins means that the pre- and post-harvest control measures recommended for aflatoxins may also provide indirect control of CPA.

It is important to consider the possible production of CPA when selecting cultures of *P. camembertii* for cheese manufacture. Although many strains appear to have the potential to produce the toxin, not all are reported to do so on cheese, and it is important to choose a non-toxin-producing strain.

Some mould species that cause mould spoilage of stored foods such as dry cured hams and fermented sausage products may be capable of producing CPA. For this reason it is preferable to control mould growth on the surface of these foods.

Legislation

There are no current specific regulations setting mandatory or recommended maximum limits for CPA in food or feed.

Sources of Further Information

Published

The mycotoxin factbook: Food & feed topics. ed. Barug, D. *et al.* Wageningen, Wageningen Academic Publishers, 2006.

Mycotoxins. Bennett, J.W. and Klich, M. *Clinical Microbiology Reviews*, 2003, 16(3), 497–516.

On the Web

European Mycotoxin Awareness Network (EMAN). http://www.mycotoxins.org/

2.1.1.4 Deoxynivalenol

Hazard Identification

What is Deoxynivalenol?

Deoxynivalenol (DON) is a toxic fungal metabolite (mycotoxin) produced by certain plant pathogenic moulds, especially *Fusarium* species, infecting cereal crops in the field. DON is also known as vomitoxin and is one of a large group of chemically related mycotoxins called the trichothecenes. DON is toxic to humans and livestock, is quite common in some food commodities and can occur at high levels. For these reasons it is of concern from a food safety point of view.

The trichothecenes are a group of around 150 compounds characterised as tetracyclic sesquiterpenes. DON ($C_{15}H_{20}O_6$) belongs to the B group of trichothecenes and has a molecular weight of 296.32. It is soluble in water and extremely stable.

Occurrence in Foods

DON is almost exclusively associated with cereals, particularly in temperate regions, and it is a common contaminant in wheat, barley, oats, rye, maize and rice. The level of contamination varies widely between regions and from year to year, but where cereals become infected with DON-producing *Fusarium* species, more than 50% of grain samples may show contamination and levels have been reported to reach approximately 9000 μg/kg for barley, 6000 μg/kg for wheat, 5000 μg/kg for rice and 4000 μg/kg for maize.

DON has also been found in processed foods, especially those produced from cereals. Foods reported to be contaminated have included flour, bread, breakfast cereals, noodles, infant foods, malt and beer. DON contamination does not seem to be a problem in foods of animal origin, despite the fact that it is frequent contaminant of animal feed. One reason for this may be that the presence of high levels of DON in feed tends to lead to feed refusal by livestock. Furthermore, lower levels are metabolised and eliminated rapidly in food animal species. Only trace amounts have been reported in eggs and milk.

The main contributor to DON in the diet in Europe is wheat (~80%), whereas in the Far East, rice is equally important.

Hazard Characterisation

Effects on Health

DON is associated with acute toxicity in both animals and humans, but its effects are difficult to quantify because it often co-occurs with other *Fusarium* mycotoxins, especially nivalenol and zearalenone. Trichothecenes in general are known to interfere with protein synthesis, but the main effects of DON now appear to be related to its role as a pro-inflammatory agent.

Pigs are particularly sensitive to DON in feed and acute toxicity is characterised by vomiting – the source of the synonym vomitoxin. At lower levels, a variety of symptoms have been reported, including feed refusal and reduced weight gain. Acute toxicity in humans has also been reported, with symptoms including vomiting, diarrhoea, abdominal pain, headache and fever. These symptoms can develop within 30 min and are difficult to distinguish from some types of bacterial food poisoning, particularly that caused by pre-formed emetic toxins of *Bacillus cereus*. However, it should be noted that the role of DON in these cases is uncertain, as other mycotoxins are almost always present. Recovery is usually quite rapid and no deaths have been reported.

Long-term chronic toxicity from low levels of DON in the diet has been investigated in animals. Studies show changes in some blood parameters and suggest adverse effects on the immune system. However, there is no evidence of carcinogenicity, or of mutagenic or teratogenic effects. Based on the data available from animal studies, the EU Scientific Committee on Food established a temporary tolerable daily intake (TDI) for DON of 1 μg/kg bodyweight for humans in 2002. Although this is in line with TDIs established by other authorities, dietary surveys suggest that some European consumers could have an intake quite close to this figure.

Incidence and Outbreaks

There are a number of documented outbreaks of food poisoning caused by foods contaminated with DON. For example, in India in 1987, approximately 50 000 people were ill with mild gastrointestinal symptoms after eating bread made from rain-damaged wheat. Samples of the wheat showed that DON was present at levels from 340–8400 μg/kg, but several other trichothecenes were also present at lower concentrations.

Sources

The principal sources of DON in cereals are the *Fusarium* species *F. graminearum* and *F. culmorum*. Both of these species are considered to be field fungi and are pathogenic to cereals, causing *Fusarium* head blight in wheat and *Gibberella* ear rot in maize. Distribution of the two species is influenced by temperature, and *F. graminearum* is found mainly in warmer regions.

DON is produced in the crop prior to harvest, rather than during storage, and contamination in wheat is directly related to the incidence of *Fusarium* head blight, which is itself related to moisture levels at flowering. Rainfall at this time is a critical factor for the incidence of the disease, but the amount of rainfall does not appear to be important. The disease causes shrivelling of the wheat seeds and DON is typically produced on the surface of infected grains. However, where high levels are produced, it may be more evenly distributed in the wheat kernel itself.

Stability in Foods

DON is extremely heat stable and is not destroyed by temperatures of 120 °C. It therefore survives most cooking processes and significant quantities are reported to remain even in baked products cooked at 200 °C. It has also been shown to survive autoclaving and extrusion processes.

The toxin is unstable under alkaline conditions. Production of maize flour for tortillas by first boiling maize in calcium hydroxide (nixtamalisation) has been shown to reduce DON levels by approximately 80%.

Control Options

Since DON production occurs mainly in the field, the most successful controls are applied at the pre-harvest stage.

Pre-harvest

Good agricultural practice (GAP) measures designed to reduce *Fusarium* infection in cereal crops are also effective in limiting the formation of DON. Control measures include the following.

- land preparation, crop rotation and crop debris removal to reduce the inoculum of *Fusarium* in the field;
- use of fungus-resistant crop varieties;
- control of infection by appropriately timed application of effective fungicides;
- harvesting at the correct moisture level and stage of maturity.

Post-harvest Handling and Storage

Further production of DON after harvest can be prevented by rapid drying to a water activity value of 0.8, and by implementing good storage practice.

Decontamination

Physical decontamination methods can be an effective means of reducing DON levels in contaminated grain. These include gravity separation and grain washing, although this process produces large amounts of effluent. The milling process also reduces DON concentrations in wheat flour by removing the generally more heavily contaminated bran, but the effectiveness of this depends on the distribution of the toxin in the grain.

Chemical decontamination methods, such as treatment with sodium bisulfite, have been investigated, but are not yet developed for commercial use.

Heat treatments are not usually effective.

Testing

Some countries monitor cereals for DON contamination by sampling and testing using analytical methods, such as liquid chromatography (LC) with UV detection, or gas chromatography (GC) with mass spectroscopic detection. Sensitive ELISA methods have also been developed for screening purposes and commercial kits are available. However, as with other mycotoxins, the distribution of DON in bulk commodities may be highly heterogeneous and it is essential to ensure that an adequate representative sampling plan is used.

Legislation

At least 40 countries around the world have introduced mandatory or guideline levels for DON in foods, mostly since the late 1990s when the toxin became a cause for concern.

European Union

The EU sets a maximum level for DON of 1250 μg/kg in most unprocessed cereals, but the permitted level in unprocessed durum wheat, oats and maize is 1750 μg/kg. Up to 750 μg/kg is allowed in pasta and in cereals, flour and bran for direct human consumption. The limit for bread, biscuits, breakfast cereals and cereal snacks is 500 μg/kg. A limit of 200 μg/kg has been set for foods intended for babies and young children.

USA

US food safety regulations include a limit of 1000 μg/kg for DON in finished wheat products for human consumption. Higher limits apply in animal feeds.

Others

The Canadian authorities have introduced a limit of 2000 μg/kg for DON in domestic raw soft wheat and 1200 μg/kg for soft wheat flour. The limit for flour used in infant food is 600 μg/kg.

A number of other countries, such as China, have introduced a limit of 1000 μg/kg for DON in wheat and maize flour.

More information can be found at the FAO web link below.

Sources of Further Information

Published

The mycotoxin factbook: Food & feed topics. ed. Barug, D. *et al.* Wageningen, Wageningen Academic Publishers, 2006.

Mycotoxins. Bennett, J.W. and Klich, M. *Clinical Microbiology Reviews*, 2003, 16(3), 497–516.

On the Web

European Mycotoxin Awareness Network (EMAN). http://www.mycotoxins.org/

JECFA monograph on deoxynivalenol. http://www.inchem.org/documents/jecfa/jecmono/v47je05.htm

FAO Food and Nutrition Paper 81 – Worldwide regulations for mycotoxins in food and feed 2003. http://www.fao.org/docrep/007/y5499e/y5499e00.htm

2.1.1.5 Ergot

Hazard Identification

What is Ergot?

The term ergot refers to fungal structures (sclerotia) produced by certain species of *Claviceps* fungi that infect cereals and wild grasses. These sclerotia are hard black masses of fungal hyphae that act as a resistant resting stage for the fungus and are visible on the grain ears of infected cereals. Ergots contain a number of different types of alkaloids, which can produce toxic effects in animals and humans. The effects of these alkaloids have been known for hundreds of years and they were the main cause of outbreaks of a toxic condition known as "St Anthony's Fire", which occurred regularly in Europe during the Middle Ages.

There are at least 40 different ergot alkaloids, but the most important are ergotamine, ergometrine, ergosine, ergocristine, ergocryptine, ergocornine and their related -inines. These compounds are derivatives of the hallucinogenic drug lysergic acid (LSD), or of isolysergic acid (-inines). In addition, some *Claviceps* species produce clavine alkaloids, such as agroclavine, which are also toxic and are derivatives of dimethylergoline.

Occurrence in Foods

Ergot can occur in all common cereals, including wheat, barley, oats, rye, millet, sorghum, maize and rice, but rye is more susceptible to infection than other cultivated crops. Ergot contamination in cereals is usually expressed in terms of the percentage, by weight, of sclerotia present in the grain, rather than as ergot alkaloids. However, some studies have measured the levels of individual and total ergot alkaloids in contaminated grain. For example, the concentrations of total ergot alkaloids in sclerotia from rye and wheat have been reported to be 700 mg/kg and 920 mg/kg, respectively. A survey of cereal products on the market in Switzerland showed levels of total ergot alkaloids between 4.2 μg/kg (wheat flour) and 139.7 μg/kg (rye flour). The daily intake of total ergot alkaloids by human beings in Switzerland was estimated to be 5.1 μg/person.

There is no evidence that ergot alkaloids transfer and accumulate in the tissues of animals fed on contaminated cereals and they have not been found in milk or eggs.

Hazard Characterisation

Effects on Health

Acute toxicity (ergotism) in humans is now rare, but it is still occasionally reported in livestock.

There is little information on the toxicity of individual ergot alkaloids, but in practice affected humans and animals are likely to be exposed to complex mixtures of varying composition. For this reason, the range of toxic effects and symptoms is quite broad, and different animals display widely different symptoms.

In general, two main types of ergotism, gangrenous and convulsive, can occur in animals. In the first type ergot alkaloids affect blood circulation, causing vasoconstriction, which may lead to a dry gangrene in the extremities, especially the limbs. Cattle affected in this way tend to become lame and may develop gangrene in their ears and tail, as well as the feet. Convulsive ergotism results from the neurotoxic activity of ergot alkaloids and symptoms include feed refusal and dizziness, as well as convulsions.

Symptoms of St Anthony's Fire in humans have been documented for centuries, and include gangrene, burning sensations (hence the name) and hallucinations. The disease was often fatal.

Little is known about the long-term effects of low levels of ergot alkaloids in the diet, or the potential carcinogenicity of these compounds.

Incidence and Outbreaks

Outbreaks or ergotism are rare in recent times and no documented outbreaks have occurred in Europe since 1928. However, a serious outbreak of gangrenous ergotism did occur in Ethiopia in 1978, when 93 cases were reported, along with a further 47 related deaths. The outbreak was caused by a high level of ergot-infected wild oats in the local barley crop, and 0.75% ergot was reported in the implicated grain.

Outbreaks of ergotism have also been reported in India, most recently in 1975, caused by consumption of infected millet, but the symptoms were mainly nausea and vomiting followed by drowsiness, and no deaths occurred. These outbreaks were found to be related to clavine alkaloids, such as agroclavine, present in implicated grain at levels of 15–199 mg/kg.

Sources

The principal source of ergot alkaloids in cereals is the ascomycete fungus *Claviceps purpurea.* The clavine alkaloids are produced mainly by a different species, identified as *Claviceps fusiformis*, which is primarily a parasite of pearl millet in tropical regions. Ergot alkaloids have been isolated from other fungi, including some *Penicillium* and *Aspergillus* species, but their significance for human health is unknown.

When *Claviceps purpurea* spores infect a susceptible host, the fungus invades the developing grains in the floret, then destroys and replaces them. Eventually the hard, dark sclerotia, or ergots, are formed and are easily visible as dark purple bodies up to 20 × 6 mm in size. At this stage, the ergot alkaloids begin to accumulate in the sclerotia.

Cereals are more susceptible to infection in wet weather, which favours the germination of sclerotia in the soil and the production of fruiting bodies and airborne ascospores. Cool, wet conditions during flowering of cereals and grasses favour the invasion of the florets, whereas hot dry conditions inhibit infection. If weather conditions or other factors result in prolonged flowering periods, infection becomes more likely.

Stability in Foods

The heat stability of the ergot alkaloids is quite variable, but the most pharmacologically active forms tend to be less stable than the inactive isomers. Heat processes such as baking produce a significant reduction (50% or more) in the concentration of the most important ergot alkaloids.

Beer made from ergot-contaminated grain has been reported to contain only low levels (10 ng/ml) of ergot alkaloids.

Control Options

Ergot infection occurs entirely in the field, and there are control options that can be applied at the pre-harvest stage. Decontamination is also an important control.

Pre-harvest

Control measures include the following:

- land preparation (*e.g.* deep ploughing), crop rotation with non-susceptible crops and crop debris removal to reduce the inoculum of ergot;
- plant only ergot-free seed;
- effective control of wild grasses in and around the crop.

Decontamination

Physical decontamination methods, such as grain cleaning, can achieve considerable reduction in ergot contamination. However, where small pieces of sclerotia, similar in size to individual grains, are present, they may not be removed effectively.

Testing

The presence of ergot in foodstuffs can be detected by analysis for ricinoleic acid, which is diagnostic for ergot in the absence of other sources, such as castor oil. This marker compound can be detected using a gas-liquid chromatography method.

Methods for detection of specific ergot alkaloids in cereals have also been developed, using a variety of techniques, including HPLC, GC–MS (Gas chromatography – mass spectometry) and TLC.

Legislation

Most regulations for the control of ergot in foods specify a recommended, or mandatory limit based on the percentage by weight, or number, of ergots in grain, rather than ergot alkaloid concentration. These limits are most often applied to animal feed. In general, feed containing >0.1% of ergot is not suitable for livestock, but many countries have developed higher voluntary standards.

Australia and New Zealand have set a maximum level of 0.5% ergot kernels by weight for cereal grains used in human food and Canada has also set various tolerances for different cereal foods and pulses based on percentage by weight. Canada is also unusual in having specific limits for ergot alkaloids in animal feed.

More information can be found at the FAO web link below.

Sources of Further Information

Published

The mycotoxin factbook: Food & feed topics. ed. Barug, D. *et al.* Wageningen, Wageningen Academic Publishers, 2006.

Mycotoxins. Bennett, J.W. and Klich, M. *Clinical Microbiology Reviews*, 2003, 16(3), 497–516.

On the Web

European Mycotoxin Awareness Network (EMAN). http://www.mycotoxins.org/

International Programme on Chemical Safety report on selected mycotoxins (1990). http://www.inchem.org/documents/ehc/ehc/ehc105.htm

FAO Food and Nutrition Paper 81 – Worldwide regulations for mycotoxins in food and feed 2003. http://www.fao.org/docrep/007/y5499e/y5499e00.htm

2.1.1.6 Fumonisins

Hazard Identification

What are Fumonisins?

The fumonisins are a group of at least 15 chemically related toxic fungal metabolites (mycotoxins) produced by certain mould species of the genus *Fusarium*, which may colonise cereals, especially maize, in the field. They were first identified as recently as 1988, although their effects had been noted many years before. Fumonisins are known to cause adverse health effects in livestock and other animals and are considered to be potentially toxic to humans. They have been found in maize and maize products worldwide. For these reasons they are of concern from a food safety point of view.

The fumonisins are polar compounds based on a long hydroxylated hydrocarbon chain containing methyl and amino groups. They are quite stable compounds and are divided into five groups, A, B, C, P and H, according to their chemical structure. The most widespread fumonisins in nature are the B group, and of these the most important and probably the most toxic is fumonisin B_1 (FB_1), although fumonisins B_2, B_3 and B_4 have also been found in food commodities. The chemical formula of FB_1 is $C_{34}H_{59}NO_{15}$ and its molecular weight is 721.

Occurrence in Foods

The fumonisins were initially thought to be confined to maize and maize products, but more recently they have also been found in other food commodities, such as rice, sorghum, asparagus and mung beans. Contamination levels in maize can vary considerably from year to year, and are strongly influenced by climatic conditions. High levels of fumonisins are associated with hot and dry weather, followed by a period of high humidity. Surveys of maize harvested in Iowa, in the USA, showed that the average concentration of FB_1 from 1988–91 was $>2000\,\mu g/kg$, but from 1992–6 it was $<450\,\mu g/kg$. Similar variation has been found elsewhere. The mean level of FB_1 in sound maize traded around the world in any given year has been estimated to vary from 200 to 2500 μg/kg. However, it should be noted that much higher levels may be present in visibly mouldy maize. For example, a sample tested in Italy in 1994 recorded a level FB_1 and FB_2 of 300 000 μg/kg. Detectable levels in other crops are much less common.

Fumonisins have also been found in processed foods, especially those produced from maize, such as maize meal and cornstarch, popcorn, maize-based breakfast cereals and snack products, polenta and beer. Levels are usually much lower than those found in unprocessed maize. Foods of animal origin do not seem to be a significant source of fumonisins.

A mean daily intake for FB_1 in the European diet has been estimated at 0.2 μg/kg bodyweight. By far the main contributors to fumonisins in the diet worldwide are maize and maize products.

Hazard Characterisation

Effects on Health

The acute toxicity of fumonisins in animals is relatively low in comparison with other mycotoxins, such as aflatoxins, although it is important to note that they may be present at very high levels in mouldy maize. Exposure to fumonisins in mouldy feed is associated with diseases in some livestock, especially horses and pigs. Horses exposed to fumonisins in feed over a period can develop a fatal disease known as equine leucoencephalomalacia (ELEM), which causes neurotoxic effects, liver damage and degeneration in the brain. A minimum dose of FB_1 of 200–440 μg/kg bodyweight per day is reported to be sufficient to cause ELEM in horses. Pigs may suffer from pulmonary oedema and develop respiratory problems. An outbreak of human gastrointestinal disease in India was reported to be associated with mouldy sorghum or maize containing FB_1 at a level of 64 000 μg/kg, but other mycotoxins were probably also present.

Toxicity testing in animals shows that the liver and kidneys are the main targets for fumonisin toxicity, especially in rodents. Cardiovascular effects have also been reported. The basis for the toxicity of fumonisins is thought to be interference with the synthesis of complex glycol-sphingolipids, which has effects on cell growth, development and function. The long-term chronic toxicity and carcinogenicity of FB_1 has been investigated in animals. Studies show adverse effects on the liver and kidneys of rats and mice and the development of cancers at higher levels (2500–7000 μg/kg bodyweight).

Epidemiological studies have suggested links between consumption of fumonisin-contaminated maize and high incidences of oesophageal cancer in humans, but these studies are considered inconclusive. FB_1 is classified as "possibly carcinogenic to humans. Based on the data available from animal studies, the EU Scientific Committee on Food established a tolerable daily intake (TDI) for FB_1 of 2 μg/kg bodyweight for humans.

Sources

The only known source of fumonisins in maize and other crops are *Fusarium* species fungi. The two species most associated with FB_1 and FB_2 production in maize are *F. verticillioides* (older synonym *F. moniliforme*) and *F. proliferatum*. However, other species, such as *F. nygamai, F. napiforme, F. anthophilum* and *F. dlamini* are also reported to produce fumonisins and are associated with food grains.

F. verticillioides is considered to be the main cause of Fusarium kernel rot in maize, a disease that occurs predominately in warm dry weather. High levels of FB_1 can accumulate in infected maize grains under these conditions, especially in maize that has been damaged by insects. *F. verticillioides* is very common in tropical and subtropical regions, but less so in cooler climates. It is able to grow over a fairly wide temperature range (2–35 °C) and FB_1 and FB_2 production occur at water activity levels down to about 0.90. Most toxin production in maize occurs in the field, or during the early stages of drying, rather than during storage.

Stability in Foods

Fumonisins are fairly heat stable and significant destruction occurs only when temperatures above 150 °C are reached. They therefore survive many cooking processes, but are less heat stable under alkaline conditions. Production of maize flour for tortillas by first boiling maize in calcium hydroxide (nixtamalisation) has been shown to reduce fumonisin levels considerably, but the hydrolysed breakdown products formed are also thought to be toxic.

FB_1 has been shown to survive fermentation and brewing processes and has been detected in beer.

Control Options

Since fumonisin production occurs almost entirely in the field, the most effective controls are applied at the pre-harvest stage.

Pre-harvest

Good agricultural practice (GAP) measures designed to reduce *Fusarium* infection in cereal crops are also effective in limiting the formation of fumonisins. Control measures include the following.

- land preparation, crop rotation and crop debris removal to reduce the inoculum of *fusarium* in the field;
- use of fungus-resistant crop varieties;
- control of infection by appropriately timed application of effective fungicides;
- effective control of insect crop pests;
- harvesting at the correct moisture level and stage of maturity.

Post-harvest Handling and Storage

Further production of fumonisins during storage can be prevented by rapid drying to a water activity value of 0.8 immediately after harvest, and by implementing good storage practice.

Decontamination

Physical decontamination methods, such as separation of screenings, can be an effective means of reducing fumonisin levels in contaminated maize. However, fumonisins also occur in whole undamaged grains. Milling processes also reduce fumonisin concentrations in maize flour by removing the generally more heavily contaminated bran and germ, but the effectiveness of this depends on the distribution of the toxin in the grain. In wet milling processes, significant quantities of fumonisins leach out of the grain into the steep water.

Chemical decontamination methods for FB_1, such as a modified nixtamalisation process and ammoniation, have been investigated, but are not yet developed for commercial use.

Heat treatments are not usually effective, unless high temperatures (>150 °C) are used.

Testing

Some countries monitor cereals for FB_1 and FB_2 contamination in maize and maize products by sampling and testing using analytical methods, usually based on HPLC. ELISA methods for FB_1 and FB_2 have been developed for screening purposes and commercial kits are available. However, as with other mycotoxins, the distribution of fumonisins in bulk commodities may be highly heterogeneous and it is essential to ensure that an adequate representative sampling plan is used.

Legislation

Very few countries outside Europe and North America have introduced mandatory or guideline levels for fumonisins in foods.

European Union

The EU has set maximum levels for FB_1 and FB_2 in combination. The maximum level for unprocessed maize (other than maize intended to be processed by wet milling) is 4000 µg/kg, for maize and maize-based foods intended for direct human consumption it is 1000 µg/kg, and for maize-based breakfast cereals and snacks it is 800 µg/kg. The limit for maize-based foods for infants and young children is 200 µg/kg.

USA

US food safety regulations include maximum guidance levels for FB_1, FB_2 and FB_3 in combination for maize products. These vary from 2000 to 4000 µg/kg depending on the product. Much higher levels are allowed in animal feeds.

Others

Switzerland has set a maximum level for FB_1 and FB_2 in maize of 1000 µg/kg.

More information can be found at the FAO web link below.

Sources of Further Information

Published

The mycotoxin factbook: Food & feed topics. ed. Barug, D. *et al.* Wageningen, Wageningen Academic Publishers, 2006.

Mycotoxins. Bennett, J.W. and Klich, M. *Clinical Microbiology Reviews*, 2003, 16(3), 497–516.

On the Web

European Mycotoxin Awareness Network (EMAN). http://www.mycotoxins.org/

JECFA monograph on fumonisins (2001). http://www.inchem.org/documents/jecfa/jecmono/v47je03.htm

FAO Food and Nutrition Paper 81 – Worldwide regulations for mycotoxins in food and feed 2003. http://www.fao.org/docrep/007/y5499e/y5499e00.htm

2.1.1.7 Moniliformin

Hazard Identification

What is Moniliformin?

Moniliformin is a toxic fungal metabolite (mycotoxin) produced by some moulds of the genus *Fusarium* growing on certain food commodities, especially cereals. It was originally reported to be produced by *Fusarium moniliforme* (now reclassified as *F. verticillioides*), which also produces fumonisins, but these reports are now discounted. Although comparatively little is known about the occurrence of moniliformin, it exhibits a number of toxic effects in animals and its presence in food is undesirable.

Moniliformin is an ionic compound with a four-carbon ring structure and occurs as sodium or potassium salts of 1-hydroxycyclobut-1-ene-3,4-dione. It is soluble in water.

Occurrence in Foods

Moniliformin appears to be relatively uncommon in food commodities, but it has been reported in cereals, including wheat, rye, rice and especially maize. Levels of up to 12 mg/kg were reported in maize intended for human consumption in South Africa, and up to 4.6 mg/kg moniliformin was reported in 60% of samples of milled maize imported into the UK for use in animal feed. Moniliformin has also been found in Polish cereals showing mould damage.

Little is known about the occurrence of moniliformin is processed foods, but it was detected in corn tortillas at levels of up to 0.1 mg/kg. Similar levels have also been reported in other maize-based foods, such as polenta.

Natural occurrence in foods and the potential for human exposure from the diet appear to be quite low.

Hazard Characterisation

Effects on Health

Most of the information on the toxicity of citrinin is derived from a limited number of animal studies and there is little or no experimental, or epidemiological, data on acute or chronic toxicity in humans. Its significance for human health is therefore still unclear.

The toxicity of moniliformin is based on its ability to inhibit mitochondrial pyruvate and ketoglutarate oxidation. But relatively high doses appear to be necessary to cause significant toxic effects on mammals, and an oral LD_{50} (lethal dose) of 25–50 mg/kg bodyweight has been reported for rodents. Birds are reported to be more sensitive to moniliformin (LD_{50} of 4 mg/kg for day-old chicks). The main effect of acute toxicity is intestinal haemorrhage, but chronic toxicity mainly affects the heart. The interpretation of animal studies based on the feeding of contaminated maize is complicated by the likely presence of other *Fusarium* mycotoxins.

There is no significant evidence for carcinogenicity, but the amount of reported data is quite limited.

It has been proposed that moniliformin may be implicated in human disease, notably Keshan disease, a cardiomyopathy endemic in certain parts of China. However, it is likely that other factors, such as selenium deficiency, are also involved in this condition.

Sources

Moniliformin is reportedly produced by several species of *Fusarium*, including *F. avenaceum, F. subglutinans* and some strains of *F. proliferatum* and *F. oxyporum*, at least in laboratory culture. Erroneous reports of production by *F. moniliforme* are now thought to be the result of working with mixed cultures of more than one species.

F. subglutinans is thought to be a producer of moniliformin in the field and this species has a global distribution. It has been isolated from maize in Europe, North and South America, Asia and Australia and is also a pathogen of pineapples and bananas. *F. avenaceum* is also found worldwide, but is rarely isolated from food commodities and is not regarded as a major pathogen of cereals. It has been reported to cause occasional spoilage in fruits and vegetables, such as apples and tomatoes. It is able to grow in a temperature range of –3 to 35 °C and at water activity values as low as 0.90.

Stability in Foods

Relatively little is known about the stability of moniliformin during food processing, but like many mycotoxins, it is thought to be quite heat stable. It has been reported to survive autoclaving of creamed corn at 121 °C for 65 min, and roasting corn meal at 218 °C for 15 min gave a 45% reduction. Significant concentrations have also been shown to survive in the manufacture of corn chips from spiked maize. Moniliformin is less stable at alkaline pH, and production of tortillas using nixtamalisation processes gave a 70% reduction.

Control Options

There are few specific documented control measures for moniliformin, but its co-occurrence with other *Fusarium* mycotoxins in cereals means that the pre- and post-harvest control measures recommended for fumonisins may also provide indirect control of moniliformin.

Legislation

There are no current specific regulations setting mandatory or recommended maximum limits for moniliformin in food or feed.

Sources of Further Information

Published

The mycotoxin factbook: Food & feed topics. ed. Barug, D. *et al.* Wageningen, Wageningen Academic Publishers, 2006.

Mycotoxins. Bennett, J.W. and Klich, M. *Clinical Microbiology Reviews*, 2003, 16(3), 497–516.

On the Web

European Mycotoxin Awareness Network (EMAN). http://www.mycotoxins.org/

2.1.1.8 Ochratoxins

Hazard Identification

What are Ochratoxins?

Ochratoxins are a small group of chemically related toxic fungal metabolites (mycotoxins) produced by certain moulds of the genera *Aspergillus* and *Penicillium* growing on a wide range of raw-food commodities. Some ochratoxins are potent toxins and their presence in food is undesirable.

The ochratoxins are pentaketides made up of dihydro-isocoumarin linked to β-phenylalanine. The most important and most toxic ochratoxin found naturally in food is ochratoxin A (OTA). The only other ochratoxin found in food is ochratoxin B, which is rare and much less toxic. Other structurally related ochratoxins include ochratoxin C, α and β. These have been isolated from fungal cultures, but are not normally found in foods. The remainder of this section therefore refers specifically to OTA.

Occurrence in Foods

In surveys, OTA has been found in a very wide range of raw and processed food commodities all over the world. It was first reported in cereals, but has since been found in other products, including coffee, dried fruits, wine, beer, cocoa, nuts, beans, peas, bread and rice. It has also been detected in meat, especially pork and poultry, following transfer from contaminated feed.

OTA levels in different food products vary, but are generally low in properly stored commodities (mean value <1 μg/kg for cereals from temperate regions). However, much higher concentrations can develop under inadequate storage conditions. Levels of up to 6000 μg/kg and 5000 μg/kg have been reported in Canadian wheat and UK barley, respectively, but the concentrations found are usually below 50 μg/kg. The major contributors to OTA in the diet in Europe are cereals and wine. Coffee was thought to be important in this respect, but is now considered less significant. Pork products have also been suggested as a significant dietary source.

Hazard Characterisation

Effects on Health

OTA is a potent nephrotoxin and causes both acute and chronic effects in the kidneys of all mammalian species tested. The sensitivity of different species varies, but a level of 200 μg/kg in feed over three months is sufficient to cause acute damage to the kidneys of pigs and rats. There are no documented cases of acute OTA toxicity in humans.

OTA is also genotoxic (damages DNA) and teratogenic (damages the foetus) and is considered a probable carcinogen, causing renal carcinoma and other cancers in a number of animal species, although the mechanism for this is uncertain. It is also reported to have adverse effects on the immune system in some species. The evidence for carcinogenicity in humans is not conclusive, but in view of the evidence for other mammalian species, the presence of OTA in food and feed must be considered undesirable. Some toxicologists suspect that OTA may be a very significant food contaminant from a public health point of view.

OTA has been detected in human blood and breast milk, demonstrating dietary exposure. Daily intakes have been estimated at between 0.2 and 4.7 ng/kg bodyweight. In 2006, the European Food Safety Authority (EFSA) derived a tolerable weekly intake (TWI) of 120 ng/kg bodyweight for OTA in the diet, based on the latest scientific evidence.

Sources

In tropical and subtropical regions, OTA is produced mainly by *Aspergillus* species, particularly the widespread *A. ochraceus*. But in temperate climates (Canada, northern Europe and parts of South America), the main producer is *Penicillium verrucosum*.

OTA production by *A. ochraceus* is favoured by relatively high temperatures (13 °C to 37 °C), but *P. verrucosum* grows and produces the toxin at temperatures as low as 0 °C. *A. ochraceus* is able to produce OTA at water activities down to 0.80, while the lower limit for significant toxin production by *P. verrucosum* is thought to be about 0.86, although growth can occur at lower values. Both are considered to be storage fungi, rather than field contaminants or plant pathogens, and toxin production occurs mainly when susceptible commodities are stored under inappropriate conditions, particularly at high moisture levels.

Stability in Foods

OTA is a relatively heat-stable molecule and survives most cooking processes to some extent, although the reduction in concentration during heating depends on factors such as temperature, pH and other components in the product. For example, heating wet wheat at 100 °C for 2.3 h gave a 50% reduction in OTA concentration, but in dry wheat, the same reduction took 12 h.

Processes such as coffee roasting and baking of cereal products and biscuits can produce significant losses in OTA levels, but processes like pasta manufacture produce little reduction. OTA also survives brewing and winemaking and can be found in a variety of processed consumer food products.

OTA is destroyed by acid and alkaline hydrolysis and by the action of some oxidising agents.

Control Options

The ability of OTA-producing fungi to grow on a wide range of food commodities and the persistence and ubiquity of OTA in the food chain mean that control is best achieved by measures designed to prevent the contamination of foods using HACCP-type techniques. Detection and removal of OTA-contaminated material from the food supply chain is also important for imported products.

Pre-harvest

Both *A. ochraceus* and *P. verrucosum* are considered to be storage fungi rather than field fungi. Pre-Harvest controls are therefore limited to harvesting susceptible crops at the correct moisture level and stage of maturity.

Post-harvest Handling and Storage

For cereals, the most important and effective control measure in post-harvest handling and storage is the control of moisture content and hence, the water activity of the crop. Ensuring that susceptible crops are harvested at a safe moisture level, or are dried to a safe level immediately after harvest is vital to prevent mould growth and OTA production during storage. In tropical and subtropical climates stored grains must be dried rapidly to an a_w value of below 0.8 and this level must be maintained throughout storage to prevent *A. ochraceus* growth. In temperate regions a target moisture content of 18% for grain drying is recommended, together with rapid cooling of grain if hot-air drying is used. This should be followed by further drying down to a moisture level of 15% (UK Code of Good Storage Practice).

Other important cereal storage factors are effective cleaning of grain stores and handling equipment between crops, and fumigation to prevent insect infestation. In tropical regions, the use of controlled atmosphere storage to control insects may also help to inhibit mould growth.

Rapid and effective drying is also important in the control of OTA production in other commodities, especially coffee. For dried fruits, minimising mechanical and insect damage during handling and storage helps to prevent the entry of moulds into the fruit before drying.

Monitoring raw material quality is the most effective control for processed foods. Any ingredient that displays visible mould growth should not be used. Testing for the presence of OTA in susceptible materials, such as barley for brewing, may be necessary in some cases.

Decontamination

Physical separation of contaminated material can be an effective means of reducing OTA levels in contaminated commodities. Mouldy grain should not be used for food, or for animal feed.

There has been little practical evaluation of chemical decontamination methods for OTA to date, but an ammoniation process has been shown to be effective for cereals.

Testing

Some countries monitor imported commodities that are susceptible to OTA contamination, such as grains and coffee beans, by sampling and analysis. A number of analytical methods have been developed based on TLC, HPLC and ELISA and there are also rapid-screening kits available. However, moulds and mycotoxins in bulk food shipments tend to be highly heterogeneous in their distribution and it is essential to ensure that an adequate sampling plan is used to monitor imported materials.

Legislation

A number of countries, particularly in Europe, have regulations governing OTA in food and feed and most include maximum permitted, or recommended levels for specific commodities.

European Union

The EU sets limits for OTA in cereals, dried vine fruits, roasted coffee beans and ground coffee, soluble coffee, wine and grape juice. Limits vary according to the commodity, but range from 2–10 μg/kg. The limit for unprocessed cereals is 5.0 μg/kg, but for processed cereal products intended for direct human consumption it is 3.0 μg/kg. The limit for dried vine fruits is 10 μg/kg. There is also a limit of 0.50 μg/kg for OTA in processed cereal-based foods for infants and young children.

Others

Switzerland applies a limit of 5.0 μg/kg for all foods except cereal based infant foods, where the limit is 0.5 μg/kg, and Turkey has set limits of between 3.0 and 10 μg/kg for various food commodities.

Few other countries outside Europe have imposed limits for OTA, but a number have proposals to do so. Uruguay sets a limit of 50 μg/kg for rice, cereals and dried fruits and Canada sets a limit of 2000 μg/kg for OTA in pig and poultry feed.

More information can be found at the FAO web link below.

Sources of Further Information

Published

Bayman, P. and Baker, J.L. Ochratoxins: a global perspective. *Mycopathologia*, 2006, 162(3), 215–23.

The mycotoxin factbook: Food & feed topics. ed. Barug, D. *et al.* Wageningen, Wageningen Academic Publishers, 2006.

Mycotoxins. Bennett, J.W. and Klich, M. *Clinical Microbiology Reviews*, 2003, 16(3), 497–516.

On the Web

Food-Info.net: Overview of foodborne toxins – mycotoxins (ochratoxins). http://www.food-info.net/uk/tox/ochra.htm

European Mycotoxin Awareness Network (EMAN). http://www.mycotoxins.org/

JECFA monograph on ochratoxin A (2001). http://www.inchem.org/documents/jecfa/jecmono/v47je04.htm

FAO Food and Nutrition Paper 81 – Worldwide regulations for mycotoxins in food and feed 2003. http://www.fao.org/docrep/007/y5499e/y5499e00.htm

2.1.1.9 Patulin

Hazard Identification

What is Patulin?

Patulin is a toxic fungal metabolite (mycotoxin) produced by certain moulds of the genera *Penicillium, Aspergillus* and *Byssochlamys* growing on certain food commodities, especially fruit. Patulin exhibits a number of toxic effects in animals and its presence in food is undesirable.

Chemically, patulin is a polyketide lactone. It is a relatively small molecule ($C_7H_6O_4$) and is soluble in water.

Occurrence in Foods

Patulin occurs most often in apples that have been spoiled by mould growth, or in products made from spoiled apples, such as apple juice, pies and conserves. It has also been found in other fruits, including pears and grapes, in vegetables and in cereal grains and cheese.

Apples and apple products are considered to be by far the most significant contributor to patulin in the diet. Contaminated apple juice usually contains patulin at levels below 50 μg/litre, but much higher levels (up to 4000 μg/litre) have been reported occasionally.

Hazard Characterisation

Effects on Health

Most of the information on the toxicity of patulin is derived from animal studies and there is little or no experimental, or epidemiological, data on acute or chronic toxicity in humans.

At relatively high doses, patulin is acutely toxic in mice and rats, causing gastrointestinal lesions, distension and haemorrhage in the stomach and small intestine. However, it is possible that these effects are due to the selective antibiotic action of patulin against gram-positive bacteria, which may give gram-negative intestinal pathogens an advantage. LD_{50} values (lethal dose) of 20–100 mg/kg bodyweight have been reported for patulin administered orally to mice and rats. These levels are much higher than those likely to be encountered in human diets. Relatively high doses of patulin have also been shown to be immunotoxic and neurotoxic in animals.

Of more concern from a food safety point of view are longer-term chronic effects. It has been suggested that patulin could be a carcinogen at low levels in the diet, but the International Agency for Research on Cancer (IARC) has reviewed the available data and concluded that there is no convincing evidence of carcinogenicity in animals or in humans, other than at extremely high doses.

Data from feeding experiments have been used to derive a no observed effect level (NOEL) of 43 μg/kg bodyweight per day and a provisional maximum tolerable daily intake (PMTDI) for humans of 0.4 μg/kg bodyweight. This is well above the maximum daily intake levels estimated for adults and children (0.1 and 0.2 μg/kg bodyweight, respectively).

Sources

Patulin is produced by certain species of *Penicillium, Aspergillus* and *Byssochlamys*, notably *Penicillium expansum* and *Aspergillus clavatus. P. expansum* is the most significant producer of patulin, as it is a common cause of rots in apples. Patulin production by *P. expansum* has been reported over a temperature range from 0–25 °C and over a pH range in apple juice of 3.2–3.8.

Stability in Foods

Patulin is relatively heat stable and is not destroyed by pasteurisation of apple juice at 90 °C for 10 s. However, it is broken down in fruit juice and other foods in the presence of sulfur dioxide used as a preservative. It does not appear to survive fermentation processes and is not usually found in alcoholic drinks, such as cider, but the toxicity of its breakdown products is uncertain.

Patulin produced by mould growth on cheese is inactivated by interaction with high cysteine levels.

Control Options

Patulin is only considered to be a significant problem in apples and apple products, especially apple juice.

Pre-harvest

Good agricultural practice (GAP) measures designed to minimise insect and bird damage to apples can help to prevent mould infection and patulin production before harvest.

At harvest, rotten and damaged apples should be discarded, as these are much more likely to contain patulin.

Post-harvest

Control in harvested apples is best achieved by good storage practice designed to ensure hygienic conditions in apple stores and to minimise physical damage that might promote fungal infection and rotting. Storage at temperatures of less than 10 °C is also a useful control measure.

Processing

Physical separation of mouldy and damaged apples before processing will help to reduce patulin levels in apple juice and other apple products. This can be done by hand, or by using water flumes or high-pressure water jets. Washing of apples can also help to reduce patulin levels.

Testing

Monitoring of patulin levels in susceptible products, such as apple juice, by sampling and analysis can be valuable – the test method of choice being HPLC with UV detection. In the UK, significant reductions in patulin levels in apple juice have been achieved since regular monitoring was implemented in 1992.

Legislation

Although patulin is now considered to be a less significant food safety hazard than previously, a number of countries have introduced regulations specifying maximum permitted levels in susceptible products.

European Union

The EU has set a maximum limit for patulin of 50 μg/kg in fruit juices and in drinks containing apple juice or derived from apples. For solid apple products, such as apple puree, the limit is 25 μg/kg. A lower limit of 10 μg/kg has been set for certain foods intended for infants.

USA

The FDA has set an upper limit of 50 μg/kg for patulin in apple juice and apple juice concentrates.

Others

The Codex Alimentarius Commission has also set a recommended upper limit of 50 μg/kg for patulin in apple juice and apple ingredients in other beverages.

More information can be found at the FAO web link below.

Sources of Further Information

Published

Drusch, S. and Ragab, W. Mycotoxins in fruits, fruit juices and dried fruits, *Journal of Food Protection*, 2003, 66(8), 1514–27.

The mycotoxin factbook: Food & feed topics, ed. Barug, D. *et al.* Wageningen, Wageningen Academic Publishers, 2006.

Mycotoxins. Bennett, J.W. and Klich, M., *Clinical Microbiology Reviews*, 2003, 16(3), 497–516.

On the Web

JECFA Monograph on patulin (WHO Food Additives Series 35). http://www.inchem.org/documents/jecfa/jecmono/v35je16.htm

FDA background paper on patulin. http://www.cfsan.fda.gov/~dms/patubckg.html

European Mycotoxin Awareness Network (EMAN). http://www.mycotoxins.org/

FAO Food and Nutrition Paper 81 – Worldwide regulations for mycotoxins in food and feed 2003. http://www.fao.org/docrep/007/y5499e/y5499e00.htm

2.1.1.10 Sterigmatocystin

Hazard Identification

What is Sterigmatocystin?

Sterigmatocystin is a toxic fungal metabolite (mycotoxin) produced by some moulds of the genus *Aspergillus* growing on certain food commodities, such as maize. Sterigmatocystin is a potent carcinogen in animals and its presence in food is undesirable.

Chemically, sterigmatocystin is closely related to, and is a precursor of, the aflatoxins. It consists of a xanthone nucleus attached to a bifuran structure. It is only slightly soluble in water. It is one of a group of at least seven related compounds, others of which may also occur naturally.

Occurrence in Foods

Sterigmatocystin has been reported in mouldy cereals, particularly maize, peanuts and pecans, green coffee beans and cheese. It appears to be much less common and less widely distributed than aflatoxins, although low levels in foods may be under-reported because sensitive analytical techniques have only recently been available. However, it has hardly ever been detected in surveys of good-quality food commodities, even with the use of reliable analytical methods.

Sterigmatocystin has very rarely been detected in naturally contaminated processed foods, but it has been reported to be present in quite high levels in bread and cured meats inoculated with toxin-producing mould cultures.

Hazard Characterisation

Effects on Health

Most of the information on the toxicity of sterigmatocystin is derived from animal studies and there is little or no experimental, or epidemiological, data on acute or chronic toxicity in humans.

The toxicity of sterigmatocystin is very similar to that of aflatoxin B_1, causing liver and kidney damage and diarrhoea, although its acute toxicity is lower for most species. Cattle ingesting feed containing about 8 mg/kg sterigmatocystin were reported to have developed bloody diarrhoea and loss of milk production.

Chronic toxicity is probably more important from a food safety point of view. Sterigmatocystin is a potent carcinogen, mutagen and teratogen in many animals, and therefore potentially in humans, and the liver is again the main target organ. However, it is considered a less potent carcinogen than aflatoxin B_1, although levels as low as 15 μg/day caused liver cancer when fed to rats.

Based on data from animals, the California Department of Health has derived a "no significant risk" intake level for humans of 8 μg/kg bodyweight/day.

Sources

Sterigmatocystin is produced mainly by the *Aspergillus* species *A. versicolor* and *A. nidulans*. The toxin is produced primarily on stored products that undergo mould spoilage rather than on crops in the field.

A. versicolor is quite widely dispersed and has been isolated from a number of foods, such as fruits and dried meats, in which sterigmatocystin itself has not been found. It is able to grow in a temperature range of 9–39 °C and at water activity values as low as 0.80.

Stability in Foods

There is little published information on the stability of sterigmatocystin in foods, but its chemical similarity to the aflatoxins suggests that it likely to be similarly heat stable and persistent.

Control Options

As sterigmatocystin is produced mainly in stored cereals and other foods that undergo mould spoilage, effective control can be achieved by applying good storage practice and by ensuring that moisture levels in cereals are low enough to prevent mould growth.

Most of the sterigmatocystin in contaminated rice is reported to be removed during the milling stage.

Legislation

There are no current specific regulations setting mandatory or recommended maximum limits for sterigmatocystin in food or feed. However, some Eastern European countries did set limits in legislation prior to becoming members of the EU. For example, the Czech Republic set maximum limits of 5 or 20 μg/kg, depending on the nature of the product.

Sources of Further Information

Published

The mycotoxin factbook: Food & feed topics, ed. Barug, D. *et al.* Wageningen, Wageningen Academic Publishers, 2006.

Mycotoxins. Bennett, J.W. and Klich, M., *Clinical Microbiology Reviews*, 2003, 16(3), 497–516.

On the Web

European Mycotoxin Awareness Network (EMAN). http://www.mycotoxins.org/

2.1.1.11 Trichothecenes

Hazard Identification

What are Trichothecenes?

The trichothecenes are a large group of around 150 chemically related toxic fungal metabolites (mycotoxins) produced by moulds, especially *Fusarium* species, which may colonise cereals and other crops in the field. Several of the trichothecenes are known to be acutely toxic to humans and livestock. They have been found in a number of food commodities and can be present at high levels. For these reasons they are of concern from a food safety point of view.

The trichothecenes are characterised as tetracyclic sesquiterpenes. They are chemically stable and persistent compounds and are divided into two groups, A and B, according to their chemical structure. The most commonly reported group A trichothecenes in foods are T2 toxin ($C_{24}H_{34}O_9$) and HT-2 toxin ($C_{22}H_{32}O_8$), while group B trichothecenes include deoxynivalenol (DON), which is covered elsewhere (see 2.1.1.4), and nivalenol ($C_{15}H_{20}O_7$). The remainder of this section refers to T-2 and HT-2 toxins and, to a lesser extent, nivalenol.

Occurrence in Foods

Trichothecenes are mainly associated with cereals, and have been found to occur in wheat, barley, oats, rye, maize and rice. Their presence has also been reported in other commodities, such as soya beans, potatoes, sunflower seeds, peanuts and bananas. The frequency of contamination in cereals varies from year to year, but surveys in Europe have shown that T-2 toxin was present in 11% of cereal samples, while HT-2 toxin occurred in 14% of samples. The level of contamination found for T-2 and HT-2 toxins in cereals is usually low (<100 μg/kg), but high levels do occur in a small number of samples. For T-2 toxin, levels have been reported to reach 820 μg/kg in wheat from Asia, 1700 μg/kg in European oats and 2400 μg/kg in American maize. A level of 2000 μg/kg of HT-2 toxin has been reported in European oats. High levels of both toxins may occasionally be present in the same samples.

Trichothecenes have also been found in processed foods, especially those produced from cereals. Foods reported to be contaminated have included bread, breakfast cereals, noodles, and beer. Foods of animal origin do not seem to be a significant source of trichothecenes in the human diet.

Daily intakes for T-2 and HT-2 toxins in Europe have been estimated at 7.6 ng/kg bodyweight and 8.7 ng/kg bodyweight, respectively. The main contributors to trichothecenes in the diet in Europe are wheat and barley, but it is probable that other crops, such as rice and maize are more significant in other regions.

Hazard Characterisation

Effects on Health

Trichothecenes are associated with acute toxicity in both animals and humans and T-2 toxin, HT-2 toxin and nivalenol are all acutely toxic to mice at much lower concentrations than DON. The toxicities of T-2 and HT-2 toxins are generally considered in combination, largely because T-2 toxin is rapidly converted to HT-2 toxin and other metabolites in the gut. Trichothecenes in general are known to inhibit protein synthesis, and are immunosuppressive at low concentrations.

Acute toxicity in animals is characterised by haemorrhaging in the gastrointestinal tract and severe gastroenteritis, which may eventually be fatal. Other symptoms include necrotic lesions in the mouth and on the skin and degeneration of the bone marrow and lymph nodes. Acute toxicity in humans has also been reported, with symptoms including nausea and vomiting, dizziness, diarrhoea, abdominal pain and distension, throat irritation and chills. In some suspected outbreaks a high mortality rate was recorded, but in others no deaths occurred. It should be noted that the role of individual toxins in these cases is usually uncertain, as other mycotoxins are almost always present. T-2 and HT-2 toxins are thought to be the most significant in most cases, but the role of other trichothecenes, such as DON and nivalenol may also be important.

Long-term chronic toxicity from low levels of T-2 and HT-2 toxins in the diet has been investigated in animals. Studies show adverse effects to the immune system, leading to changes in the white blood cell count and, in some cases, decreased resistance to microbial infection. Other effects in animals include reduced feed intake and weight gain. However, there is little evidence of carcinogenicity, and T-2 and HT-2 toxins are not considered likely to be potent carcinogens. Based on the data available from animal studies, the EU Scientific Committee on Food established a temporary tolerable daily intake (TDI) for T-2 and HT-2 toxins (alone or in combination) of 0.06 μg/kg bodyweight for humans. A temporary TDI of 0–0.7 μg/kg bodyweight was established for nivalenol.

Incidence and Outbreaks

There are a number of documented outbreaks of food poisoning-like illness caused by foods contaminated with trichothecenes. For example, a series of outbreaks of a condition termed alimentary toxic aleukia were reported in the former Soviet Union during the 1940s and 1950s and caused widespread disease with many deaths. These outbreaks were associated with consumption of overwintered wheat and subsequent analysis of fungi isolated from wheat samples showed that some could produce T-2 toxin and other trichothecenes.

There have also been reported outbreaks affecting hundreds of people in China and India. These were associated with eating contaminated rice. T-2 toxin at concentrations of 180–420 μg/kg was found in rice from one Chinese outbreak, but it seems likely that other trichothecenes were involved in some of these cases.

Sources

The principal sources of trichothecenes in cereals and other crops are *Fusarium* species fungi. Group A trichothecenes are produced by mainly saprophytic species such as *F. poae, F. sporotrichioides* and *F. acuminatum*, whereas group B trichothecenes are produced by cereal pathogens such as *F. graminearum* and *F. culmorum*. All of these are common soil fungi and may colonise or infect cereals in the field.

F. sporotrichioides is the most important producer of T-2 and HT-2 toxins in cereals in temperate regions and it is able to grow at low temperatures (–2 °C to 35 °C). However, it cannot grow at water activities of below 0.88. Most toxin production by this species occurs in water-damaged grains that have either remained in the field for long periods, especially in cold weather, or become damp during storage. T-2 and HT-2 toxins are typically produced on the surface of infected grains. However, where high levels are produced, it may be more evenly distributed in the kernel.

Stability in Foods

Trichothecenes are extremely heat stable and are not destroyed by temperatures of 120 °C. They therefore survive most cooking processes and T-2 and HT-2 toxins are reported to be relatively stable even in baking processes. Some natural degradation seems to occur in grain in the field or during storage, but the mechanism for this is uncertain.

Control Options

Since trichothecene production occurs in the field and during storage, controls are applied at both the pre-harvest and post-harvest stages.

Pre-harvest

Good agricultural practice (GAP) measures designed to reduce *Fusarium* infection in cereal crops are also effective in limiting the formation of trichothecenes. Control measures include the following.

- land preparation, crop rotation and crop debris removal to reduce the inoculum of *Fusarium* in the field;
- use of fungus-resistant crop varieties;
- control of infection by appropriately timed application of effective fungicides;
- harvesting at the correct moisture level and stage of maturity.

Post-harvest Handling and Storage

Further production of trichothecenes after harvest can be prevented by rapid drying to a water activity value of 0.8, and by implementing good storage practice.

Decontamination

Physical decontamination methods, such as gravity separation, can be an effective means of reducing trichothecene levels in contaminated grain. The milling process also reduces trichothecene concentrations in wheat flour by removing the generally more heavily contaminated bran, but the effectiveness of this depends on the distribution of the toxin in the grain.

Chemical decontamination methods for T-2 toxin, such as treatment with calcium hydroxide monomethylamine, have been investigated, but are not yet developed for commercial use.

Heat treatments are not usually effective.

Testing

Some countries monitor cereals for T-2 and HT-2 toxin contamination by sampling and testing using analytical methods, such as liquid chromatography (LC), or gas chromatography (GC), with mass spectroscopic detection. HPLC methods have also been developed for some group B trichothecenes. Sensitive ELISA methods for T-2 and HT-2 toxins are available for screening purposes and commercial kits are available. However, as with other mycotoxins, the distribution of trichothecenes in bulk commodities may be highly heterogeneous and it is essential to ensure that an adequate representative sampling plan is used.

Legislation

Very few countries around the world have introduced mandatory or guideline levels for trichothecenes, other than DON, in foods.

European Union

The EU has not yet set maximum levels for T-2 and HT-2 toxins. However, the appropriateness of a combined maximum level for cereals and cereal products is due for review by July 2008. A number of Eastern European countries did set limits for T-2 toxin in cereals (typically 100 μg/kg) in national legislation prior to EU accession.

USA

US food safety regulations include a limit for DON in finished wheat products for human consumption, but not for other trichothecenes.

Others

The Russian Federation and the Ukraine both set a limit of 100 μg/kg for T-2 toxin in cereals.

The Canadian authorities have introduced a limit of 1000 μg/kg for T-2 toxin in pig and poultry feed and 100 μg/kg for HT-2 toxin in cattle and poultry feed.

More information can be found at the FAO web link below.

Sources of Further Information

Published

Aldred, D. and Magan, N., Prevention strategies for trichothecenes, *Toxicology Letters*, 2004, 153(1), 165–71.

The mycotoxin factbook: Food & feed topics, ed. Barug, D. *et al*. Wageningen, Wageningen Academic Publishers, 2006.

Mycotoxins. Bennett, J.W. and Klich, M., *Clinical Microbiology Reviews*, 2003, 16(3), 497–516.

On the Web

JECFA monograph on T-2 and HT-2 toxins. http://www.inchem.org/documents/jecfa/jecmono/v47je06.htm

Food-Info.net: Overview of foodborne toxins – mycotoxins (trichothecenes). http://www.food-info.net/uk/tox/trich.htm

European Mycotoxin Awareness Network (EMAN). http://www.mycotoxins.org/

FAO Food and Nutrition Paper 81 – Worldwide regulations for mycotoxins in food and feed 2003. http://www.fao.org/docrep/007/y5499e/y5499e00.htm

2.1.1.12 Zearalenone

Hazard Identification

What is Zearalenone?

Zearalenone is a toxic fungal metabolite (mycotoxin) produced by certain mould species of the genus *Fusarium* colonising cereal crops in the field and during storage. Zearalenone is an oestrogenic mycotoxin well known as a cause of hormonal effects in livestock, especially pigs and sheep. It is also commonly found in a wide range of food commodities and can be found in processed, ready-to-eat foods. For these reasons it is of concern from a food safety point of view.

Zearalenone ($C_{18}H_{22}O_5$) is characterised chemically as a phenolic resorcyclic acid lactone and has a molecular weight of 318.36. It is only slightly soluble in water and is quite stable. Several closely related metabolites of zearalenone have been identified in fungal cultures, notably α- and β-zearalenols, but the presence and significance of these compounds in foods is uncertain.

Occurrence in Foods

Zearalenone has been found worldwide in a range of cereals and other crops, including wheat, barley, maize, rice, oats, sorghum and some legumes. High levels have also been reported in bananas grown in India. The level of contamination in cereal crops varies widely depending on climatic conditions. For example, zearalenone was found in 11–80% of wheat samples collected randomly in Germany between 1987 and 1993. The mean yearly contents were 3–180 μg/kg and the highest level found was 8000 μg/kg. There is evidence that cereal crops produced by "alternative" or "ecological" cultivation methods may develop higher levels of contamination than those produced by conventional methods.

Zearalenone has also been found in processed foods, especially those produced from cereals, although levels are usually low. Foods reported to be contaminated have included wheat and corn flour, bread, breakfast cereals, noodles, biscuits, snacks and corn beer. The metabolite β-zearalenol may be produced from zearalenone by yeast fermentation and so may occur in beer. Contamination with zearalenone does not seem to be a major problem in foods of animal origin. It has been found to be excreted into the milk of lactating cows, along with α- and β-zearalenols, but only when very high oral doses (6000 mg) were used.

Average dietary intakes of zearalenone in humans have been estimated at 1.5 μg per day for the European diet and 3.5 μg per day for the Middle Eastern diet. Cereals are the major contributor of zearalenone in the diet.

Hazard Characterisation

Effects on Health

The acute toxicity of zearalenone is low and its toxic effects are related to the potent oestrogenic activity of the toxin itself and its metabolites.

Zearalenone is metabolised in the gut of animals, especially pigs and potentially humans, forming α- and β-zearalenols. These metabolites are then conjugated with glucuronic acids and may be more potent oestrogens than zearalenone itself.

Zearalenone has been shown to cause hormonal effects on the reproductive systems of pigs and sheep, which appear to be more sensitive than other animal species. Feeding zearalenone to female pigs at levels of up to 0.25 mg/kg produced slight inflammation of external sexual organs. Effects of higher doses (50 mg/kg) in the diet of pigs included abortion and stillbirths, while more moderate doses (10 mg/kg) caused reduced litter sizes and birth weights. Sheep are similarly affected and zearalenone is reported to be a cause of infertility in flocks in New Zealand. Dairy cows are also reported to develop reproductive abnormalities when the toxin is present in the diet.

There is some evidence for similar effects in humans. Zearalenone was suspected as a cause of an outbreak of early secondary breast development affecting girls from six months to eight years old in Puerto Rico between 1978 and 1981. A similar incident was reported in Hungary in 1997.

There is only very limited evidence for the carcinogenicity of zearalenone. It has been evaluated by the International Agency for Research on Cancer (IARC) and was judged a Group 3 compound (not classifiable as to their carcinogenicity in humans). Based on the data available from studies in pigs, the EU Scientific Committee on Food established a temporary tolerable daily intake (TDI) for zearalenone of 0.2 μg/kg bodyweight for humans.

Sources

The principal sources of zearalenone in cereals are *Fusarium* species, such as *F. graminearum, F. culmorum* and *F. crookwellense*. These species are considered to be field fungi and are pathogenic to cereals, causing diseases such as *Fusarium* head blight in wheat and *Gibberella* ear rot in maize. The same species also produce other mycotoxins, such as deoxynivalenol, and infected cereals may be contaminated with more than one *Fusarium* toxin.

Zearalenone is produced in the crop prior to harvest, and can continue to be produced during storage in moist grain. Important factors influencing the degree of mould growth and toxin production in crops in the field include high rainfall and high humidity, but toxin production appears to be particularly favoured by wet, cool weather.

Stability in Foods

Zearalenone is heat stable and is not destroyed by temperatures of 120 °C. It therefore survives most cooking processes and significant quantities (60–80%) are reported to remain even in baked bread and biscuits.

Moderate amounts of zearalenone also appear to survive fermentation processes, such as brewing.

Control Options

Since zearalenone production occurs both in the field and during storage, controls should be applied pre-harvest and post-harvest.

Pre-harvest

Good agricultural practice (GAP) measures designed to reduce *Fusarium* infection in cereal crops are also effective in limiting the formation of zearalenone. Control measures include the following.

- land preparation, crop rotation and crop waste removal to reduce the inoculum of *Fusarium* in the field;
- use of fungus-resistant crop varieties;
- control of infection by appropriately timed application of effective fungicides;
- harvesting at the correct moisture level and stage of maturity.

Post-harvest Handling and Storage

Further production of zearalenone after harvest can be prevented by rapid drying to a water activity value of 0.8 immediately after harvest, and by implementing good storage practice.

Decontamination

Physical decontamination methods, including gravity separation, can be effective means of reducing zearalenone levels in contaminated grain. The milling process has also been shown to reduce zearalenone concentrations in corn flour and grits by around 80–90% by removing the more heavily contaminated bran.

Heat treatments are not usually effective.

Testing

In some countries cereals are monitored for zearalenone contamination by sampling and testing using various analytical methods, such as high-performance liquid chromatography (HPLC) with UV detection. ELISA methods have also been developed for screening purposes but are less sensitive. As with other mycotoxins, the distribution of zearalenone in bulk may be highly heterogeneous and it is essential to ensure that an adequate representative sampling plan is used.

Legislation

Few countries outside Europe have yet introduced mandatory or guideline levels for zearalenone in foods.

European Union

The EU sets a maximum level for zearalenone of 100 µg/kg in most unprocessed cereals, but the permitted level in unprocessed maize is 350 µg/kg. Maize intended for direct human consumption and maize-based snacks and cereals are permitted to contain a maximum of 100 µg/kg, and the limit for other cereals, flour and bran for direct human consumption is 75 µg/kg. The limit for bread, cereal snacks, biscuits, pastries and breakfast cereals (excluding maize-based products) is 50 µg/kg. A limit of 20 µg/kg has been set for foods intended for babies and young children.

Others

Chile has set a limit for zearalenone of 200 µg/kg for all foods, while Indonesia requires the toxin to be "not detectable" in maize and Iran has a limit of 200 µg/kg for most cereals.

The Canadian authorities have introduced a limit of 3000 µg/kg for zearalenone in pig feed.

More information can be found at the FAO web link below.

Sources of Further Information

Published

Opinion of the Scientific Panel on Contaminants in the Food Chain on a request from the Commission related to zearalenone as an undesirable substance in animal feed. *The EFSA Journal*, 2004, 89, 1–35. http://www.efsa.europa.eu/EFSA/efsa_locale-1178620753812_1178620763118.htm

The mycotoxin factbook: Food & feed topics, ed. Barug, D. *et al.* Wageningen, Wageningen Academic Publishers, 2006.

Mycotoxins. Bennett, J.W. and Klich, M., *Clinical Microbiology Reviews*, 2003, 16(3), 497–516.

On the Web

WHO Food Additives Series 44 – JECFA monograph on zearalenone (2000). http://www.inchem.org/documents/jecfa/jecmono/v44jec14.htm

Food-Info.net: Overview of foodborne toxins – mycotoxins (zearalenone). http://www.food-info.net/uk/tox/zear.htm

European Mycotoxin Awareness Network (EMAN). http://www.mycotoxins.org/

FAO Food and Nutrition Paper 81 – Worldwide regulations for mycotoxins in food and feed 2003. http://www.fao.org/docrep/007/y5499e/y5499e00.htm

2.1.1.13 Other Mycotoxins

Many toxic fungal metabolites (mycotoxins) have been identified and characterised, but relatively few of these are currently thought to be important from a food safety perspective. The preceding sections have dealt with the most significant foodborne mycotoxins, but there are a number of others that may be relevant to food safety. Some are very uncommon, or usually co-occur with other mycotoxins, and others have been very little studied, so that their public-health significance is uncertain.

Brief details are given below of some mycotoxins that may have food safety significance. Most of these are thought to cause toxic effects in animals and may occur naturally in certain food commodities. They therefore have the potential to affect human health.

Aflatrem

Aflatrem is one of a group of related mycotoxins known as tremorgens. These compounds can cause a range of neurological symptoms in animals, including tremors, seizures and even death. Their presence in mould-contaminated feed has been implicated in a disease of cattle known as "staggers syndrome".

Chemically, aflatrem is an indole-diterpene with a molecular weight of 502. It is a potent tremorgen, and is of importance in food safety because it is produced by *Aspergillus flavus*, which also produces aflatoxins. It may therefore co-occur with aflatoxins in a wide range of food commodities. Aflatrem probably contributes to the overall toxicity of aflatoxins, but its precise significance to human health is uncertain. Control measures designed to prevent aflatoxin formation are also likely to be effective against aflatrem.

Alternaria *Toxins*

Mould species belonging to the genus *Alternaria*, notably *Alternaria alternata*, are able to attack a range of fruit and vegetable crops at the pre- and post-harvest stages. They also produce a number of toxic metabolites under certain conditions, but most do not seem to occur naturally in foods. Those that do include alternariol, alternariol monomethyl ether, altenuene, altertoxin I, and tenuazonic acid, of which tenuazonic acid is probably the most important and the most toxic. A few rare isolates also produce Alternaria alternata toxin (AAT), a highly toxic compound related to the fumonisins.

Alternaria toxins exhibit a range of acute and chronic toxic effects in animals, especially poultry and rabbits, and have also been implicated in human illness. Tenuazonic acid inhibits protein synthesis and most alternaria toxins are cytotoxic. The altertoxins are also mutagenic.

A. alternata and its toxins have been isolated from cereals, sunflower seeds, olives and a number of fruits and vegetables. It is an important pathogen of tomatoes and also attacks peppers and apples. Alternaria toxins are normally only found in visibly mouldy food commodities and rarely occur naturally in human food. Therefore, the potential for human exposure is thought to be very limited.

Aspergillus Clavatus *Toxins*

Aspergillus clavatus is a mould species normally found in soil. It is capable of producing a number of toxins in culture, including agroclavine (an ergot alkaloid), cytochalasin E and K and several tremorgens. *A. clavatus* grows well in malting barley and is the cause of condition known as "malt workers lung", but it does not seem to produce significant quantities of mycotoxins naturally in barley. Nevertheless, it has been implicated in the intoxication of cattle consuming mouldy grain.

Citreoviridin

Citreoviridin consists of a lactone ring conjugated to a furan ring and has a molecular weight of 402. It is produced by some species of *Penicillium*, notably *P. citreognigrum* and *P. ochrosalmoneum.* It is a neurotoxin and causes a number of severe symptoms in mice and other animals, including vomiting, convulsions, paralysis and respiratory arrest. Historically, citreoviridin was recognised as the cause of a condition known as "acute cardiac beriberi" in Japan, which was linked to the consumption of mouldy "yellow rice". The banning of this food in 1910 has eradicated the disease from Japan.

P. citreonigrum is not common, but is widespread, especially in the temperate rice growing regions. It grows in rice after harvest, but only dominates within a narrow moisture range around 15%. *P. ochrosalmoneum* is also rare, but has been isolated from unharvested maize in the USA and may produce citreoviridin naturally in maize under certain conditions.

Other Fusarium Toxins

In addition to the important mycotoxins described elsewhere, species of the genus *Fusarium* produce a number of other less well known and less studied toxic metabolites. Some of these have the potential to affect human health.

Beauvericin is cyclic hexadepsipeptide produced by *F. subglutinans, F. proliferatum* and several other species and has been shown to be toxic to human cells in culture. It has been detected in wheat infected with Fusarium head blight and also in maize, but the extent of human exposure is not known.

Enniatin is also a cyclic hexadepsipeptide and is produced by *F. avenaceum.* It too has been found in wheat infected with Fusarium head blight, but its toxicity and the potential for human exposure are uncertain.

Fusaproliferin is a sesterterpene produced by *F. subglutinans* and *F. proliferatum.* It has been shown to be cytotoxic to some human and animal cell lines and may occur in infected maize.

Gliotoxin

Gliotoxin is a potent immunosuppressive agent produced by the pathogenic mould species *Aspergillus fumigatus* and some other *Aspergillus* and *Penicillium* species. It may have a role in the development of human aspergillosis infections, but there is limited evidence that it is occasionally produced in mould-infected cereals.

Mycophenolic Acid

Mycophenolic acid is another immunosuppressant produced by some species of *Penicillium*, including *P. roqueforti*. It has been detected in mould-ripened cheese. It has been demonstrated to be toxic at quite high concentrations in rodents and primates and may also be mutagenic.

b-Nitropropionic Acid

This toxin is a toxic metabolite of *Aspergillus orysae* used in the production of soy sauce. *A. orysae* has been shown to produce b-nitropropionic acid in cooked potatoes and in ripe bananas. It is a neurotoxin and can cause toxic effects in livestock fed with contaminated feed. It has also been implicated in cases of human illness in China.

Penicillic Acid

Penicillic acid is a toxic metabolite of several species of *Penicillium* and of *Aspergillus* species, including *A. ochraceus*, which also produces ochratoxin A. It can cause liver cancers in some animal species and has been isolated from maize, dried beans and tobacco. It has also been reported to have been detected in fermented sausage.

Phomopsins

Phomopsins are produced by the fungus *Phomopsis leptostromiphoris*, which is an important pathogen of lupins. These toxins may be present at significant levels in lupin seeds used to produce animal feed, but also now increasingly used as an ingredient in human foods. The phomopsins are potent liver toxins and carcinogens in rats and other animals. Their significance in human health is not known, but their presence in foods is considered undesirable and of concern, especially as they are stable compounds likely to survive cooking processes. Australian legislation sets a maximum level of 5 μg/kg for phomopsins in lupin seeds and lupin-seed products.

PR-Toxin

PR-toxin is a toxic metabolite of *Penicillium roqueforti*. It is lethal to rats, mice and cats and is reported to cause toxic effects in the lungs, brain, liver and kidney. It has been detected at low levels in blue cheeses and mouldy cereal grains. It is not particularly stable in cheese and degrades to other less toxic compounds quite rapidly. Adverse health effects associated with consumption of blue cheese containing PR-toxin have not been reported.

Penitrem A

Penitrem A is a potent neurotoxin produced primarily by *Penicillium crustosum*, which is a common and widespread food and feed spoilage mould. It is a tremorgen and has been associated with outbreaks of tremorgenic disease in cattle, sheep and horses. Its significance for human health is so far uncertain. *P. crustosum* can cause spoilage in a variety of foods, including maize, nuts, cheese, cured and processed meat products, cakes and biscuits, and fruit. Most strains can potentially produce penitrem A at high levels, but only at high moisture levels. This may explain the comparatively few reports of animal and human poisoning caused by this toxin.

Roquefortines

Roquefortines A, B and C are reported to be produced by several *Penicillium* species, including *Penicillium roqueforti*, used in the production of some blue cheeses. They are indole compounds and have been reported to be toxic to rats, mice and poultry at relatively high levels. Their significance for human health is so far uncertain. Roquefortines have been detected in blue cheese, but only at low levels, and adverse health effects associated with consumption of blue cheese containing these compounds have not been reported.

Satratoxins

The satratoxins are trichothecene mycotoxins produced by fungi of the genus *Stachybotrys*, notably *Stachybotrys chartarum*. These fungi are widespread and have been isolated mainly from environmental samples, especially from water-damaged buildings, but also from mouldy cereals. The satratoxins are potent toxins that inhibit protein synthesis in mammalian cells. They have been linked with a disease of horses associated with consumption of mouldy hay and straw and also with illness in other animals. Their food safety significance is uncertain.

Viomellein, Vioxanthin and Xanthomegnin

These toxins are produced by some *Penicillium* species, such as *P. cyclopium* and *P. viridicatum* and also by *Aspergillus* species, including *A. ochraceus*. They are known to co-occur with other mycotoxins, especially ochratoxin A, and are nephrotoxic. They may be involved in kidney disease of animals, such as pigs, caused by ochratoxin A, but their food safety significance is not known.

Walleminol A

Walleminol A is a toxic metabolite of the xerophilic mould species *Wallemia sebi*, which is known to grow on a wide range of foods, including cereals, pulses, dried fruits, cakes, confectionary and conserves. Walleminol A has been shown to be toxic to animal cells, but its significance for human health and its occurrence in foods have not yet been investigated.

Sources of Further Information

Published

Valdes, J.J., Cameron, J.E. and Cole, R.J. Aflatrem: A tremorgenic mycotoxin with acute neurotoxic effects, *Environmental Health Perspectives*, 1985, 62, 459–63.

The mycotoxin factbook: Food & feed topics. ed. Barug, D. *et al.* Wageningen, Wageningen Academic Publishers, 2006.

Mycotoxins. Bennett, J.W. and Klich, M. *Clinical Microbiology Reviews*, 2003, 16(3), 497–516.

On the Web

Australia New Zealand Food Authority – Report on phomopsins in Food (2001). http://www.foodstandards.gov.au/_srcfiles/TR1.pdf

European Mycotoxin Awareness Network (EMAN). http://www.mycotoxins.org/

2.1.2 PLANT TOXINS

2.1.2.1 Cucurbitacins

Hazard Identification

What are Cucurbitacins?

Courgettes (zucchini), together with many closely related species of the *Cucurbitacea* family, including cucumber and squash, produce an intensely bitter group of compounds known as cucurbitacins. Some wild-type squashes are so bitter that they become almost inedible to humans and most animals. Some can even kill small animals.

The cucurbitacins are highly oxygenated triterpenoid compounds and are divided into twelve different categories according to their structure. They are potent toxins with natural insecticidal and/or fungicidal properties.

Occurrence in Foods

Natural production of cucurbitacins occurs in members of the cucumber family. As well as cucumbers, these include courgettes, marrows, melons and squashes. The compounds are responsible for the bitter taste that is sometimes evident in cucumbers and courgettes.

The varieties of courgette and squash that are grown commercially and domestically in the garden have been selected for low levels of these bitter compounds, although one notable exception to this is bitter melon, which is used in Asian cuisine, where the bitterness is a prized part of the flavour. Larger courgettes and marrows will have higher levels of cucurbitacins than smaller fruit. Natural cross-pollination with wild varieties may also increase the bitterness of cultivated varieties.

Hazard Characterisation

Effects on Health

Cucurbitacins are toxic at high levels, but they are so bitter that it is almost impossible for anyone to eat sufficient quantity of the toxins to cause significant harm. Cucurbitacin-B, for example, has an oral LD_{50} in the mouse of 5 mg/kg bodyweight. Theoretically, this means that a dose of 300 mg could be sufficient to kill a human.

In New Zealand, in the early summer of 2001, there was a series of cases of severe stomach cramps associated with eating courgettes. So many cases were reported that the health authorities instigated an official investigation. Many of those who became ill reported eating bitter-tasting courgettes. The summer had been unusually wet, which favoured fungal infection, and it is likely that increased fungal infection led to up-regulation of the genes involved in cucurbitacin production, thus increasing the toxin levels in the courgettes.

Because of their extreme bitter taste, ingestion of cucurbitacins is usually limited and symptoms of intoxication are generally mild. Stomach cramps, nausea, vomiting and diarrhoea have all been reported.

Sources

The natural production of cucurbitacins, which occurs in members of the *Cucurbitacea* family, is controlled by the plants so that they are produced only when they are needed. The gene that codes for curcubitacin production is switched on only when climatic conditions are favourable for insect infestation or fungal infection. Their concentration therefore varies according to weather and the potential for fungal infestation or insect attack.

Commercially grown cucumbers, courgettes and related vegetables have been selected for low levels of the bitter cucurbitacins. However, even carefully selected varieties will produce high levels of the toxins when environmentally stressed, or when conditions are ripe for fungal infection or insect infestation.

Stability in Foods

Cucurbitacins are heat resistant and only slightly soluble. They are therefore neither destroyed nor removed by cooking of courgettes and food plants.

Control Options

There is little that can be done to reduce the level of cucurbitacins once the plant has started to produce them. Their heat stability and poor solubility mean that cooking the vegetables in water has little effect. It is thought that cutting off the end of the courgette, nearest to the blossom, can reduce some of the bitterness. The preferred control options are to ensure that the plants are watered carefully during growth, and to harvest the crop as early as possible.

Legislation

There is no specific legislation governing cucurbitacins levels in foods.

Sources of Further Information

Fenwick, G.R., Curl, C.L., Griffiths, N.M., Heaney, R.K. and Price, K.R. Bitter principles in food plants. In: Rouseff, RL., ed. Bitterness in Foods and Beverages; Developments in Food Science 25. Amsterdam: Elsevier, 1990:205–50.

2.1.2.2 Cyanogenic Glycosides

Hazard Identification

What are Cyanogenic Glycosides?

Cyanogenic glycosides are chemical compounds that occur naturally in many plants, including species of *Prunus* (wild cherry), *Sambucus* (elderberry), *Manihot* (cassava), *Linum* (flax), *Bambusa* (bamboo) and *Sorghum* (sorghum). Chemically, they are defined as glycosides of the α-hydroxynitriles. These compounds are potentially toxic as they are readily broken down by enzymic hydrolysis to liberate hydrogen cyanide when the plant suffers physical damage.

Occurrence in Foods

There are approximately 25 known cyanogenic glycosides, and a number of these can be found in the edible parts of some important food plants. These include amygdalin (almonds), dhurrin (sorghum), lotaustralin (cassava), linamarin (cassava, lima beans), prunasin (stone fruit) and taxiphyllin (bamboo shoots). Table 2.1.1 summarises some of the main food sources of cyanogenic glycosides and their estimated potential yield of hydrogen cyanide released on hydrolysis.

Bitter apricot kernels have been marketed as a health food in the UK and elsewhere. They can contain high levels of the cyanogenic glycoside amygdalin.

Table 2.1.1 Cyanogenic food sources and their approximate hydrogen cyanide yield.

Food source	*Cyanogenic glycoside*	*Hydrogen cyanide yield (mg/100 g fresh weight)*
Almond bitter seed	Amygdalin	290
Apricot kernel	Amygdalin	60
Bamboo stem (unripe)	Taxiphyllin	300
Bamboo sprout tops (unripe)	Taxiphyllin	800
Cassava tuber bark (less toxic clones)	Linamarin and Lotaustralin	69
Cassava inner tuber (less toxic clones)	Linamarin and Lotaustralin	7
Cassava tuber bark (very toxic clones)	Linamarin and Lotaustralin	84
Cassava inner tuber (very toxic clones)	Linamarin and Lotaustralin	33
Flax seedling tops	Linamarin, Linustatin and Neolinustatin	91
Black Lima bean, Puerto Rico (mature seed)	Linamarin	400
Peach kernel	Prunasin	160
Sorghum shoot tips	Dhurrin	240
Wild cherry leaves	Amygdalin	90–360

(Adapted from: Frehner *et al.*, 1990, *Plant Physiol.* 94, 28–34).

Analytical data indicates that the bitter apricot kernels currently on sale have a cyanide content of 1450 mg/kg (approximately 0.5 mg/kernel). While swallowing of apricot kernels whole may not release much cyanide, grinding or chewing them significantly increases its release.

Hazard Characterisation

Effects on Health

The toxicity of a cyanogenic plant depends largely on the amount of hydrogen cyanide that could be released on consumption of the plant. Adequate processing or preparation is required to ensure that detoxification of the food is complete before consumption. However, if the processing or preparation is insufficient to ensure detoxification, the potential hydrogen cyanide concentration released during consumption can be high. Upon consumption of the food, the enzyme β-glycosidase will be released and hydrolysis of the cyanogenic glycoside will commence, resulting in hydrogen cyanide formation. Certain gut microflora also produce β-glycosidases, which can contribute to the breakdown of cyanogenic glycosides to hydrogen cyanide.

Hydrogen cyanide is cytotoxic and blocks the activity of cytochrome oxidase – an enzyme critical for cellular respiration. When cytochrome oxidase is blocked, ATP production stops and cellular organelles cease to function. However, cyanide is readily detoxified in animals as all animal tissues contain the enzyme rhodanese – a thiosulfate sulfurtransferase enzyme that converts cyanide to thiocyanate, which is then excreted in urine. Acute poisoning only occurs when this detoxification mechanism is overwhelmed.

The symptoms of acute cyanide poisoning include rapid breathing, drop in blood pressure, raised pulse rate, dizziness, headache, stomach pains, vomiting, diarrhoea, confusion, twitching and convulsions. In extreme cases, death may occur. The minimum lethal dose of hydrogen cyanide taken orally is approximately 0.5–3.5 mg/kg bodyweight, or 35–245 mg for a person weighing 75 kg.

The chronic effects of cyanide consumption are associated with regular long-term consumption of foods containing cyanogenic glycosides in individuals with poor nutrition. These effects are most notable in the tropics, where cassava, and to a lesser extent, sorghum, bamboo shoots and lima beans are staple components of human diets. Malnutrition, growth retardation, diabetes, congenital malformations, neurological disorders and myelopathy are all associated with cassava-eating populations subject to chronic cyanide intake.

There are a number of documented cases of poisoning caused by consumption of apricot kernels. One report concerned a 41-year-old female found comatose after eating approximately 30 bitter apricot kernels, who eventually recovered after treatment. There are also case reports of children being poisoned after consumption of wild apricot kernels and where the kernels were made into sweets without proper processing. The UK Committee on Toxicity recommended in March 2006 that a tolerable daily intake (TDI) of 20 μg cyanide/kg bw/day be applied, which is the equivalent of 1–2 bitter apricot kernels per day.

Sources

There are over 2500 known species of plants that produce cyanogenic glycosides, usually in combination with a corresponding hydrolytic enzyme – a beta-glycosidase. When the cell structure of the plant is disrupted in some way, for example by predation, the beta-glycosidase is brought into contact with its substrate – the cyanogenic glycoside. This leads to the breakdown of the glycoside to sugar and a cyanohydrin, which rapidly decomposes to release hydrogen cyanide. The purpose of the reaction is to protect the plant from predation.

Stability in Foods

Cyanogenic glycosides break down when the cells of the plant are damaged, for example during preparation and processing, and release hydrogen cyanide. Hydrogen cyanide itself is not heat stable and does not survive boiling and cooking processes. It can also be eliminated by fermentation.

Control Options

Processing

Adequate processing of cyanogenic glycoside-containing plants should be sufficient to significantly reduce or remove the toxic agents prior to consumption. Processing procedures, such as peeling and slicing disrupt the cell structure of the plant so that β-glycosidases are released and the cyanogenic glycosides are hydrolysed. Hydrogen cyanide is thus released and can be removed by cooking processes such as baking, boiling or roasting. Fermentation is also used to remove hydrogen cyanide. These methods are particularly suitable for products such as cassava and bamboo shoots. There are two main types of cassava – bitter cassava and sweet cassava. The sweet variety contains a significantly lower concentration of cyanogenic glycosides than the bitter variety, and it is the sweet variety that is used commercially. Cassava is consumed largely as cassava flour, cassava chips and tapioca pearls, all of which are processed products with a long history of safe consumption.

Treatments for removing cyanogenic compounds from flaxseed include boiling in water, dry and wet autoclaving and acid treatment followed by autoclaving. Solvent extraction has also been used to remove cyanogenic glycosides from flaxseed and oil.

Legislation

A safe level of cyanide in cassava flour for human consumption has been set by the WHO at 10 ppm.

Low levels of cyanide are also present in almonds, sweet apricot kernels and in the stones of other fruit such as cherries, as well as in bitter apricot kernels.

In the UK, the maximum level of cyanide that can be present as a result of using such materials as flavourings is regulated under the terms of the Flavourings in Food Regulations 1992 (as amended).

Sources of Further Information

Published

Oke, O.L., Some aspects of the role of cyanogenic glycosides in nutrition. *World Review of Nutrition and Diet*, 1979, 33, 70–103.

Vetter, J., Plant cyanogenic glycosides. *Toxicon*, 2000, 38, 11–36.

World Health Organization (WHO). Toxicological evaluation of certain food additives and naturally occurring toxicants. *WHO Food Additive Series: 30*. 1993, World Health Organization, Geneva.

On the Web

UK Committee on Toxicity Background Paper. www.food.gov.uk/multimedia/pdfs/TOX-2006-13.pdf

Food Standards Australia New Zealand Technical Report. www.foodstandards.gov.au/_srcfiles/28_Cyanogenic_glycosides.pdf

JECFA Monograph – Cyanogenic Glycosides. http://www.inchem.org/documents/jecfa/jecmono/v30je18.htm

2.1.2.3 Furocoumarins

Hazard Identification

What are Furocoumarins?

The furocoumarins are a group of naturally occurring chemicals that are found in a wide variety of plants, but which are present at their highest concentrations in members of the *Umbelliferae* family, particularly parsnips, celery and parsley. They are also present in lower concentrations in other foods such as citrus fruit, celeriac and figs. There are many different furocoumarins, but they all have similar molecular structures. Examples include psoralen, bergapten, xanthotoxin and isoimperatorin. The furocoumarins all have insecticidal and/or fungicidal activity, but they are also photoactivated carcinogens and are therefore significant from a food safety point of view.

Occurrence in Foods

The highest concentrations of furocoumarins are found in parsnips, celery and parsley (see Table 2.1.2).

Organically grown vegetables often have higher levels of furocoumarins. This may be because conventional cultivation involves the use of pesticides, and conventionally grown plants have less need to produce natural chemical defences in response to the threat of predation by insects. Damaged vegetables also contain significantly higher levels of furocoumarins than intact produce.

Furocoumarins have also been detected in some processed foods, particularly purees and soups, with the highest levels being found in soups containing celery. Other sources include citrus fruits, marmalade and sweet fennel.

Hazard Characterisation

Effects on Health

Furocoumarins are photoactivated carcinogens. This means that they absorb long-wave ultraviolet radiation upon exposure of the skin to sunlight and are activated by the light to form carcinogens. Prolonged exposure can result in cell damage, by binding pyrimidine bases and nucleic acids and thus inhibiting DNA synthesis. The oral LD_{50} for psoralen in rats has been reported to be 791 mg/kg bodyweight.

Table 2.1.2 Furocoumarins in commonly eaten foods (MAFF survey 1996).

Plant	*Main furocoumarin*	*Concentration (mg/kg)*
Celery	Bergapten	1.3–47
Parsnip	Bergapten	40–1740
Parsley	Isoimperatorin	11–112

They can also cause skin sensitisation to UV light, resulting in skin rashes after prolonged skin exposure to the sun. A fairly high intake is required to cause photosensitisation. The main symptom is peeling and blistering of the light-exposed parts of the skin of someone who has consumed a fairly large quantity of parsnips or celery, particularly damaged produce that has been organically produced. There have been two reported cases of phototoxic reactions after consumption of celery. Both involved extreme intakes of celery and strong UVA exposure.

A condition known as "celery dermatitis" has also been noted. The symptoms include blistering of the arms of farm workers handling celery when the celery is diseased with pink rot (*Sclerotinia sclerotiorum*) and produces xanthotoxin and trisoralen.

In 1996, the UK Committee on Toxicity carried out a risk assessment on toxic compounds in plants, including furocoumarins. Overall, the Committee concluded that the likelihood of any risk to health from dietary intakes of furocoumarins was very small.

Sources

Furocoumarins are produced by many plants in response to stresses such as bruising or injury caused by predation. The plants respond to damage by up-regulating natural pesticide production to prevent insect attack or fungal infection.

Stability in Foods

Furocoumarins are quite heat stable and cooking does not reduce their concentration significantly.

Control Options

There are few effective controls for furocoumarins in celery and parsnips, although it has been recommended that new cultivated varieties be monitored for furocoumarin content before widespread planting. Avoiding damage to crops in the field and during harvesting may help to reduce furocoumarin levels.

Processing

Although furocoumarins are not inactivated by heating, they are water soluble. Therefore, if furocoumarin-containing vegetables are cooked in water, the levels in the vegetable can be appreciably reduced.

Product Use

Consumers with high dietary exposure to vegetables with potentially high furocoumarin levels may benefit by avoiding produce showing evidence of physical damage.

Legislation

The content of furocoumarins in vegetables is not generally regulated by legislation.

Sources of Further Information

Published

MAFF (1996). Inherent Natural Toxicants in Food – the 51st Report of the Steering Group on Chemical Aspects of Food Surveillance. The Stationery Office, London.

On the Web

Committee on Toxicity Report. http://www.archive.official-documents.co.uk/document/doh/toxicity/chap-1c.htm

MAFF furocoumarin surveillance sheet. http://archive.food.gov.uk/maff/archive/food/infsheet/1993/no09/09furo.htm

Cornell University fact sheet. http://www.ansci.cornell.edu/plants/toxicagents/coumarin.html#furo

2.1.2.4 Glycoalkaloids

Hazard Identification

What are Glycoalkaloids?

Many plants in the *Solanaceae* family contain glycoalkaloids, and they are considered to be natural toxins. They are active as pesticides and fungicides and are produced by the plants as a natural defence against animals, insects and fungi that might attack them.

The plant glycoalkaloids are toxic steroidal glycosides and the commonest types found in food plants are α-solanine and α-chaconine, with α-solanine ($C_{45}H_{73}NO_{15}$) being the more toxic of the two.

Occurrence in Foods

Amongst the most widely cultivated food crops, aubergines, tomatoes and potatoes are in the *Solanaceae* family; however, the levels of glycoalkaloids in tomatoes and aubergines are generally quite low and are therefore not a concern. The glycoalkaloids of most relevance to food safety are those occurring in the potato, since even in commercially available tubers destined for human consumption a residual level of these compounds is always present.

The predominant toxic steroidal glycosides in potato are α-solanine and α-chaconine. They occur in potato tubers, peel, sprouts and blossoms and their concentration in tubers depends on a number of factors, such as cultivar, maturity, environmental factors and stress conditions.

In the UK, the total glycoalkaloid level in tubers destined for human consumption is generally in the range 25–150 mg/kg fresh weight, but considerably higher levels have been recorded for certain commercial varieties. As an example, the *Lenape* potato variety was withdrawn from commercial growing in Canada and the USA as it contained unacceptably high levels of glycoalkaloids. In Sweden, a conditional sales ban had to be imposed on potato tubers of the commercially established variety *Magnum Bonum* harvested in 1986, as they contained potentially toxic levels of glycoalkaloids.

Hazard Characterisation

Effects on Health

Most cases of suspected potato poisoning involve only mild gastrointestinal effects, which generally begin within 8–12 h after ingestion and resolve within one or two days. However, reported symptoms have included nausea and vomiting, diarrhoea, stomach cramps and headache. More serious cases have experienced neurological problems, including hallucinations and paralysis, and fatalities have also been recorded.

Although suspected potato poisoning is rare, a number of incidents have been documented, and a few of the more recent ones are tabulated below (taken partly from a review by McMillan and Thompson, 1979; *Quart. J. Med.*, **48**, 227–243):

Year	*Details*	*Effects*
1925	7 family members ate greened potatoes.	Extreme exhaustion, restlessness, rapid breathing, loss of consciousness. Death of 2 family members.
1933	In Cyprus, 60 people consumed young potato shoots and leaves as a vegetable.	Headache, nausea, vomiting, diarrhoea, fever, throat irritations. One death.
1952–53	382 North Koreans affected following consumption of rotten potatoes.	Pain, nausea, vomiting, facial oedema, respiratory failure, cardiac arrest. 52 hospitalised and 22 deaths.
1979	78 London schoolboys consumed potatoes left over from a previous term.	Diarrhoea, vomiting, circulatory, neurological, dermatological problems. 17 hospitalised.
1986	11 people in Sweden consuming Magnum Bonum variety potatoes.	Nausea, vomiting, pain, headache.

Although glycoalkaloids are suspected to be the cause of these symptoms, there is little data to confirm this. One study examined case reports of poisoning incidents and estimated that glycoalkaloid doses of 2–5 mg/kg bodyweight would be enough to cause symptoms in humans and that 3–6 mg/kg bodyweight could be fatal. However, a toxicological monograph produced by the Joint FAO/WHO Expert Committee on Food Additives (JECFA) in 1992 states that "Glycoalkaloids are not acutely toxic by the oral route in laboratory animals even at very high doses (up to 1 g/kg bodyweight) in some species." The Committee considered that the evidence implicating glycoalkaloids in potato poisoning cases was not convincing. JECFA concluded that levels of α-solanine and α-chaconine normally found in potatoes (20–100 mg/kg) were not of toxicological concern.

Nevertheless, JECFA and others have expressed concern about glycoalkaloids in skin-on potato products, such as crisps, that became widely available in the mid-1990s. Glycoalkaloid concentrations of up to 720 mg/kg were found in "green-skinned" crisps, compared with a maximum of 150 mg/kg in normal crisps.

Apart from their toxicity, glycoalkaloids are also associated with a bitter taste and burning sensation in the throat.

Sources

Although glycoalkaloids in potatoes are produced naturally by the plant, certain factors can have a significant effect on the levels present:

Maturity

The highest concentrations of glycoalkaloids are usually associated with areas that are undergoing high metabolic activity, such as potato flowers, young leaves, sprouts, peels and the area around the potato "eyes". Small immature tubers are normally high in glycoalkaloids since they are still metabolically active.

Exposure to Light

Exposure to light has a significant effect on the concentration of both total and individual glycoalkaloids. Potatoes that become sunburned during growth and start to "green", owing to lack of soil cover, tend to taste very bitter as a result of their high glycoalkaloid content.

In retail outlets, tubers may be displayed under fluorescent lighting and this can increase glycoalkaloid concentration. Studies have indicated that replacing fluorescent lights with mercury lighting for potatoes on display would significantly reduce glycoalkaloid content and improve food safety.

Storage Temperature

Storage at very low temperatures (0–5 °C) results in more bitter-tasting potatoes and thus more glycoalkaloids than storage at higher temperatures (up to 20 °C). On the whole, storage at lower temperatures will prolong potato quality, but at very low temperatures (0–5 °C), stress becomes a factor and glycoalkaloid accumulation starts to occur.

Injury/Damage

Any type of injury or damage to the tuber will result in the accumulation of glycoalkaloids. Disease, insect attack or rough handling, during or after harvest, will all initiate glycoalkaloid synthesis (as it is a defence response). Damaged potatoes from retail generally contain elevated levels of glycoalkaloids.

Stability in Foods

Glycoalkaloids are relatively stable in potatoes and levels are not affected by boiling, freeze-drying, or dehydration. Microwave cooking has only a limited effect, but cooking at temperatures at or above 170 °C is more effective at lowering levels.

Control Options

Cultivar Selection

The amounts of total and individual potato glycoalkaloids are genetically controlled. The most effective way of obtaining low levels is to select breed varieties that are initially very low in glycoalkaloids.

Processing

Peeling

In normal tubers, potato glycoalkaloids appear to be concentrated in a small 1.5-mm layer immediately under the skin, therefore, with normal tubers, peeling will remove between 60–95% of the glycoalkaloids present. However, if the tubers are very high in glycoalkaloids, peeling will remove only up to 35%, as, in potatoes with a high level, diffusion into the deeper tissues occurs. Unfortunately, peeling or slicing also elicits a stress response in the tubers and causes a slow rise in glycoalkaloid levels. If long delays occur before subsequent processing, glycoalkaloids can accumulate.

Cooking

The heat stability of glycoalkaloids means that only high-temperature processing, such as deep frying has any significant effect on levels in potatoes. Other processes give little or no reduction in the concentration of these compounds.

Physical/Chemical Treatments

Gamma irradiation has been shown to control glycoalkaloid levels, particularly in damaged tubers. Treatment with certain chemicals, most of which function as sprout inhibitors, has also been shown to control glycoalkaloid accumulation.

Legislation

Although there is no specific legislation governing glycoalkaloid levels in potatoes, the generally accepted safe upper limit is considered to be 200 mg glycoalkaloids per kg of fresh potato.

Sources of Further Information

Published

Friedman, M. and McDonald, G.M., Potato glycoalkaloids: chemistry, analysis, safety and plant physiology. *Critical Reviews in Plant Sciences*, 1997, 16 (1), 55–132.

Cantwell, M., Glycoalkaloids in Solanaceae. *Food Reviews International*, 1994, 10 (4), 385–418.

On the Web

JECFA review 1992. http://www.inchem.org/documents/jecfa/jecmono/v30je19.htm

Potato glycoalkaloid toxicity article (Cornell University). http://www.ansci.cornell.edu/courses/as625/1999term/andrew/index.htm

US National Toxicology Program literature review (α-solanine and α-chaconine) 1998. http://ntp.niehs.nih.gov/ntpweb/index.cfm?objectid=6F5E933B-F1F6-975E-7B1D19DE73F21505

2.1.2.5 Grayanotoxin

Hazard Identification

What is Grayanotoxin?

Grayanotoxins are natural plant toxins found in rhododendrons and other plants of the family *Ericaceae*. Specific grayanotoxins vary according to the plant species in which they are found. They can be found in honey made from the nectar produced by the flowers of these plants, and can cause a very rare poisonous reaction.

Grayanotoxin compounds are diterpenes – polyhydroxylated cyclic hydrocarbons that do not contain nitrogen. Alternative names for grayanotoxin include andromedotoxin, acetylandromedol, and rhodotoxin.

Occurrence in Foods

Honeys originating from Japan, the United States, British Colombia, Brazil and Nepal are those most likely to be contaminated with grayanotoxin. Honey obtained locally from farmers who may have only a few hives is at increased risk, particularly in regions where plants of the *Ericaceae* family dominate the vegetation. The pooling of massive quantities of honey during commercial processing generally dilutes any toxic substances.

Hazard Characterisation

Effects on Health

Grayanotoxins elicit their effects by binding to sodium channels in cell membranes. All of the observed responses of skeletal and heart muscles, nerves, and the central nervous system are related to these membrane-binding effects.

Grayanotoxin intoxication is rarely fatal. Symptoms include dizziness, weakness, excessive perspiration, nausea, and vomiting shortly after the toxic honey is ingested. Other symptoms may include low blood pressure or shock, bradyarrhythmia (slowness of the heart beat associated with an irregularity in the heart rhythm) and other cardiac abnormalities. Despite the potential cardiac problems, the condition is rarely fatal and generally lasts less than a day.

Several cases of grayanotoxin poisoning have been documented, many associated with honey originating in Turkey. Between 1984 and 1986, 16 patients in Turkey had to be treated for honey intoxication. One case in Austria, which resulted in cardiac arrhythmia, was attributed to honey brought back from a holiday in Turkey. In this case, the patient needed a temporary cardiac pacemaker to deal with the decrease in heart rate. Grayanotoxin poisoning has also been reported in goats in the UK.

Sources

Rhododendrons are the main documented source of grayanotoxins, but not all rhododendrons produce them. *Rhododendron ponticum*, which grows extensively in the mountains of the eastern Black Sea area of Turkey has been associated with honey poisoning since 401 BC (according to the writings of Pliny the Elder). Other species known to produce the toxins grow over large areas of the USA. In the eastern part of the country, grayanotoxin-contaminated honey may be derived from other members of the family *Ericaceae*.

Control Options

Most honey contaminated with grayanotoxin originates in areas of the world where the vegetation is dominated by *Ericaceae*, particularly areas of Turkey, Japan, Brazil, the United States, Nepal, and British Columbia. Extra care should be taken with honeys originating from these parts.

Chemical Analysis

The grayanotoxins can be isolated from the suspect product by the typical extraction procedures used for naturally occurring terpenes, and the toxins can be identified by thin layer chromatography.

Sources of Further Information

Published

Gunduz, A., Turedi, S., Uzun, H. and Topbas, M. Mad Honey Poisoning. *Am. J. Emerg. Med.* 2006, Sep, 24 (5), 595–8.

Lampe, K.F., Rhododendrons, mountain laurel, and mad honey. *JAMA*. 1988, Apr 1, 259 (13), 2009.

Yavuz, H., Ozel, A., Akkus, I. and Erkul, I. Honey poisoning in Turkey. *Lancet*. 1991 Mar 30, 337 (8744), 789–90.

2.1.2.6 Lectins

Hazard Identification

What are Lectins?

Lectins are proteins that are widely distributed in nature and occur in many plants commonly consumed in the diets of humans and animals. They are toxic to humans and animals but their toxicity varies depending on their source. They were originally discovered in the 19th century, when it was found that the extreme toxicity of castor beans could be attributed to a protein fraction capable of agglutinating erythrocytes (red blood cells). This protein fraction was given the name *ricin*, as it was derived from *Ricinus communis* (the castor oil plant). Since then, many other lectins similar to ricin have been discovered. For example, lectins are found in common edible legumes such as kidney beans, soya beans, lentils, peas and peanuts. Lectins are also commonly known as phyto-haemagglutinins, owing to their ability to agglutinate red blood cells.

Lectins are characterised by their highly specific carbohydrate-binding activity, and it was this high degree of specificity that led Boyd and Shapleigh in the 1950s to coin the term "lectins" from the Latin word *legere*, meaning to choose.

Most lectins are actually glycoproteins containing 2 or 4 subunits, each of which has a sugar-binding site. Lectins are generally identified by the plant species that they are derived from.

Occurrence in Foods

As can be seen from Table 2.1.3, leguminous vegetables are the most frequently encountered food sources of lectins, although other sources have been reported, such as dry cereals and wheat germ. The amounts and specificity of the lectins obtained from different sources vary widely, but the highest concentration is found in red kidney beans (*Phaseolus vulgaris*). The unit of toxin measure is the hemagglutinating unit (hau). Raw kidney beans contain from 20 000 to 70 000 hau, while fully cooked beans contain from 200 to 400 hau. White kidney beans,

Table 2.1.3 Properties of some common lectins.

Common name	*Botanical name*	*Molecular weight*	*Number of subunits*
Peanut	*Arachis hypogeae*	110 000	4
Kidney bean	*Phaseolus vulgaris*	126 000	4
Fava bean	*Vicia faba*	52 500	4
Soya bean	*Glycine max*	120 000	4
Lentil	*Lens esculenta*	46 000	4
Winged bean	*Psophocarpus tetragonolobus*	58 000	2
Garden pea	*Pisium sativum*	49 000	4
Horse gram	*Dolichos biflorus*	110 000	4
Lima bean	*Phaseolus lunatus*	60 000	2
Navy bean	*Phaseolus vulgaris*	128 000	4
Jack bean	*Canavalia ensiformis*	110 000	4

another variety of *Phaseolus vulgaris*, contain about one-third the amount of toxin as the red variety; broad beans (*Vicia faba*) contain 5 to 10% the amount that red kidney beans contain.

Despite the fact that most food-derived lectins are inactivated by heat processing, lectin activity has been detected in processed food items such as dry cereals and peanuts, dry-roasted beans and processed wheat germ.

Hazard Characterisation

Effects on Health

One of the most important structural features of lectins is the fact that they consist of 2 or 4 subunits, each having a sugar-binding site. This feature of multivalency enables the lectins to agglutinate red blood cells by binding to one cell via its surface proteins and attaching another cell to a different part of the protein molecule, effectively sticking the red blood cells together to form a clot, which can block blood vessels.

It has been shown that kidney-bean lectins are able to bind specific receptor sites on the surface of the epithelial cells lining the intestine. This is accompanied by the appearance of lesions and disruption of the microvilli lining the digestive tract, which then leads to a severe impairment in the absorption of nutrients across the intestinal wall. Some lectins are highly toxic, for example, phasin from red kidney beans can lead to death at a concentration as low as 5 μg/kg bodyweight.

Onset of symptoms usually starts within 1–3 h of consumption of raw or undercooked lectins. Symptoms include acute gastroenteritis, sickness and abdominal pain, which may be severe enough to require hospitalisation. The symptoms generally clear within 3–4 h and recovery is usually rapid and complete.

A number of incidents of human intoxication by lectins have been documented in the literature. In 1948, the population of West Berlin suffered a serious bout of gastroenteritis caused by the consumption of partially cooked beans that had been air-lifted into the city during the Russian blockade. Illness has been reported in countries such as Tanzania, where a mixture of beans and maize is cooked as porridge for infants. The mixture often retains lectin activity owing to insufficient cooking, possibly caused by poor heat transfer to the beans through the viscous food mass. In 1976, an acute outbreak of sickness and diarrhoea occurred in a group of schoolboys in the UK and was attributed to the consumption of kidney beans that had been soaked in water but not cooked. An intake of 4–5 beans was sufficient to elicit the response, and two of the boys were hospitalised and required intravenous infusion. Following this incident, the Ministry of Health asked the public to report any similar experiences, which resulted in over 800 reports of illness.

Several UK outbreaks have been associated with "slow-cooking" devices, or casseroles that had not reached a high enough internal temperature to destroy the glycoprotein lectin. It has been shown that heating beans to 80 °C may potentiate toxicity five-fold, so that these beans are more toxic than if eaten raw.

In studies of casseroles cooked in slow cookers, internal temperatures often did not exceed 75 °C, and were probably insufficient to destroy all of the lectin activity, even though the beans were deemed acceptable in terms of texture and palatability.

Sources

Although many different lectins have now been identified in a wide range of plant species as detailed above (see Table 2.1.3), their role in plants is still uncertain. It seems likely that they do perform a physiological function connected with their ability to bind to carbohydrate-containing molecules. However, in some plants they are also thought to play a role in protecting the plant against attack by insects and fungi, and physical damage or fungal invasion may result in elevated lectin levels.

Stability in Foods

Lectins are proteins, and are denatured and inactivated by an adequate heat process. Boiling or autoclaving lectin-containing beans has been found to be effective, although preliminary soaking in water may be required. Dry heat is much less effective and lectin activity in some beans may remain after heating for several hours if they have not been soaked in water.

Control Options

Toxic lectins in edible legume species can be inactivated by adequate preparation and cooking procedures.

The following procedure is recommended by the UK Health Protection Agency and other authorities for the safe cooking of red kidney beans:

1. Soak in water for at least 5 h.
2. Pour away the soaking water.
3. Boil briskly in fresh water, with occasional stirring, for at least 10 min.

Food processors should be aware of these guidelines when using lectin-containing bean species as ingredients. Canning processes will inactivate lectins and canned beans can be used without further treatment.

Legislation

There is no specific legislation governing lectin levels in foods.

Sources of Further Information

Published

Liener, I., Sharon, N. and Goldstein, I. The Lectins. Properties, Functions and Applications in Biology and Medicine. Academic Press, New York, 1986.

Nachbar, M. and Oppenheim, J., Lectins in the US diet: A survey of lectins in commonly consumed foods and a review of the literature. *American Journal of Clinical Nutrition*, 1980, 33, 2338.

On the Web

Paper in Livestock Research for Rural Development (Vol 3, issue 3, December 1991). http://www.fao.org/ag/AGA/AGAP/FRG/lrrd/lrrd3/3/tropap.htm

Cornell University fact sheet. http://www.ansci.cornell.edu/plants/toxicagents/lectins/lectins.html

2.1.3 FISH TOXINS

2.1.3.1 Amnesic Shellfish Poisoning (ASP)

Hazard Characterisation

What is Amnesic Shellfish Poisoning?

Amnesic shellfish poisoning (ASP) is a foodborne intoxication associated with the consumption of contaminated shellfish harvested from waters affected by growth of certain types of toxic algae. It is also sometimes referred to as domoic acid poisoning because amnesia is not a symptom in every case. ASP was first identified in 1987 following a shellfish-related food poisoning incident in Canada.

ASP is caused by the ingestion of toxins that accumulate in certain types of shellfish that have been feeding on the algae that produce the toxins, or have preyed on contaminated species. ASP is an acute form of human poisoning, which causes a wide range of symptoms and can sometimes be fatal.

ASP is caused by domoic acid (DA), a water-soluble acidic amino acid that has been isolated from a number of marine macro- and micro-algae species. DA is a powerful neurotoxin and belongs to the kainoid class of compounds.

Occurrence in Foods

Most human cases of ASP are related to bivalve molluscs, especially mussels, but DA has also been isolated from scallops, oysters and razor clams. DA has also been found at levels high enough to cause human illness in Dungeness crabs, carnivorous gastropods, and anchovies. Mussels and other bivalves are filter feeders and accumulate toxins when the water contains sufficient levels of toxin-producing algae. It is thought that some small finfish, such as anchovies, may also feed directly on high densities of algae when other food sources are limited. There have been instances of other marine predators, notably pelicans and sealions, dying in large numbers after feeding on contaminated fish.

DA has been shown to accumulate in several bivalve species. Most of the toxin is concentrated in the viscera, especially in the digestive gland (hepatopancreas). Different species accumulate DA at different rates and variation has been observed in individuals of the same species growing in the same area. A toxin level of $>3000\,\mu g/g$ in the digestive gland of scallops has been reported, but negligible amounts were found in muscle tissue.

DA levels in shellfish do reduce naturally after they stop feeding on toxic algae, but retention times vary greatly between species. For example, mussels accumulate DA quite quickly, but it is also lost quickly from their tissues. Razor clams, by contrast lose DA from their tissues only slowly, and the toxin can remain in the edible muscle for a considerable time.

Hazard Characterisation

Effects on Health

DA is a potent neurotoxin, which can affect both central and peripheral nervous systems in humans and is also an emetic. It acts as an excitatory neurotransmitter that binds to receptor proteins on nerve cells. This causes a repeated depolarisation of the cell and results in its eventual destruction. In the first documented ASP outbreak in Canada, consumption of 60–110 mg DA (0.9–2.0 mg/kg bodyweight) was sufficient to cause mild symptoms.

The onset of symptoms of ASP in the Canadian outbreak occurred between 15 min and 38 h after ingestion of toxic shellfish. The main symptoms of ASP include nausea, vomiting, abdominal cramps headache, diarrhoea and memory loss. Memory loss is usually temporary and is more common in older people. The severity of symptoms depends on the amount of DA ingested, and a wide variety of more severe neurological symptoms can occur, including coma, disorientation, seizures, uncontrolled weeping or aggressive behaviour, eye problems and unstable blood pressure and pulse. Patients falling into a coma may not recover and may eventually die.

The effect of long-term exposure to small concentrations of DA is unknown.

Incidence and Outbreaks

Documented ASP outbreaks in humans are known only from Canada and the USA, but DA has been found in shellfish taken from European waters. This has resulted in the closure of fisheries in several countries, including Scotland, Ireland and Spain. High levels of the toxin have also been isolated from shellfish harvested in New Zealand, but no outbreaks in humans are recorded. Algal species known to be capable of producing DA have been found over a much wider geographical area, including the Pacific Ocean.

The first documented outbreak in 1987 affected over 100 people in Prince Edward Island off the east Canadian coast. Three deaths were reported during the outbreak. The toxin was traced to blue mussels produced locally by aquaculture. Since then, dangerous levels of DA have been found in shellfish on a number of occasions. In 1991, 24 people in the US state of Washington were taken ill suffering from gastrointestinal symptoms and memory loss. Although ASP was not confirmed, the outbreak coincided with high DA levels being identified in razor clams and a ban on harvesting the shellfish. There have been repeated incidents of DA being found in shellfish from US waters, especially on the west coast, and a number of examples of fisheries being closed.

Sources

DA is unusual among shellfish toxins, as it is not produced by species of dinoflagellates. It was first isolated from the red macroalga *Chondria armata* in the 1950s, but the source of DA implicated in the first documented ASP outbreak was identified as a microalga, the diatom *Pseudo-nitzschia pungens* forma *multiseries* (now recognised as two separate species, *P. pungens* and *P. multiseries*). DA production has been reported in at least nine species of *Pseudo-nitzschia*: *P. australis, P. delicatissima, P. pseudodelicatissima, P. multiseries, P. pungens, P. seriata, P. multistriata, P. turgidula* and *P. fraudulenta*. Another species, *Nitzschia navis-varingica*, isolated from shrimp ponds in Vietnam, has also been shown to produce DA. These species are widely distributed around the world's oceans, although certain species tend to be found more often in a specific region.

Production of DA by the different species is very variable and seems to be affected by environmental conditions, although the relationship with factors such as temperature and nutrient availability is unclear. Generally, DA is produced when rapid growth of *Pseudo-nitzschia* spp. occurs, forming an algal bloom. Toxin production has been observed during exponential and stationary growth phases. Reports suggest that cell densities of at least 3×10^5 cells per litre are required before feeding shellfish accumulate sufficient toxin to cause ASP.

Stability in Foods

DA is relatively heat stable and is not destroyed by practical cooking processes, or by frozen storage. In scallops, DA has been shown to spread from the digestive gland into other tissues during frozen storage and even a canning process was found to be ineffective in reducing DA levels, although migration from flesh to canning brine was observed. The meat of Dungeness crabs can also become contaminated during cooking if they are not eviscerated before processing.

Natural detoxification (depuration) in shellfish does occur, but the rate of this process varies greatly with the species, being rapid in mussels, but very slow in razor clams.

Control Options

The stability of DA and the variability of natural detoxification mean that neither depuration in clean water nor cooking processes are effective or economically viable methods of reducing the toxicity of affected shellfish to safe levels.

The only effective controls available currently are the monitoring of the marine environment and the testing of shellfish for DA when contamination is suspected. Regular inspection of the waters where shellfish are harvested, or produced by aquaculture, for the presence of toxic algae can be a useful source of data and indicate when a risk of toxicity is present. *Pseudo-nitzschia* diatoms are quite easy to identify under the microscope, but distinguishing between species is very difficult. As species vary in their ability to produce DA it is important to be able to identify individual species and molecular biology methods have been developed to do this.

When potentially toxic conditions are detected, bans on harvesting shellfish have to be imposed until toxicity can be shown to have returned to safe levels and contaminated shellfish should not be allowed to enter the human food chain.

Legislation

There are regulations relating specifically to ASP toxin in shellfish in a number of countries.

In the EU the European Commission has set a guideline limit for total ASP toxin in the edible parts of molluscs of 20 mg/kg. A liquid chromatography (LC) method is specified and if levels above the guideline value are found, then the complete batch of shellfish must be destroyed. Monitoring of toxin-producing algae and DA in shellfish occurs in several European countries.

In both Canada and the USA a guideline value of 20 mg DA/kg of mussel and/or bivalves is in force and the LC-based method must be used. In the USA, a guideline value for cooked crab (viscera and hepatopancreas) of 30 mg DA/kg is in place. Some monitoring for toxin-producing algae and DA in shellfish is carried out in both countries.

Monitoring is also undertaken in Australia and New Zealand and New Zealand has set a regulatory limit of 20 mg DA/kg of shellfish meat, to be determined by LC.

Sources of Further Information

Published

Sobel, J. and Painter, J. Illnesses caused by marine biotoxins. *Clinical Infectious Diseases*, 2005, 41(9), 1290–6.

Australia New Zealand Food Authority Shellfish toxins in food: A toxicological review and risk assessment. *Technical Report Series*, No. 14, 2001.

On the Web

US Centers for Disease Control & Prevention (CDC) – Marine toxins factsheet. http://www.cdc.gov/ncidod/dbmd/diseaseinfo/marinetoxins_g.htm

FAO Food and Nutrition Paper 80 – Marine Biotoxins (2004). http://www.fao.org/docrep/007/y5486e/y5486e00.htm

2.1.3.2 Azaspiracid Shellfish Poisoning (AZP)

Hazard Characterisation

What is Azaspiracid Shellfish Poisoning?

Azaspiracid shellfish poisoning (AZP) is a foodborne intoxication associated with the consumption of contaminated shellfish harvested from waters affected by growth of certain types of toxic algae. AZP was first recognised in 1995 following an outbreak of illness in the Netherlands associated with the consumption of mussels imported from Ireland.

AZP is caused by the ingestion of toxins that accumulate in certain types of shellfish that have been feeding on the algae that produce the toxins. AZP is a form of food poisoning with symptoms typical of gastroenteritis, broadly similar to DSP.

The toxin responsible for AZP has been identified and characterised as azaspiracid (AZA). AZA is a polyether toxin with an unusual spiral ring structure. Up to 11 analogues of AZA have been identified and characterised, but some of these are thought to be shellfish metabolytes and are less toxic than AZA. Only AZA-1, AZA-2 and AZA-3 are considered to have public-health significance and AZA-1 is thought to be the main cause of illness.

Occurrence in Foods

Recorded cases of AZP have been associated with consumption of mussels, but AZAs have also been found in oysters, clams, scallops, razor clams and cockles. There have also been reports of AZA contamination in crabs.

AZAs tend to accumulate in shellfish digestive glands initially, but unlike other shellfish toxins, they can be readily transported to other tissues, though not predictably. This means that the rate of natural detoxification (depuration) in contaminated shellfish can be very slow.

Hazard Characterisation

Effects on Health

The mechanism of AZA toxicity is unknown, but evidence from AZP outbreaks suggests that a lowest observable adverse effect level (LOAEL) of AZA is 23 to 86 μg per person (mean value 51.7 μg). Mussels collected from Irish waters after outbreaks were found to contain total AZAs at levels up to 1.4 μg/g of meat.

Symptoms of AZP resemble those of DSP and include nausea, vomiting, severe diarrhoea and stomach cramps. Severity of symptoms appears to be linked to the quantity of toxin ingested.

Incidence and Outbreaks

AZP has only been associated with shellfish harvested from Irish waters to date. However, AZAs have also been isolated from shellfish harvested in UK and Norwegian waters.

The first recorded outbreak of AZP affected eight people in the Netherlands who had consumed mussels imported from Ireland. Since 1996 other incidents have been reported in Ireland, notably in 1997 when contaminated mussels from Arranmore, in Donegal, caused human cases in Ireland and elsewhere in Europe. Further incidents were reported in 2001 and 2005, resulting in mussel fisheries being closed for prolonged periods.

Sources

The source of AZAs is uncertain, but is likely to be dinoflagellates. Evidence suggests that the species *Protoperidinium crassipes* is most likely to be responsible, but as this species preys on other dinoflagellates, it may not be the only species involved. Other known toxin-producing species have not been found when AZAs have been identified in shellfish.

Incidents have not been linked to visible algal blooms and the cell density needed to produce hazardous AZA levels is not known.

Stability in Foods

There are conflicting reports on the heat stability of AZAs, but recent evidence suggests that they survive cooking processes, as do other polyether shellfish toxins.

Natural detoxification in shellfish does occur, but the rate of this process in mussels is slow, and toxicity has been reported to last for up to six months.

Control Options

The stability of AZAs and the prolonged duration of natural detoxification mean that neither depuration in clean water nor cooking processes are effective or economically viable methods of reducing the toxicity of affected shellfish to safe levels.

The only effective control available currently is the regular monitoring of shellfish samples for the presence of AZAs using a mouse bioassay or LC-MS analysis.

When toxic conditions are detected, bans on harvesting shellfish have to be imposed until toxicity can be shown to have returned to safe levels and contaminated shellfish should not be allowed to enter the human food chain.

Legislation

There are regulations relating specifically to AZP toxins in the EU where the European Commission has set a maximum level of 160 μg/kg in bivalve molluscs, echinoderms, tunicates and marine gastropods. The reference method for analysis is the mouse bioassay, although other alternative or complementary methods can be used.

The Irish authorities undertake weekly shellfish testing for several toxins, including AZAs.

Sources of Further Information

Published

Scientific Committee of the Food Safety Authority of Ireland (FSAI). Risk assessment of azaspiracids (AZAs) in shellfish *FSAI Report*, 2006.

On the Web

FAO Food and Nutrition Paper 80 – Marine Biotoxins (2004). http://www.fao.org/docrep/007/y5486e/y5486e00.htm

2.1.3.3 Ciguatera Fish Poisoning

Hazard Identification

What is Ciguatera?

Ciguatera fish poisoning (CFP) is a foodborne intoxication associated with consumption of coral reef fish from tropical and subtropical waters in the Pacific and Indian Oceans and the Caribbean sea. It was first recorded by Spanish explorers some 500 years ago. CFP is the commonest form of marine food poisoning worldwide and is considered to be a significant public health problem. CFP is caused by ingestion of toxins (ciguatoxins) that accumulate in certain fish species. They occur in fish that feed on toxic algae, or on toxic herbivorous prey fish species.

The ciguatoxins are lipid-soluble polyether compounds made up of 13 or 14 rings fused into rigid ladder-like structures. Multiple forms of ciguatoxin with small structural differences have been described and there are important geographic differences. The Pacific ciguatoxin-1 (molecular weight 1112) is the most potent and its structure is slightly different from that of the Caribbean ciguatoxin-1. These differences are also reflected in the symptoms produced.

Occurrence in Foods

Ciguatoxins are found in a broad range of fish that live in or around coral reefs in comparatively shallow tropical waters. Over 400 species have been reported to be involved in CFP outbreaks. The toxins tend to concentrate as they move up the food chain, so that large carnivorous fish are more likely to be toxic. Species such as barracuda, grouper, snapper, jack, moray eel, Spanish mackerel and some inshore tuna carry the highest risk, but herbivorous and coral-eating species such as parrot fish may also cause CFP outbreaks.

The highest concentrations of toxins in the fish are found in the viscera, particularly in the liver and kidneys, and levels can be up to 100 times higher than in other tissues. The fish themselves suffer no detectable symptoms even though the toxin is persistent and affected fish can remain toxic for long periods.

In former times, CFP was restricted to indigenous populations in areas where ciguatoxins are endemic, but this has changed in recent years with the increase in global travel and the increasing importation of exotic foodfish species into developed countries.

Hazard Characterisation

Effects on Health

Ciguatoxins cause a wide variety of neurological, gastrointestinal and cardiovascular symptoms. They are extremely powerful toxins and an oral dose of 0.1 μg may be enough to cause illness. They act by increasing the sodium ion permeability of the plasma membranes in nerve and muscle cells, causing membrane depolarisation and thus disrupting cell function. Similarly, they affect intracellular calcium transport in gut epithelial cells.

Symptoms may appear within one hour in severe cases, but onset may be delayed for 24 or even 48 h in milder cases. Gastrointestinal symptoms, including nausea, vomiting, diarrhoea and abdominal pain often occur first, followed by neurological symptoms, such as a tingling of the lips and extremities and severe localised skin irritation. However, there is geographic variation, with neurological symptoms being more common in the Pacific and gastrointestinal in the Caribbean.

Other recorded symptoms include hallucinations, depression and anxiety, fatigue and aching in the muscles and joints. Hypotension, respiratory problems and even paralysis can occur in severe cases, but death is uncommon, with a reported fatality rate of less than 1%. Gastrointestinal symptoms usually resolve within a few days, but where neurological symptoms occur they may last much longer, typically several weeks or months. Individuals can also become sensitised to ciguatoxin so that they may react to eating fish that do not affect others.

The varied nature of the symptoms can result in CFP being misdiagnosed as multiple sclerosis or chronic fatigue disorder in developed countries.

Incidence and Outbreaks

It is estimated that between 10 000 and 50 000 cases of CFP occur each year. Most of these cases occur in tropical and subtropical coastal regions adjoining the Pacific and Indian Oceans and the Caribbean. However, more cases are being reported in temperate developed countries and it is thought that under-reporting could be significant in Europe and North America because of misdiagnosis.

CFP outbreaks have been reported in France, Italy, Germany and the Netherlands. In the USA, 129 outbreaks affecting 508 people were recorded between 1983 and 1992. Most of these occurred in Hawaii and Florida, but outbreaks linked to imported fish were reported elsewhere. A number of outbreaks have occurred in Australia and an annual incidence of 30 per 100 000 has been estimated.

Sources

The principal known source of ciguatoxins is an alga, the marine dinoflagellate *Gambierdiscus toxicus*, which is associated with seaweeds, sediments and dead coral. It is distributed around the tropics within the latitudes 32 °N and 32 °S and grows in shallow waters, but its presence and numbers are unpredictable. There is also evidence that other species of dinoflagellates may sometimes be involved.

Certain strains of *G. toxicus* produce toxins referred to as gambiertoxins – less oxidised and less toxic precursors of ciguatoxins. When the algae are consumed by herbivorous fish, the gambiertoxins accumulate in the fish and a biotransformation begins to occur, in which they are converted to ciguatoxins. Over time, the toxins become transferred to carnivorous fish and the biotransformation is completed. The highest levels of ciguatoxins are found in the largest carnivorous fish. Different strains of *G. toxicus* are thought to produce different ciguatoxin precursors, which are then transformed into the various ciguatoxin types.

G. toxicus also produces another type of highly potent toxin called maitotoxins. These occur in the guts of herbivorous fish, but are not now thought to be involved in CFP.

Stability in Foods

Ciguatoxins are temperature stable and are not destroyed by cooking or by freezing. Other processes, including salting and smoking, also have little or no effect. Affected fish can remain toxic for years, even when their diet ceases to contain toxin or precursors.

Control Options

Ciguatoxins are odourless and tasteless and do not alter the appearance of the fish. They can only be detected using animal bioassays, following extraction and purification techniques, although an immunoassay method is being developed. This, plus their stability, severely limits the control options available.

The only practical control is to avoid consumption of susceptible fish species from areas where ciguatera is endemic. Large predatory reef fish, such as barracuda, present a high risk and should be particularly avoided. Parts of the fish where the highest toxin levels accumulate, such as the head, gut, liver and roe should not be eaten. Health Canada advises travellers not to eat large reef fish weighing more than 3 kg.

Legislation

There are few specific regulations for ciguatera toxins in fish.

In the EU, legislation covering fishery products states that "fishery products containing biotoxins such as ciguatera toxins" cannot be placed on the market, but no methods of analysis are given.

In the USA there are no standards or official methods as yet and no action limits have so far been established.

The most common legislative control in use around the world is the prohibition of the sale of high-risk fish taken from areas where ciguatera toxins are known to be present. Such bans have been used with success in Australia, Fiji, Hawaii and Florida.

Sources of Further Information

Published

Lewis, R.J., Ciguatera: Australian perspectives on a global problem Toxicon *(official journal of the International Society on Toxinology)*, 2006, 48(7), 799–809.

Sobel, J. and Painter, J., Illnesses caused by marine biotoxins. *Clinical Infectious Diseases*, 2005, 41(9), 1290–6.

On the Web

US Centers for Disease Control & Prevention (CDC) – Marine toxins factsheet. http://www.cdc.gov/ncidod/dbmd/diseaseinfo/marinetoxins_g.htm

FAO Food and Nutrition Paper 80 – Marine Biotoxins (2004). http://www.fao.org/docrep/007/y5486e/y5486e00.htm

CDC Ciguatera page. http://www.cdc.gov/nceh/ciguatera/

2.1.3.4 Diarrhoeic Shellfish Poisoning (DSP)

Hazard Characterisation

What is Diarrhoeic Shellfish Poisoning?

Diarrhoeic shellfish poisoning (DSP) is a foodborne intoxication associated with the consumption of contaminated shellfish harvested from waters affected by growth of certain types of toxic algae. DSP has been known for around 30 years and is most common in Europe and Japan, but DSP toxins are being increasingly reported in shellfish from previously unaffected areas.

DSP is caused by the ingestion of toxins that accumulate in certain types of shellfish that have been feeding on the algae that produce the toxins. DSP is a non-lethal form of food poisoning with symptoms typical of gastroenteritis, especially diarrhoea.

There are a number of chemically different toxins associated with DSP. They are lipophilic and polyether compounds and can be divided into three main groups:

1. acidic toxins – okadaic acid (OA) and its derivatives named dinophysistoxins (DTXs);
2. neutral toxins – pectenotoxin group (PTXs);
3. other toxins – yessotoxin (YTX) and a derivative.

The DTXs are the most important group in causing DSP symptoms and it is possible that the other two groups are not involved in illness, but are often found in association with DTXs.

Occurrence in Foods

Most cases of DSP are related to bivalve molluscs, especially mussels, but also scallops, oysters and clams. These species are filter feeders and accumulate toxins when the water contains sufficient levels of toxin-producing algae. Toxicity is seasonal and tends to be highest during the summer months in Europe and Japan, although DSP cases in Scandinavia have been reported in February and in October. Predatory fish and other marine animals that prey on toxic shellfish may also accumulate DSP toxins, especially in liver tissue, but the significance of this for human health is uncertain.

DSP toxins are fat soluble and so tend to accumulate in the fatty tissue of affected shellfish. The highest levels are normally found in the viscera and shellfish can accumulate enough toxin to cause illness within h when large populations of toxic algae are present in the water. DTXs may also be metabolised in the digestive gland (hepatopancreas) of contaminated shellfish, producing related toxic by-products. Toxin levels as high as 10 mg OA/g hepatopancreas have been reported in mussels grown in Japanese waters.

DSP toxin levels in shellfish do reduce naturally after they stop feeding on toxic algae, but there is little definite information on how this process occurs or on toxin retention times in different species. It is likely that some toxin is excreted in faeces before it can be assimilated.

Hazard Characterisation

Effects on Health

DSP toxins are powerful phosphatase inhibitors and this property is associated with inflammation of the gut in humans. This leads to fluid loss from intestinal cells resulting in diarrhoea. A minimum dose of OA for toxic effects to occur is estimated to be 48 μg, whereas for its derivative DTX1 the minimum it is 38.4 μg.

Levels of DSP toxins are commonly expressed as toxic equivalents of OA (mg OA eq/kg) or as mouse units (MU/kg) relating to a standard mouse bioassay method.

The onset of symptoms of DSP may occur between 30 min and 12 h after ingestion of toxic shellfish. The main symptoms of DSP include diarrhoea, nausea and vomiting and abdominal pain. The severity of symptoms depends on the amount of DSP toxins ingested, but complete recovery typically occurs within three days. No fatalities caused by DSP have been reported to date and hospital treatment is not usually needed.

DSP toxins have also been shown to have other effects in animals and in cell cultures. For example, OA and DTX1 are probable carcinogens, but the significance of this for human health is unknown. PTXs and YTX are lethal in mice and are certainly toxic, but the effects of oral doses in humans are not known.

Incidence and Outbreaks

DSP mainly affects Western Europe and Japan, but DSP toxin-contaminated shellfish and toxin-producing algae have been found in more widespread locations, including Canada, Mexico, South America, India, Thailand, China and Australia, and incidences of DSP toxin seem to be increasing.

There have been a number of major outbreaks of DSP in Europe. Mussels imported from Denmark caused 415 cases of illness in France in 1990. In 1984, 10 000 people in France were affected by DSP symptoms caused by domestically produced mussels and a further 2000 became ill the following year in a similar outbreak. 1984 also saw a major outbreak in Norway affecting at least 300 people. Over 5000 cases of DSP-related gastroenteritis were reported in Spain in 1981, and DSP toxins have repeatedly been found at high levels in shellfish from the Galician region, resulting in prolonged disruption to local fisheries from 1993 onward. DSP cases were not reported in the UK until 1997, when 49 people were made ill after eating mussels in a London restaurant. Since then, the frequency of DSP events in UK waters has increased and shellfish harvesting has been restricted in several areas on a regular basis.

In Japan, cases of DSP were first reported in 1976 and 1977 when more than 150 people were affected by vomiting and diarrhoea. A total of at least 1300 cases were reported between 1976 and 1981.

Elsewhere, outbreaks have been reported in Australia, Canada, Chile and the USA. Although the Northeast USA, especially New York and New Jersey, experienced large outbreaks of DSP-like illness between 1980 and 1985, outbreaks of human illness have not been reported since then, although DSP toxins have been found occasionally in US waters.

Sources

DSP toxins are produced by dinoflagellates of the genus *Dinophysis*. Seven species have been shown to produce the toxins. These are *D. acuminata* (Europe) *D. acuta*, *D. fortii* (Japan), *D. mitra*, *D. norvegica* (Scandinavia), *D. rotundata* and *D. tripos*. Three other species are also suspected of being able to produce toxins. Certain *Prorocentrum* species (*P. concavum*, *P. lima* and *P. redfieldi*) also produce DSP toxins.

If conditions are favourable, exponential growth of these species may occur resulting in an algal bloom. However, it is not necessary for visible blooms to occur for DSP toxins to be present at harmful levels. The production of toxins by different dinoflagellate species is highly variable and the same species may produce widely varying quantities of toxin in different locations. Some *Dinophysis* spp. can produce sufficient toxin in shellfish to cause illness in consumers at populations as low as 200 cells per litre. On other occasions much greater densities ($>20\,000$ cells per litre) may be involved.

Stability in Foods

DSP toxins are all relatively heat stable and are not destroyed by practical cooking processes.

Natural detoxification in shellfish does occur, but the rate of this process varies greatly with the species, the season (low water temperature slows toxin loss) and with the site of toxin accumulation. It has been reported that the retention time of DSP toxins in mussels can vary from one week to six months.

Control Options

The stability of DSP toxins and the variability of natural detoxification mean that neither depuration in clean water nor cooking processes are effective or economically viable methods of reducing the toxicity of affected shellfish to safe levels.

The only effective controls available currently are the monitoring of the marine environment and the testing of shellfish flesh for DSP toxins. Regular inspection of the waters where shellfish are harvested, or produced by aquaculture, for the presence of toxic algae can be a useful source of data and indicate when a risk of toxicity is present. The routine testing of shellfish, especially mussels, for DSP toxins by chemical, immunological, or bioassay methods is the key prevention measure. However, the variability of toxin production by the algae and other factors must be taken into account when designing a suitable sampling plan.

When toxic conditions are detected, bans on harvesting shellfish have to be imposed until toxicity can be shown to have returned to safe levels and contaminated shellfish should not be allowed to enter the human food chain.

Legislation

There are regulations relating specifically to PSP toxins in shellfish in a number of countries.

In the EU the European Commission has set a maximum limit for combined OA, DTXs and PTXs in molluscs, echinoderms, tunicates and marine gastropods of 160 mg STX eq/kg of edible tissues. The maximum level of YTXs in edible tissues is set at 1 mg YTX eq/kg. The mouse bioassay method is the official reference method of analysis, if required in association with a chemical detection method. Monitoring programmes for toxic dinoflagellates are in place in most European countries where shellfish are harvested.

Japan actively monitors both phytoplankton and shellfish and applies a tolerance level for DSP toxins of 5 MU/100 g whole meat, when detected by the mouse bioassay method. This equates to approximately 0.2 µg/g.

In the USA, there is no current monitoring programme or limit for DSP toxins in shellfish, although monitoring is carried out in Canada.

Sources of Further Information

Published

Sobel, J. and Painter, J., Illnesses caused by marine biotoxins *Clinical Infectious Diseases*, 2005, 41(9), 1290–6.

Australia New Zealand Food Authority Shellfish toxins in food: A toxicological review and risk assessment *Technical Report Series*, No. 14, 2001.

On the Web

US Centers for Disease Control & Prevention (CDC) – Marine toxins factsheet. http://www.cdc.gov/ncidod/dbmd/diseaseinfo/marinetoxins_g.htm

FAO Food and Nutrition Paper 80 – Marine Biotoxins (2004). http://www.fao.org/docrep/007/y5486e/y5486e00.htm

2.1.3.5 Neurologic Shellfish Poisoning (NSP)

Hazard Identification

What is Neurologic Shellfish Poisoning?

Neurologic shellfish poisoning (NSP) is a foodborne intoxication associated with the consumption of contaminated shellfish harvested from waters affected by growth of certain types of toxic algae. It is also sometimes referred to as neurotoxic shellfish poisoning. NSP-like symptoms associated with "red tides" off the Florida coast and in the Gulf of Mexico were first noted in the nineteenth century.

NSP is caused by the ingestion of toxins that accumulate in certain types of shellfish that have been feeding on the algae that produce the toxins. NSP is an acute toxic syndrome having some similarities with PSP, although PSP is usually more severe. NSP causes a wide range of symptoms, but is not reported to be fatal.

NSP is caused by brevetoxins, ten of which have been isolated from algal blooms or cultures. The brevetoxins are stable, lipid-soluble polyether neurotoxins, consisting of 10 to 11 rings and having molecular weights of around 900. In addition to the ten naturally occurring brevetoxins, a further four analogues have been found in contaminated shellfish. These are thought to arise through biotransformation of brevetoxins, probably in the digestive glands of some shellfish species.

Occurrence in Foods

Most human cases of NSP are related to bivalve molluscs, including oysters, clams and mussels, all of which can accumulate brevetoxins during feeding when the water contains sufficient levels of toxin-producing algae. Brevetoxins have also been reported in some seabirds and finfish, but most fish, birds and mammals are susceptible to the toxins and toxic algal blooms have caused extensive fish kills and the deaths of marine mammals and birds.

There is little published information on the rate or site of brevetoxin accumulation in shellfish. Toxin levels in shellfish do reduce naturally after they stop feeding on toxic algae, but little is known about this process and retention times vary greatly between species. Furthermore, biotransformation of brevetoxins in some shellfish may produce analogues that are more toxic than the natural toxins.

Hazard Characterisation

Effects on Health

Brevetoxins are neurotoxins that act by affecting the sodium channels in the membranes of nerve cells. This causes the cells to fire repeatedly, giving rise to various neurological symptoms. Brevetoxin is considered potentially toxic to humans at any detectable level in shellfish, but a residue toxicity of 20 mouse units (MU) per 100 g of shellfish flesh is commonly used for regulatory purposes.

The onset of symptoms of NSP occurs between 30 min and 3 h after ingestion of toxic shellfish. The main symptoms of NSP include nausea, vomiting, diarrhoea, chills and sweating, hypotension, numbness, pins and needles, cramps and in some severe cases, paralysis and coma, but deaths have not been reported. Symptoms usually persist only for a few days.

Brevetoxins can also cause skin and eye irritation in people swimming in waters affected by algal blooms and inhalation of toxic aerosols can cause respiratory problems.

Incidence and Outbreaks

For many years NSP was known only in Florida and the coasts around the Gulf of Mexico. However, in 1993 an outbreak of NSP-like illness was reported in New Zealand. Algal species known to produce brevetoxins have also been identified in the coastal waters of several western European countries, South Africa, Canada, the east and west coasts of the USA, Japan and Australia.

The first documented outbreak caused by shellfish harvested from waters north of Florida occurred in North Carolina in 1987. This outbreak affected 48 people and lasted for several months. In the 1993 outbreak in New Zealand, 186 cases of illness were recorded. This outbreak was identified as NSP, but it seems that PSP may also have been involved in some of the cases. Brevetoxin levels in contaminated shellfish were reported to have reached 592 MU/100 g at the height of the outbreak.

Sources

Brevetoxins are produced by the motile form of a dinoflagellate species usually referred to as *Gymnodinium breve* (also known as *Ptychodiscus brevis* and recently renamed as *Karenia brevis*). This is the species causing toxic red tides around the Florida coast, but it probably has a much wider geographical distribution. Toxins that correspond closely to brevetoxins have also been identified in four species of algae belonging to the class *Raphidophyceaea*. These species are *Chattonella antiqua* and *Chattonella marina*, *Fibrocapsa japonica* and *Heterosigma akashiwo*, and they too are widely distributed.

The presence of low numbers of these algae is probably not a health hazard, but under certain conditions rapid growth may occur, resulting in an algal bloom. When this happens the numbers of cells can become high enough to colour the water reddish brown (a red tide). Cell densities of *G. breve* of $>10^7$ cells per litre have been recorded during a red tide along the southwest coast of Florida.

Any filter-feeding shellfish in water affected by a toxic bloom are likely to accumulate high levels of toxin quite quickly as they feed on and digest the algal cells. Thus, shellfish harvested from such waters carry a high risk of toxicity.

Stability in Foods

Brevetoxins are known to be relatively heat stable, and acid stable. They have been reported to survive both cooking and freezing processes. Even retorting processes cannot be relied upon to eliminate toxin.

Natural detoxification (depuration) in shellfish does occur, but the rate of this process varies greatly between and even within species. Commercially grown shellfish are generally regarded as safe to eat after one or two months following the end of a toxic algal bloom.

Control Options

The stability of brevetoxins and the variability of natural detoxification mean that neither depuration in clean water nor cooking processes are effective or economically viable methods of reducing the toxicity of affected shellfish to safe levels. Depuration of mussels with ozonated water has been investigated and appears to enhance depuration.

The development of potentially toxic *G. breve* blooms is highly unpredictable and the only effective control is the monitoring of the marine environment for evidence of a bloom, such as large fish kills and discoloured water. Toxicity is then confirmed using chemical analysis or mouse bioassay. Monitoring of water quality using microscopy to identify and count potentially toxic algae can be of value in preventing NSP outbreaks, but it is time consuming and requires highly skilled staff. New diagnostic tests using biomarkers for *G. breve* have been investigated in the laboratory.

When potentially toxic conditions are detected, bans on harvesting shellfish have to be imposed until toxicity can be shown to have returned to safe levels and contaminated shellfish should not be allowed to enter the human food chain.

Legislation

There are regulations relating specifically to NSP toxins in shellfish in the USA and New Zealand.

In the USA a regulatory limit of 80 μg type-2 brevetoxin/100 g of shellfish tissue (equivalent to 20 MU/100 g) determined by the APHA mouse bioassay is applied. The health authorities in Florida monitor coastal waters for *G. breve* and close shellfish fisheries when cell densities exceed 5000 cells/litre.

In New Zealand, a maximum acceptable level for brevetoxin in shellfish of 20 MU/100 g has also been adopted, again determined using the APHA mouse bioassay. Water from shellfish harvesting areas is monitored every week throughout the year.

Some South American and European countries also carry out monitoring, but have not set regulatory limits.

Sources of Further Information

Published

Sobel, J. and Painter, J., Illnesses caused by marine biotoxins. *Clinical Infectious Diseases*, 2005, 41(9), 1290–6.
Australia New Zealand Food Authority Shellfish toxins in food: A toxicological review and risk assessment. *Technical Report Series*, No. 14, 2001.

On the Web

US Centers for Disease Control & Prevention (CDC) – Marine toxins factsheet. http://www.cdc.gov/ncidod/dbmd/diseaseinfo/marinetoxins_g.htm
FAO Food and Nutrition Paper 80 – Marine Biotoxins (2004). http://www.fao.org/docrep/007/y5486e/y5486e00.htm

2.1.3.6 Paralytic Shellfish Poisoning (PSP)

Hazard Characterisation

What is Paralytic Shellfish Poisoning?

Paralytic shellfish poisoning (PSP) is a foodborne intoxication associated with the consumption of contaminated marine shellfish harvested from waters affected by a sudden and rapid growth of certain types of toxic algae. PSP was recorded in Canada over 100 years ago and reports were restricted to temperate waters until the 1970s. Since then there has been an apparent increase in outbreaks and a geographical spread into more tropical southern waters.

PSP is caused by the ingestion of toxins that accumulate in certain types of shellfish that have been feeding on the algae that produce the toxins. PSP can cause a variety of neurological symptoms and severe cases can prove fatal within hours. There are at least 21 known PSP toxins. There are considered to be four subgroups, but all are tetrahydropurines and they are closely related. They vary in toxicity, but the most toxic and first to be identified is referred to as saxitoxin (STX).

Occurrence in Foods

Most cases of PSP are related to bivalve molluscs, especially mussels and clams, but also oysters and scallops. In total, at least 50 shellfish species have been reported to cause PSP. All these species are filter feeders and accumulate toxins when the water contains significant levels of toxin-producing algae. When the algae are digested PSP toxins are released into the animal's digestive tissue. PSP cases in Japan have also been associated with consumption of certain reef-dwelling crab species.

Different shellfish species vary greatly in the way that they accumulate PSP toxins and in the retention time of the toxins within the body. Some species seem to be able to detect toxins in the water and stop feeding, but others do not have this ability. Some detoxification within the body also occurs as the toxins are broken down, but the rate varies enormously between species. Generally, the viscera accumulate the highest levels of toxins, but detoxification tends to proceed more rapidly in these tissues. The variation is illustrated by a comparison between mussels and oysters. Mussels accumulate toxins much more quickly and at higher concentrations than oysters, but they also detoxify much more quickly. For these reasons, potential levels of PSP toxin in affected shellfish are almost impossible to predict.

Hazard Characterisation

Effects on Health

PSP toxins are potent neurotoxins, and operate by selectively blocking the voltage-gated sodium channel – a large protein that extends across the plasma membrane of nerve and muscle cells. This slows or stops the cells ability to generate an action potential and so affects cell function. Reports of the level needed to cause symptoms vary greatly. The Australia New Zealand Food Authority has reported that 120 to 180 μg of PSP toxin is sufficient to produce symptoms in humans, 400 to 1060 μg may prove fatal and levels above 2000 μg are likely to cause death. However, in some reported cases 300 μg of PSP toxin proved fatal, while intakes as high as 320 μg have apparently not caused symptoms. It is likely that varying sensitivity between individuals may be partly responsible for these observations.

Levels of PSP toxins are commonly expresses as toxic equivalents of STX (mg STX eq/kg) or as mouse units (MU/100 g) relating to a standard AOAC mouse bioassay method.

The first symptom of toxicity in mild cases is usually numbness, or tingling around the mouth, which normally appears within 30 min. This then spreads to the head and neck. Within a few h, other symptoms, including "pins and needles" in the hands and feet, headaches, nausea, vomiting and diarrhoea usually occur and vision may be affected temporarily. Muscular weakness is also common and symptoms can last for several days.

Symptoms of more severe toxicity include numbness or tingling and weakness in arms and legs, incoherent speech and dizziness, motor coordination is affected and the patient may have difficulty breathing. In very severe cases, muscle and respiratory paralysis can develop leading to death within 2 to 24 h of ingestion of toxin. Mortality rate is variable (0–14%), but if the patient can be kept alive for at least 24 h, the chances of recovery are good.

Incidence and Outbreaks

The geographical distribution of PSP appears to have been expanding since the 1970s. Before then, PSP contamination events were restricted to temperate waters off the coasts of Europe, North America and Japan. More recently, PSP toxins have been reported in shellfish all over the southern hemisphere, including South Africa, Central and South America, Australia, China, India, Malaysia and Thailand. It is not clear whether this is due to increasing awareness of toxic algae and improved diagnosis of PSP, or whether other factors are involved. There are estimated to be 1600 cases of PSP each year worldwide, with approximately 300 of these proving fatal.

PSP toxin contamination in shellfish has been recorded repeatedly in Western Europe waters, especially off the coasts of Scotland, Spain, Portugal and Norway. Harvesting of scallops, mussels and other shellfish is regularly prohibited during the summer months when contamination occurs.

In the UK, an outbreak of PSP in 1968 affected many people in north-east England, with 78 requiring hospital treatment, but no deaths. The outbreak was linked to mussels containing between 600 and 6000 μg STX eq/kg. Since then monitoring of the fishing grounds has largely prevented similar outbreaks.

In 1976, mussels exported from Spain caused PSP outbreaks in several other European countries, including France, Germany and Italy. At least 120 people were affected, but there were no deaths.

PSP outbreaks have been known in Canada for over 200 years. Between 1880 and 1995, some 106 documented outbreaks occurred affecting 538 people and killing 32. Outbreaks have also occurred repeatedly on the east coast of the USA and in Alaska. In 1980 an outbreak in the northeast USA affected 51 people who had eaten locally caught mussels and oysters containing 3000 to 40 000 μg STX eq/kg.

Sources

PSP toxins are produced mainly by dinoflagellates of the genus *Alexandrium* (previously called *Gonyaulax* spp.). Several species are involved, notably *A. catenella*, *A. cohorticula*, *A. fraterculus*, *A. fundyense*, *A. minutum* and *A. tamarensis*. Certain other dinoflagellate species, such as *Pyrodinium bahamense* and *Gymnodinium catenatum* also produce STXs. Many of these species exist as free-swimming forms and as resting cysts that are also toxic.

The presence of low numbers of these algae is not a health hazard, but if conditions are right – increasing temperature, high nutrient levels and sunlight – exponential growth may occur resulting in an algal bloom. When this happens the numbers of cells can become high enough to colour the water reddish brown (a red tide). During a bloom the cells are at their most toxic during the late exponential phase.

Any filter-feeding shellfish in water affected by a toxic bloom are likely to accumulate high levels of toxin quite quickly as they feed on and digest the algal cells. Thus shellfish harvested from such waters carry a high risk of toxicity.

Stability in Foods

PSP toxins are relatively heat stable, especially at acid pH, but are easily oxidised under alkaline pH conditions. Cooking processes reduce toxin levels, but do not eliminate the risk of toxicity. Even retorting processes cannot be relied upon to eliminate toxin. Their effectiveness depends on the initial toxin concentration and only very severe processes (120 °C for 60 min) have been shown to give complete detoxification.

Natural detoxification in shellfish does occur, but the rate of this process varies greatly between species and some may remain toxic for months or even years in the case of clams.

Control Options

The stability of PSP toxins and the variability of natural detoxification mean that neither depuration in clean water nor cooking processes are effective or economically viable methods of reducing the toxicity of affected shellfish to safe levels.

Research into better methods is ongoing, but the only effective control available currently is the monitoring of waters where bivalve molluscs are harvested or produced by aquaculture. This can be done by regular inspection of water for the presence of dinoflagellates and their cysts, using biological, or immunological detection and identification methods and, more recently, molecular-biology techniques. Regular inspection and testing of shellfish flesh for the presence of toxins using bioassay or chemical methods is also important.

When toxic conditions are detected, bans on harvesting shellfish have to be imposed until toxicity can be shown to have returned to safe levels and contaminated shellfish should not be allowed to enter the human food chain.

Legislation

There are regulations relating specifically to PSP toxins in shellfish in a number of countries.

In the EU there is a limit for bivalve molluscs of 80 μg STX eq/100 g of meat. The mouse bioassay method is the official reference method of analysis, if required in association with a chemical detection method. Monitoring programmes to check for toxic dinoflagellates are in place in most European countries where shellfish are harvested.

In the USA, Canada and Australia the limit for bivalves is also 80 μg STX eq/100 g of meat and the mouse bioassay method is used. However, in the USA and Canada, some shellfish with higher levels of PSP toxin can be harvested if they are to be canned.

Both China and Japan set a limit of 400 MU/100 g in bivalves and specify the mouse bioassay as the reference method.

Sources of Further Information

Published

Sobel, J. and Painter, J., Illnesses caused by marine biotoxins. *Clinical Infectious Diseases*, 2005, 41(9), 1290–6.

Australia New Zealand Food Authority Shellfish toxins in food: A toxicological review and risk assessment. *Technical Report Series*, No. 14, 2001.

On the Web

US Centers for Disease Control & Prevention (CDC) – Marine toxins factsheet. http://www.cdc.gov/ncidod/dbmd/diseaseinfo/marinetoxins_g.htm

FAO Food and Nutrition Paper 80 – Marine Biotoxins (2004). http://www.fao.org/docrep/007/y5486e/y5486e00.htm

2.1.3.7 Tetrodotoxin

Hazard Characterisation

What is Tetrodotoxin?

Tetrodotoxin (TTX), also known as anhydrotetrodotoxin 4-epitetrodotoxin, or tetrodonic acid, is a marine biotoxin associated with certain fish species, notably pufferfish. Consumption of these fish can cause very severe foodborne intoxication, often referred to as pufferfish poisoning, or fugu poisoning. Unlike other marine biotoxins, it is not produced by the growth of toxic algae. Pufferfish poisoning has been known for many years, especially in Japan where the fish are a delicacy. Probable cases were documented by Captain James Cook as long ago as the eighteenth century. The term tetrodotoxin was first applied to the toxin nearly 100 years ago and the TTX molecule itself was first characterised in 1964.

TTX is a potent non-proteinaceous neurotoxin belonging to a group referred to as guanidinium toxins, which also includes the PSP toxin, saxitoxin. It consists of a positively charged guanidinium group and a pyrimidine ring with five additional fused rings. A number of derivatives of TTX have also been identified.

Occurrence in Foods

TTX is mainly associated with fish of the order Tetraodontidae (pufferfish, balloon fish, fugu, globe fish, blowfish, toad fish) from the Pacific and Indian Oceans. These fish are a traditional food in Japan, where they are sold as "fugu" in specialised restaurants employing specially trained and licensed chefs who are able to remove the most toxic parts of the fish to reduce the poisoning risk. The highest levels of TTX are found in the viscera, particularly the liver and ovaries, and skin of the fish, but the muscle tissue does not usually contain dangerous levels of toxin.

TTX has also been found in a wide range of other animals, such as the blue-ringed octopus, goby, triggerfish, parrotfish, angelfish, xanthid crabs, certain marine molluscs and worms and some terrestrial amphibians, such as the Californian newt.

The trumpet shell (*Charonia sauliae*) has also been reported to contain a TTX derivative and has been implicated in some cases of foodborne intoxication.

Hazard Characterisation

Effects on Health

TTX is a very potent neurotoxin, and operates in a similar way to the PSP toxin (saxitoxin) by selectively blocking the voltage-gated sodium channel – a large protein that extends across the plasma membrane of nerve and muscle cells.

This slows or stops the cells ability to generate an action potential and so affects cell function. A minimum dose of 0.2 mg has been estimated to be sufficient to cause symptoms and an LD_{50} in man of 2 mg has been reported.

Initial symptoms appear between 20–180 min of ingestion and are similar to those of PSP. A slight numbness of the lips and tongue is then followed by increasing paraethesia (tingling, pins-and-needles) in the face, hands and feet. Those affected may also suffer dizziness, headaches, nausea and diarrhoea.

These symptoms may then develop into increasing paralysis and respiratory problems. Victims may be completely paralysed and unable to move or speak, yet remain conscious. Death usually occurs within 4–6 h but may be as rapid as 20 min in some cases. Those who have not died within 24 h generally recover completely. Mortality rates of almost 50% have been reported, but this is strongly influenced by the quantity of TTX ingested.

Incidence and Outbreaks

TTX poisoning is most frequently reported in Japan. Between 1987 and 1996, almost 300 cases involving 500 individuals were recorded, with a mortality rate averaging approximately 7%. Most of these cases are thought to be associated with home preparation of fugu. Other Pacific countries, including the United States, have reported sporadic cases. Outbreaks elsewhere are rare, although three people died in Italy in 1977 after consuming wrongly labelled imported frozen pufferfish from Taiwan.

Sources

No algal source of TTX has ever been identified, and it was thought until quite recently that the toxin was produced endogenously by pufferfish as a metabolic by-product. However, there is now considerable evidence suggesting that this is not the case. The toxicity of pufferfish is very variable and when they are grown in culture they do not become toxic unless fed material containing TTX. Furthermore, the discovery that many other unrelated animals also contain TTX suggests an exogenous source.

It is now generally accepted that the source of TTX is production by certain bacteria – notably members of the Vibrionaceae, some *Pseudomonas* spp., *Photobacterium phosphoreum* and *Alteromonas* spp. It is thought that the toxin passes up the food chain through plankton, small gastropods and flatworms and is eventually accumulated in the tissues of pufferfish species, possibly as a defence against predators. Pufferfish appear to be immune to the toxic effects of TTX, but other fish species do not accumulate it, even when fed low-dose toxic material. Some other marine animals, especially the blue-ringed octopus, are reported to accumulate the toxin in special glands and may use it as venom to subdue their prey.

Stability in Foods

TTX is reported to be relatively heat stable and is not affected by normal cooking procedures. Furthermore, it does not appear to be significantly reduced during prolonged frozen storage.

Control Options

The stability and toxicity of TTX means that the only effective control for prevention of poisoning is to avoid consuming those fish species that are known to contain the toxin. In Japan, where pufferfish are traditionally eaten, strict licensing and training of fugu chefs is required to protect the consumer. These individuals are skilled in the removal of toxin-containing tissue from the fish, but the possibility of human error remains.

TTX can be monitored in pufferfish using the same mouse bioassay developed for quantifying PSP toxin and an HPLC method has also been developed. These methods may be useful in cases where pre-prepared frozen tissues from unknown, or wrongly identified species of fish are intended for consumption.

Legislation

Neither the USA, nor the EU normally permit the importation of pufferfish products for human consumption, although exceptions may be granted under special circumstances.

In Japan, there is a strict licensing system covering the marketing and preparation of pufferfish for human consumption.

Sources of Further Information

Published

Hwang, D.F. and Noguchi, T., Tetrodotoxin poisoning. *Advances in Food and Nutrition Research*, 2007, 52, 141–236.

Sobel, J. and Painter, J., Illnesses caused by marine biotoxins. *Clinical Infectious Diseases*, 2005, 41(9), 1290–6.

On the Web

Food-Info.net–tetrodotoxin. http://www.food-info.net/uk/tox/tetrodo.htm

2.1.4 BIOGENIC AMINES

2.1.4.1 Biogenic Amines (Excluding Histamine)

Hazard Identification

What are Biogenic Amines?

Biogenic amines are produced in a variety of foods by the decarboxylation of specific free amino acids. This may occur naturally as a result of the action of endogenous decarboxylase enzymes in the food, or more importantly as a by-product of bacterial growth and the production of exogenous decarboxylases. The presence of significant amounts of biogenic amines, especially in meat and fish products, is often an indicator of bacterial spoilage.

Histamine is the best known and most studied biogenic amine in foods, but this is considered in detail in the section on scombrotoxic poisoning. Other important biogenic amines and their precursor amino acids are as follows.

Biogenic amine	*Precursor*
Tyramine	Tyrosine
Cadaverine	Lysine
Putrescine	Ornithine
Tryptamine	Tryptophan
β-phenylethylamine	Phenylalanine

In terms of chemical structure, Cadaverine and putrescine are alipathic di-amines, tyramine and β-phenylethylamine are aromatic amines and tryptamine is a heterocyclic amine.

In addition to these compounds, certain other biogenic polyamines, such as spermine and spermidine are present at significant levels in some foods, especially fish and vegetables. However, these are thought to be produced by endogenous decarboxylation pathways rather than as a result of microbial decomposition.

The presence of significant quantities of biogenic amines in foods can have adverse effects on health and is generally undesirable.

Occurrence in Foods

Biogenic amines are known to occur in a wide variety of food products, but they are of particular significance in foods that contain a high level of free amino acids and high numbers of decarboxylase-producing bacteria. These include fish products, cheese, meat products (especially fermented meats), wine, beer and fermented vegetable products, such as sauerkraut. Certain biogenic amines are also found naturally in a range of fruit juices and fresh fruit and vegetables, including cocoa beans, mushrooms and lettuce.

Different amines tend to predominate in different foods, depending on the amino acids present, the nature of the bacterial population and the nature of the processing and storage environments. Putrescine and cadaverine levels tend to increase in the tissues of fish after capture, especially under temperature-abuse conditions, and high levels indicate spoilage. In ripened cheese, tyramine, putrescine and cadaverine predominate, while tyramine is found in higher concentrations (up to 150 mg/100 g) than other amines in fermented meats.

Hazard Characterisation

Effects on Health

Although the role of histamine in scombrotoxic poisoning is well established, the food safety significance of other biogenic amines is much more uncertain.

In acute toxicity testing using rats, most biogenic amines are found to have quite low oral toxicity. Tyramine, cadaverine and putrescine all have acute oral toxicities of at least 2000 mg/kg bodyweight. Spermine and spermidine were reported to be slightly more toxic, with acute oral toxicities of 600 mg/kg bodyweight. When administered intravenously, all these amines, except tyramine, caused a drop in blood pressure. However, the levels of most biogenic amines that are toxic in humans have not been reliably determined and a wide range of figures has been suggested. Furthermore, there is evidence that individuals vary considerably in their sensitivity.

Tyramine has been associated with hypertension and headaches in sensitive individuals, especially those who suffer from migraine headaches. Tyramine also interacts with a class of drugs called monoamine oxidase inhibitors (MAOI). These drugs are antidepressants and, although largely superseded by more modern drugs, they are still prescribed for a minority of patients. MAOI inhibit monoamine oxidase enzymes in the gut that would normally inactivate tyramine in foods. This allows more tyramine to enter the circulatory system and increases the risk of dangerous rises in blood pressure. Patients taking MAOI are advised to avoid tyramine-rich foods, such as cheese.

There is evidence that some biogenic amines, particularly putrescine and cadaverine, may be indirectly involved in histamine poisoning. There have been reports of cases of scombrotoxic poisoning being caused by fish containing unusually low levels of histamine, but with high amounts of other amines. It is thought that other amines may increase histamine uptake by inhibiting intestinal enzymes, such as diamine oxidase, that would normally metabolise histamine. It has been suggested that cadaverine and putrescine may also facilitate histamine transport through the wall of the intestine, but the mechanism involved is unknown.

In foods containing nitrite, such as cured-meat products, putrescine and cadaverine may react with nitrate and produce carcinogenic compounds.

Incidence and Outbreaks

The uncertainty surrounding the public-health significance of dietary exposure to biogenic amines, other than histamine, means that there is virtually no published information on the incidence of toxic events or outbreaks.

Sources

Although biogenic enzymes such as spermine and spermidine are produced endogenously in foods as a result of cellular metabolism, it is the exogenous decarboxylation of free amino acids by bacteria that is of most significance for food safety.

Bacterial sources of biogenic amines vary with the food commodity concerned and with the environmental conditions of processing and storage. For example, putrescine and cadaverine are produced in fish tissue by a wide range of bacterial species, many of which are also involved in histamine production. Post-harvest contaminants, such as members of the Enterobacteriaceae, are particularly active amine producers, especially when temperature control is poor. Species such as *Proteus* spp., *Klebsiella* spp., *Morganella morganii* and *Hafnia alvei* are all capable of producing high levels of biogenic amines in fish. *Pseudomonas* spp. too have been reported to generate high levels of putrescine and cadaverine in fish stored at temperatures between 0 °C and 15 °C.

Tyramine in cheese is produced mainly by non-starter bacteria during the ripening process. Various *Lactobacillus* spp., enterococci and propionibacteria have been reported to produce biogenic amines during cheese ripening. In fermented meats, lactobacilli have been found to produce tyramine, while members of the Enterobacteriaceae produced cadaverine and *Pseudomonas* spp. produced putrescine. Again, non-starter contaminating bacteria are thought to be mainly responsible.

In wines, lactobacilli that perform the malolactic fermentation have also been found to produce tyramine and putrescine and contaminating lactobacilli are also thought to produce biogenic amines in some beers.

Stability in foods

Like histamine, other biogenic amines are relatively heat stable and are not destroyed by cooking or even during canning processes. However, unlike histamine, cadaverine and putrescine in particular are detectable by their unpleasant and pungent odours at high levels, especially in fish and meat.

Bacterial decarboxylase enzymes are heat labile and are destroyed by cooking, so that further biogenic amine production does not occur unless foods are recontaminated.

Control Options

In non-fermented foods, biogenic amines are produced mainly by contaminating spoilage bacteria. Therefore the many controls routinely applied to prevent microbial spoilage and extend shelf life are also helpful in preventing their production at high levels. Good hygienic practice and effective refrigeration and temperature control are especially important in minimising the contamination of foods by spoilage bacteria and in inhibiting their growth.

For fermented foods where a starter culture is used, it is recommended that a starter culture strain that has been shown not to produce biogenic amines is chosen. The initial microbiological quality of the raw materials also has a significant influence on amine production during manufacture and storage. Good hygiene is important in preventing contamination by species of non-starter species that may produce large amounts of amines, particularly for products with lengthy ripening periods. A heat treatment in processing helps to reduce non-starter bacterial populations and raw-milk cheeses and fermented-meat products made without pasteurisation are more likely to develop high amine concentrations during ripening.

Legislation

Most of the legislation relating to biogenic amines applies specifically to histamine and is dealt with separately. However, the European Commission has suggested a maximum legal limit for total biogenic amines of 30 mg/100 g in fish and fish products for future consideration.

Sources of Further Information

Published

Flick, G.J. and Ankenman Granata, L., Biogenic amines in foods. In *Toxins in Food*, Ed W. Dabrowski. Cambridge, Woodhead Publishing, 2004, 121–53.

Stratton, J.E., Hutkins, R.W. and Taylor, S.L., Biogenic amines in cheese and other fermented foods: a review *Journal of Food Protection*, 1991, 54(6), 460–470.

2.1.4.2 Scombrotoxin (Histamine)

Hazard Identification

What is Scombrotoxin?

Scombrotoxin is a foodborne toxin most often associated with the consumption of fish, particularly species belonging to the *Scombridae* and *Scomberesocidae* families (scombroid fish), such as mackerel and tuna. It can cause a mild, though sometimes distressing, form of foodborne intoxication (scombroid or scombrotoxic food poisoning) when ingested in sufficient quantities.

Scombrotoxic poisoning is also known as histamine poisoning, since histamine is considered to be the toxic component of scombrotoxin, although other compounds may be involved. Histamine ($C_5H_9N_3$) is a biogenic amine and can be produced during processing and/or storage in fish and certain other foods, usually by the action of spoilage bacteria.

Occurrence in Foods

Scombrotoxin is most often associated with scombroid fish, especially tuna, skipjack, bonito and mackerel, but other non-scombroid fish, such as sardines, herring, pilchards, marlin and mahi-mahi have also be involved in outbreaks of illness. There are also reports that scombrotoxin could occur in salmon species. Generally, fast swimming and migratory finfish species with red-coloured meat are more likely to develop high histamine levels that whitefish species.

The toxin is not limited to fresh and frozen fish. It may be present in canned and cured fish products at high enough concentrations to cause illness.

The concentration of histamine can vary considerably between different sampling sites in a single fish, or between individual cans in a single lot. Levels of $\geq$3000 ppm have been recorded in fish products implicated in outbreaks.

Histamine can also be produced at levels toxic to humans by bacterial action in other foods, notably Swiss cheese.

Hazard Characterisation

Effects on Health

Scombrotoxic (histamine) poisoning is a chemical intoxication, in which symptoms typically develop rapidly (from 10 min to 2 h) after ingestion of food containing toxic histamine levels.

The range of symptoms experienced is quite wide, but may include an oral burning or tingling sensation, skin rash and localised inflammation, hypotension, headaches and flushing. In some cases vomiting and diarrhoea may develop and elderly or sick individuals may require hospital treatment. The symptoms usually resolve themselves within 24 h.

The evidence for histamine as the active toxin in scombrotoxic poisoning is strong, but the condition is very difficult to replicate in humans using pure histamine. It is thought possible that other biogenic amines in spoiled fish, such as putrescine and cadaverine, may act as potentiators for histamine toxicity, but the mechanism for this is not known.

For this reason the threshold toxic level for histamine remains uncertain. Individuals also vary in the severity of their response to histamine in fish. Analysis of outbreaks suggests that levels of histamine above 200 ppm are potentially toxic. Although histamine occurs naturally in the human body, exposure to large doses can rapidly produce the symptoms of toxicity.

Incidence and Outbreaks

The symptoms of histamine poisoning resemble an allergic reaction and there is potential for misdiagnosis. Furthermore, since symptoms are usually mild, it is likely that the illness is considerably under-reported. Nevertheless, it is thought that histamine poisoning is probably the commonest form of fish-related toxicity.

The highest numbers of cases are reported in the USA, Japan and the UK, but this may be a reflection of reporting systems rather than incidence. Between 1992 and 2004 England and Wales reported 56 outbreaks affecting 296 people. Outbreaks were more common in summer than in winter. In the USA, between 1968 and 1980, 103 outbreaks involving 827 people were reported and in Japan over the same period 42 outbreaks affecting 4122 people.

Large outbreaks also occur. In 1973, at least 200 US consumers became ill after eating domestic canned tuna.

In the first six months of 2005 an unusual increase in incidence was reported in England and Wales, with 16 outbreaks affecting 38 people. This was thought to be associated with poor temperature control and hygiene in certain catering premises.

Sources

Histamine in fish and other foods is produced by the decarboxylation of the amino acid histidine and fish species that have high levels of free histidine in their tissues are most likely to develop toxic histamine levels. This is usually the result of the action of the enzyme histidine decarboxylase, which is found in a number of bacterial species that may occur on fish.

Species such as *Vibrio* spp. *Pseudomonas* spp. and *Photobacterium* spp. are found in the marine environment and occur naturally on fish. Others, especially the Enterobacteriaceae, are contaminants that are introduced post-harvest. It is this second group that is considered most important in the development of histamine. Species such as *Morganella morganii, Klebsiella pneumoniae* and *Hafnia alvei* are able to produce high levels of histamine very rapidly at mesophilic temperatures (20–30 °C). For this reason, histamine is more often produced during spoilage in this temperature range, although high levels can also develop at lower temperatures over time.

In tropical waters the indigenous microflora may be more important histamine-producing organisms, particularly when fishing methods such as longlining are used, where the fish may die before landing. Under these conditions, it is possible for histamine to be formed before the fish is landed and chilled.

There is evidence that histidine decarboxylase remains active at chill temperatures, even though the bacteria themselves are not active. Therefore once the enzyme has been formed at higher temperatures, it may continue to produce histamine even when the fish is properly chilled.

It is also possible for histamine to form after cooking or canning if the fish subsequently becomes contaminated with histidine decarboxylase producing bacteria. This can happen when canned fish is handled under conditions of poor hygiene.

Stability in Foods

Histamine is extremely stable once formed and is not affected by cooking. It can survive canning and retorting processes and is not reduced during freezing or frozen storage. Furthermore, high histamine levels may not be accompanied by other signs of spoilage and may be undetectable other than by chemical analysis.

The enzyme histidine decarboxylase is inactivated by cooking and further histamine will not then be produced unless recontamination occurs.

Control Options

Temperature Control

Chilling

The key measure for the control of histamine production in fish is rapid chilling as soon as possible after death, particularly where the fish has been exposed to warm water. This will inhibit the formation of bacterial histidine decarboxylase. Once the enzyme is present control options are very limited.

Accepted guidelines (FAO/FDA) recommend that fish should be placed in ice, or chilled seawater or brine at <4.4 °C within 12 h of death, or placed in chilled seawater or brine at <10 °C within 9 h of death. If the fish have been exposed to air or water temperatures above 28.3 °C they should be chilled to <4.4 °C within 6 h and very large fish such as tuna that are eviscerated before chilling also should have the body cavity packed with ice.

Further chilling to a temperature as close to the freezing point as possible is desirable to prevent less rapid formation of histidine decarboxylase at lower temperatures. Even rapid chilling to <4.4 °C may only give a safe shelf life of 5–7 days.

Once frozen, the fish can be stored safely for extended periods and further histidine decarboxylase will not be formed. However, enzyme produced before freezing will not be destroyed and will continue to produce histamine after thawing.

Cooking

Cooking will destroy both histamine-producing bacteria and bacterial decarboxylases, but not histamine itself. Cooked fish therefore can be stored safely for longer periods and canned fish can be kept almost indefinitely.

It is important to note that once cooked or canned fish becomes recontaminated with histamine-producing bacteria, temperature control again becomes critical to prevent a hazard. For example, canned tuna that is not consumed immediately after opening should be stored at $<5\,^{\circ}\mathrm{C}$ as soon as possible.

Good Hygienic Practice

Good hygienic practice on board fishing vessels, especially during landing and processing, is important to minimise contamination with non-indigenous histamine-producing bacterial species.

Careful handling of fish to avoid damage to muscle tissue is also important in preventing contamination. For example, puncture wounds in fish can introduce contaminating bacteria into deep tissue where large concentrations of histidine are available. Histamine production may then happen much more quickly.

Good hygiene at processing and preparation stages further along the supply chain, such as cutting and packing or in catering operations, is also important to prevent contamination of fresh fish, or recontamination of frozen and cooked fish.

Chemical Testing

Histamine is only detectable by chemical analysis and affected fish may appear otherwise satisfactory. Chemical testing can provide some assurance that toxic levels of histamine are not present, but the variability in histamine levels in a single fish mean that very large numbers of samples must be taken. For this reason, chemical testing cannot be relied upon to demonstrate adequate control of the hazard, but can be useful as a HACCP verification tool.

Legislation

European legislation states that fish species belonging to families known to contain large amounts of histidine (*e.g.* Scombridae, Clupeidae, *etc.*) in their tissues should be tested for the presence of histamine. Nine samples should be tested from each lot and the mean value should be ≤ 100 ppm. The lot is considered unsatisfactory if more than two samples give results of between 100 and 200 ppm, or if any sample gives a result of ≥ 200 ppm. A mean level of 200 ppm and a maximum limit of 400 ppm are permitted for fish that have undergone enzyme maturation in brine.

In the USA the Food and Drug Administration has issued guidelines for tuna and related fish establishing a "defect action level" of 50 ppm in any sample. This is said to be indicative of spoilage and may mean that toxic levels are present in other samples. A separate toxicity level of 500 ppm is also given.

The international Codex standard for fish also includes histamine levels as indicators of decomposition and hygiene and handling. A maximum average level of not more than 100 ppm is considered satisfactory in relation to decomposition, while an upper limit of 200 ppm in any one sample is applied for hygiene and handling.

Australia and New Zealand also apply a maximum limit of 200 ppm for histamine in fish or fish products.

Sources of Further Information

Published

McLauchlin, J., *et al.* Scombrotoxic fish poisoning. *Journal of Public Health*, 2006, 28, 61–2.

Bremer, P.J., Fletcher, G.C. and Osborne, C., Scombrotoxin in Seafood Report for the New Zealand Institute for Crop & Food Research, 2003.

On the Web

Seafood Network Information Center-scombrotoxin formation. http://seafood.ucdavis.edu/haccp/compendium/Chapt27.htm

NZFSA datasheet. http://www.nzfsa.govt.nz/science/data-sheets/scombroid-poisoning.pdf

CHAPTER 2.2

Non-Biological Contaminants

2.2.1 CONTAMINANTS PRODUCED DURING PROCESSING

2.2.1.1 Acrylamide

Hazard Identification

What is Acrylamide?

Acrylamide ($CH_2{=}CH{-}CONH_2$) is a synthetic vinyl compound produced by the chemical industry mainly as a building block for various polymers, particularly polyacrylamide. Polyacrylamide is widely used in various applications, such as in the treatment of wastewater, in textile and paper processing and in mining and mineral production. Acrylamide is also present in cigarette smoke.

The wide use of polyacrylamide in industry means that human exposure to acrylamide is likely and a number of toxicological studies have been carried out. The results of these studies suggest that acrylamide may have adverse effects on human health under some circumstances.

Occurrence in Foods

The possibility of acrylamide contamination of foods did not become widely known until April 2002, when a report from the Swedish National Food Administration was published. This report revealed that acrylamide could be produced in significant concentrations in certain carbohydrate-rich foods processed at relatively high temperatures, such as fried potato and baked cereal products. The work on which this report was based was done following an earlier study into the adverse health effects of polyacrylamide used by

The Food Safety Hazard Guidebook
By Richard Lawley, Laurie Curtis & Judy Davis

construction workers in the building of a tunnel. The discovery that control subjects showed unexplained evidence of exposure to acrylamide gave rise to the idea that food could be a source of the chemical.

Since 2002 a very wide range of foods around the world have been surveyed for the presence of acrylamide and the contaminant has been found to occur widely in many different food categories. Fried potato products, such as French fries and crisps, and baked cereal products, such as biscuits, bread, toasted breakfast cereals and pastries are the main foods affected, but roasted and ground coffee has also been found to be an important source. Animal-based foods and plant foods that are eaten raw, or cooked at lower temperatures, tend not to contain significant levels of acrylamide.

Acrylamide is not confined to commercially processed foods. It can also be found in home-baked or fried foods at relatively high levels. It seems certain that acrylamide has been present and gone undetected in cooked foods for centuries. It has been found in such diverse products as olives, prune juice and chocolate confectionery and many countries have published survey data covering a wide range of foods. Much of this data is available at the Acrylamide Infonet web site, maintained on behalf of the World Health Organisation and the Food and Agriculture Organisation (link provided below).

The amount of acrylamide found in foods varies widely, both with the food category and with the process applied. Some approximate examples of recorded levels in different food groups are given below.

Approximate observed ranges of acrylamide concentration by food group[a]

Food Group	*Acrylamide (ppb)*
Breakfast cereals	20–250
Bread	10–130
Roast and ground coffee	100–400
Crackers	50–600
Potato crisps and snacks	100–2500
Chocolate products	10–100

[a]Source: Acrylamide Infonet Analytical Database.

Hazard Characterisation

Effects on Health

Acrylamide is a neurotoxin at high levels of exposure and may cause a range of symptoms such as numbness in the hands and feet. It has also been shown to be genotoxic in animal studies. However, it is considered unlikely that the levels found in foods could result in sufficient exposure to cause neurological damage or reproductive toxicity.

Of more concern to the food industry is the finding that acrylamide is also carcinogenic in animal studies. The International Agency on Research on Cancer (IARC) classifies it as "probably carcinogenic to humans (IARC

Group 2A)." Epidemiological studies on humans have so far found no evidence of a link between acrylamide in the diet and the development of certain common cancers, but these findings are not considered conclusive.

The Joint FAO/WHO Expert Committee on Food Additives (JECFA) reviewed all the available toxicity and likely intake data for acrylamide in 2005 and carried out a risk assessment for the effect on human health. They found that the average person ingests enough acrylamide in the diet each day to equate to 1/300th of the dose required to cause a 10% increase in the risk of breast cancer in rats, with high consumers ingesting as much as 1/75th of that dose. The Committee considered this to be a low safety margin in comparison with other carcinogens in the diet. They concluded that, although there was considerable uncertainty in estimating the risk to human health, exposure to acrylamide in the diet might indeed be a concern. A number of long-term carcinogenicity and toxicological studies are currently in progress and these should help to reduce the level of uncertainty.

The JECFA review acknowledged that acrylamide is an inadvertent contaminant introduced during cooking and unlikely ever to be eliminated from foods. Nevertheless, the Committee recommended that the food industry should work towards lowering acrylamide levels in critical food groups, such as potato crisps and chips, coffee, bakery products and biscuits and that guidance should be developed to help consumers reduce the levels produced in home-cooked foods.

Sources

The original Swedish report into acrylamide in food in 2002 indicated that the contaminant is produced as a result of heating certain foods, especially those containing high levels of carbohydrate, at temperatures above 120 °C. It is therefore a contaminant generated during processing. Since then considerable research has been carried out into the mechanism by which acrylamide is generated during frying, baking or roasting.

The major mechanism for the formation of acrylamide during cooking is now acknowledged to be the reaction of the free amino acid asparagine with reducing sugars, such as glucose or fructose, during the Maillard browning reactions that occur during cooking at high temperatures. Other mechanisms have since been suggested, including formation via acrolein, produced during the degradation of lipids from frying oil. However, most attention has been focused on Maillard browning as the main source of acrylamide. The key factors that affect the quantity of acrylamide produced appear to be the amount of free asparagine and sugars present in the food and the cooking time and temperature.

Stability in Foods

The large amount of data collected from food surveys suggests that acrylamide is relatively stable in food, but this has not been widely studied to date. Nevertheless, acrylamide levels have been found not to decrease significantly in crisps or baked cereal products during shelf life, while levels in roast and ground coffee do decrease significantly.

Control Options

A considerable amount of research has been initiated since 2002 to investigate possible strategies for minimising the formation of acrylamide during the cooking of food products. Much of this work has been published and many of the most useful and practical techniques have been brought together by the Confederation of the Food and Drink Industries of the EU (CIAA) in an "Acrylamide Toolbox" available on the Internet (link provided below).

Product Formulation

One obvious strategy for the control of acrylamide formation is to minimise the amount of free asparagine and reducing sugars in food prior to cooking. The development of low-asparagine varieties of potato is one approach that is receiving attention.

The modification of product recipes also shows some promise. For example, replacing ammonium bicarbonate with other raising agents in baked products can reduce acrylamide formation significantly, as can a reduction in pH. However, care must be taken to ensure that unacceptable textural and flavour changes do not result from such modifications.

Processing

The main factors that can be modified to minimise acrylamide formation are cooking time and temperature. The "thermal input" to a cooking process has been shown to be directly linked to the amount of acrylamide produced. As a general rule, higher thermal input results in higher levels, with the exception of coffee production, where acrylamide levels decrease with longer roasting times and "darker" roasts.

Frying, baking and roasting at lower temperatures and for shorter times reduce the amount of browning of the product and also reduce the amount of acrylamide produced. For example, consumers have been advised to cook French fries only until golden, rather than brown, and some crisp manufacturers have altered frying times and temperatures to reduce acrylamide production. While this may be successful, it must be recognised that the browning of baked and fried foods is an essential component in their sensory acceptability. Also, frying at lower temperatures may allow foods to take up higher levels of fat, which may be undesirable from a nutritional point of view. Reducing acrylamide by changing processing times and temperatures results in a compromise between product quality and safety.

The industry has already made significant progress in reducing acrylamide in processed foods and it is likely that improved strategies and techniques will be developed in the near future.

Legislation

Acrylamide is not yet covered specifically by legislation in Europe or North America and no permitted limits have been set. At present, national and international food safety and public health authorities request that the food industry continue to work to minimise the levels of acrylamide in critical food groups.

Sources of Further Information

Published

Friedman, M. and Mottram, D., American Chemical Society Chemistry and safety of acrylamide in food: proceedings of a symposium, Anaheim 2004. New York, Springer Science, 2005.

Taeymans, D. *et al.* A review of acrylamide: an industry perspective on research, analysis, formation, and control. *Critical Reviews in Food Science and Nutrition*, 2004, 44 (5), 323–347.

Friedman, M. Chemistry, biochemistry and safety of acrylamide. A review. *Journal of Agricultural and Food Chemistry*, 2003, 51 (16), 4504–26.

On the Web

Acrylamide Infonet. http://www.acrylamide-food.org/index.htm

JECFA monograph on acrylamide (2006). http://whqlibdoc.who.int/publications/2006/9241660554_ACR_eng.pdf

CIAA "Acrylamide Toolbox." http://www.ciaa.be/documents/brochures/CIAA_Acrylamide_Toolbox_Oct2006.pdf

European Commission acrylamide pages. http://ec.europa.eu/comm/food/food/chemicalsafety/contaminants/acrylamide_en.htm

Swedish National Food Administration. http://www.slv.se/templates/SLV_DocumentList.aspx?id=4089

US Food and Drug Administration acrylamide pages. http://www.cfsan.fda.gov/~lrd/pestadd.html#acrylamide

2.2.1.2 Benzene

Hazard Identification

What is Benzene?

Benzene (C_6H_6, CAS No. 71-43-2) is an aromatic hydrocarbon compound used extensively in the chemical industry as an intermediate in the manufacture of polymers and other products. It is also a common atmospheric contaminant and is present in motor vehicle exhaust emissions and cigarette smoke.

In 1990, it was discovered by the US soft drinks industry that benzene could be produced at low levels in certain soft drinks containing a benzoate preservative and ascorbic acid. Since benzene is a known human carcinogen, its presence in food and beverages is clearly undesirable.

Occurrence in Foods

Detectable levels of benzene have been found in a number of soft drinks that contain either a sodium or potassium benzoate preservative and ascorbic acid, and "diet"-type products containing no added sugar are reported to be particularly likely to contain benzene at detectable levels. Recent surveys carried out in the USA, the UK and Canada have all confirmed that a small proportion of these products may contain low levels of benzene. For example, in a survey of 86 samples analysed by the US Food and Drug Administration (FDA) between April 2006 and March 2007, only five products were found to contain benzene at concentrations above 5 μg/kg. The levels found were in a range from approximately 10–90 μg/kg. A survey of 150 UK-produced soft drinks by the Food Standards Agency published in 2006 showed that four products contained benzene at levels above 10 μg/kg, and the highest level recorded was 28 μg/kg. However, it has been reported that higher levels may develop in these products during prolonged storage, especially if they are exposed to daylight.

Benzene may also be formed in some mango and cranberry drinks in the absence of added preservatives, because these fruits contain natural benzoates.

Hazard Characterisation

Effects on Health

Although benzene can cause acute toxicity, especially when inhaled at high levels, it is its carcinogenicity that is of concern in foods and beverages. Benzene is a proven carcinogen and has been shown to cause cancers in industrial workers exposed to high airborne levels. Much less is known about its effects when ingested at low levels over long periods, but current risk assessments suggest that the contribution of soft drinks to benzene exposure levels is negligible, as is any additional risk to human health. Nevertheless, the soft drinks industry has been requested to take action to eliminate benzene from its products and product recalls have been initiated in the UK following the discovery of benzene contamination.

Sources

It has been established that the source of benzene in soft drinks is the decarboxylation of benzoic acid when ascorbic acid and trace amounts of a suitable metal catalyst (copper or iron) are present. Elevated temperature and light are both reported to stimulate this reaction, whereas it is inhibited by sugars and by EDTA salts. This may be why benzene is most likely to be found in diet drinks containing low sugar levels. Benzene levels may continue to rise during storage if the product is kept in the light and the storage temperature is high.

Stability in Foods

There is little information available on the stability of benzene in soft drinks during storage.

Control Options

The preferred approach for controlling the production of benzene in soft drinks is to reformulate the product. Once a specific soft-drink formulation has been shown to be capable of generating benzene during storage, alternatives to benzoate preservatives, such as potassium sorbate, should be evaluated. Benzene generation may be effectively prevented by the removal of benzoates from the product. However, it should be noted that the majority of soft drinks containing benzoates and ascorbic acid have not been shown to produce benzene and may not need to be reformulated in this way.

Legislation

Current US and European legislation does not set maximum limits for benzene in soft drinks. However, the FDA has adopted the Environmental Protection Agency (EPA) maximum contaminant level (MCL) for drinking water of 5 parts per billion (ppb) as a quality standard for bottled water. This MCL has been used to evaluate the significance of benzene contamination in the soft drinks tested in recent surveys.

The UK Food Standards Agency has used the World Health Organization (WHO) guideline level for benzene in water of 10 μg/kg as a point of reference for its own survey results.

Sources of Further Information

Published

Gardner, L.K. and Lawrence, G.D. Benzene production from decarboxylation of benzoic acid in the presence of ascorbic acid and a transition metal catalyst *Journal of Agricultural and Food Chemistry*, 1993, 41(5), 693–5.

On the Web

FDA benzene documents and data. http://www.cfsan.fda.gov/~lrd/pestadd.html#benzene

UK Food Standards Agency survey of benzene in soft drinks. http://www.food.gov.uk/news/newsarchive/2006/mar/benzenesurvey

2.2.1.3 Chloropropanols

Hazard Identification

What are Chloropropanols?

The chloropropanols are a group of related chemical contaminants that may be produced in certain foods during processing. They first became a concern to the food industry in the late 1970s when small concentrations were found to be generated during the manufacture of acid-hydrolysed vegetable protein (acid-HVP) used as a savoury ingredient in soups, sauces, especially soy sauce, snacks, stock cubes and ready-meals. Chloropropanols are potentially carcinogenic and their presence in food, even at low levels is therefore undesirable.

Several different chloropropanols have been identified in food. The most common and the best studied is 3-monochloropropane-1,2-diol (3-MCPD), but other foodborne chloropropanols include 2-monochloro-1,3-propandiol (2-MCPD), 1,3-dichloro-2-propanol (1,3-DCP) and 2,3-dichloro-2-propanol (2,3-DCP). Chloropropanols are probably produced by a number of different mechanisms during food processing, but these are not yet fully understood.

Occurrence in Foods

The highest levels of chloropropanols (mainly 3-MCPD and 1,3-DCP) have been found in acid-HVP and in soy sauce and related products. A UK survey of 3-MCPD in acid-HVP in 1980 showed levels of up to 100 mg/kg and surveys of soy sauce products in Europe and North America in 1999–2000 showed levels varying from undetectable <0.01 mg/kg) to a highest concentration of 330 mg/kg in a sample tested in Canada. High levels were shown to be produced during the manufacturing process of acid-HVP, which is a major ingredient of soy sauce. Changes in acid-HVP manufacturing methods have produced a dramatic reduction in levels of 3-MCPD in products on the market in the UK since 1990, and typical levels in 1998 were in the range 0.01–0.02 mg/kg.

Since chloropropanols were first identified in acid-HVP and soy sauce, they have also been found in a variety of other food products that do not contain acid-HVP as an ingredient. For example, 3-MCPD has been found in bread, biscuits and other baked products, coffee, roasted barley malt, certain cured and fermented-meat products, cheeses, salted fish and smoked foods. Levels of 3-MCPD are generally low in these foods. For example, a concentration of 0.5 mg/kg is not unusual in malt used as a food ingredient, and maximum concentrations of 3-MCPD found in surveys of bakery products, meat, fish and cheese range from 0.01–0.1 mg/kg. It is thought that the contaminant is usually produced during the manufacturing process, especially at high temperatures, but the mechanism is not known in all cases.

Foodborne chloropropanols may also be derived from migration from food-contact materials, such as sausage casings and teabags, and they can also be produced during domestic cooking of such foods as grilled cheese and meats.

Hazard Characterisation

Effects on Health

Although chloropropanols can cause acute toxicity at high concentrations, it is extremely unlikely that this could occur through consumption of contaminated food, and it is the effect of low doses over a long time that is of most concern from a food safety point of view. Both 3-MCPD and 1,3-DCP have been shown to be carcinogenic in animal studies and are therefore potential human carcinogens.

3-MCPD was formerly considered to be genotoxic, but recent studies suggest that there is little solid evidence for this. The Joint FAO/WHO Expert Committee on Food Additives (JECFA) has recently reviewed the toxicity of 3-MCPD and concluded that a threshold-based approach for deriving a tolerable daily intake could be used. A provisional maximum tolerable daily intake (PMTDI) of 2 μg/kg bodyweight has thus been set to replace the previous recommendation that levels in foods should be reduced as far as technically possible. For 1,3-DCP, JECFA was unable to rule out the possibility of genotoxicity and so no PMTDI has been set.

Sources

The mechanism for chloropropanol production in acid-HVP is known to be a reaction between hydrochloric acid and lipids that occurs more rapidly at the high temperatures used in processing. 3-MCPD and other chloropropanols then contaminate other foods, for which acid-HVP is a key flavour ingredient.

In bread and other baked products, chloropropanols are thought to be formed by a reaction during the baking process between the chloride in added salt and glycerol from flour and yeast. In other foods, the mechanisms of chloropropanol production are unclear. One proposed mechanism for 3-MCPD production in meat, fish and cheese at relatively low temperatures suggests that hydrolytic enzymes (lipases) may be involved, but this has yet to be confirmed.

Stability in Foods

Chloropropanols are relatively non-volatile and may be quite persistent in foods once formed. However, degradation does occur during storage, and 3-MCPD has been shown to be lost more rapidly from foods at higher pH values and at higher temperatures.

Control Options

The control of chloropropanols in foods focuses on limiting their production during processing.

Processing

The production of chloropropanols during the manufacture of acid-HVP is well understood and control strategies have been successful in reducing the level of contamination significantly. This has been achieved by a number of changes to the manufacturing process.

- replacing acid hydrolysis with an enzymatic process;
- reducing lipid concentrations in the raw materials;
- effective control of the acid hydrolysis process;
- use of an over-neutralisation treatment with NaOH to remove chlorohydrins after acid hydrolysis.

The mechanism of formation of chloropropanols in other foods is less well known and it is therefore more difficult to design effective control strategies. However, in many cases common salt is a source of chloride ions and a precursor for chloropropanol production. Therefore, reducing salt levels without compromising sensory properties or microbiological stability may be an effective control, especially in bread and other bakery products. Reducing processing temperatures and avoiding excessive browning of these products may also be useful controls.

For meat, fish and cheese, there is little information on how chloropropanols are formed at lower temperatures. However, salt concentration is again likely to be a factor and the inactivation of lipases may also be helpful. Fortunately, levels of 3-MCPD and other contaminants are usually very low in these foods.

Legislation

In the EU, permitted levels of 3-MCPD in hydrolysed vegetable protein and soy sauce are prescribed by a European Commission Regulation (EC) 1881/2006, which sets a maximum level of 20 μg/kg. This is based on the PMTDI for 3-MCPD of 2 μg/kg bodyweight. For other chloropropanols, manufacturers are requested to reduce levels as far as is technically possible.

Sources of Further Information

Published

Studer, A., Blank, I. and Stadler, R.H. Thermal processing contaminants in foodstuffs and potential control strategies *Czech Journal of Food Science*, 2004, 22 (special issue), 1–10.

Hamlet, C.G., *et al.* Occurrence of 3-chloro-propane-1,2-diol (3-MCPD) and related compounds in foods: a review *Food Additives and Contaminants*, 2002, 19(7), 619–31.

On the Web

JECFA monograph on 3-MCPD (2001). http://www.inchem.org/documents/jecfa/jecmono/v48je18.htmSCOOP report on chloropropanols (2004) http://ec.europa.eu/food/food/chemicalsafety/contaminants/scoop_3-2-9_final_report_chloropropanols_en.pdf

IFST Information Statement – 3-MCPD in Foods. http://www.ifst.org/uploadedfiles/cms/store/ATTACHMENTS/3mcpd.pdf

2.2.1.4 Furan

Hazard Identification

What is Furan?

Furan (C_4H_4O, CAS No. 110-00-9) is a volatile heterocyclic organic chemical often found as an intermediate in industrial processes for producing synthetic polymer materials. It is a very different compound from the diverse group of chemicals sometimes referred to collectively as "furans", which includes various antimicrobials (nitrofurans) and dioxin-like toxins.

Concern over furan in foods dates back only to 2004, when a Food and Drug Administration (FDA) survey of heat-processed foods in the USA revealed that low levels of furan could be found in an unexpectedly large proportion of products processed in closed containers, such as cans and jars. Furan is a possible human carcinogen, and therefore, even low levels in foods are undesirable.

Occurrence in Foods

Furan has been recognised as a food flavour volatile for a considerable time, and quite high levels (up to 4000 μg/kg) were reported in canned meat as long ago as 1979, it was not known to occur widely in heat-processed foods until the 2004 FDA survey. This found furan at concentrations of up to 125 μg/kg in a variety of heat-processed foods, including baby foods, canned beans, soups, sauces and pasta meals. Since then, a wide range of foods has been surveyed in the USA and in Europe, notably by the FDA, the Swiss Federal Office of Public Health and by certain food manufacturers. Detectable levels of furan have now been found in savoury snacks, coffee, canned fruits and juices, preserves, canned vegetables, ready-to-use gravies and breakfast cereals.

Most of the positive samples recorded levels of furan of less than 100 μg/kg, but the Swiss survey found much higher concentrations (up to 5900 μg/kg) in some ground, roasted coffee samples. Both the FDA and the European Food Safety Authority have appealed for the submission of more data on furan levels in foods so that a valid risk assessment can be carried out.

Hazard Characterisation

Effects on Health

Furan is cytotoxic and the liver is the target organ for acute toxic effects. However, it is the effect of prolonged dietary exposure to furan and its possible carcinogenic potential that is of concern for food safety. Furan has been shown to be carcinogenic in rats and mice and is probably genotoxic. For this reason, it has been classified by the International Agency for Research on Cancer (IARC) as "possibly carcinogenic to humans." The EFSA Scientific Panel on Contaminants in the Food Chain concluded in 2004 that the difference between human exposure to furan and doses causing carcinogenic effects in animals was "relatively small." However, this conclusion was based on limited data, and the extent of the health risk presented by furan in food will not be properly established until more toxicity and exposure data are available for evaluation.

Sources

It is thought probable that furan is a by-product of the high temperatures involved in the heat processing of foods, but the means by which it is produced is not known. In view of the wide variety of heat-processed foods that may contain furan, it is considered likely that a number of different mechanisms are involved. Proposed sources of furan formation include the thermal degradation of reducing sugars alone, or in combination with amino acids, thermal degradation of some amino acids, and thermal oxidation of ascorbic acid, poly unsaturated fatty acids and carotenoids. The presence of furan residues in canned foods, and products in sealed jars and other containers, is probably a consequence of the volatile compound being trapped in the container.

Stability in Foods

There is little data as yet on the stability of furan in food, although it is a highly volatile compound and is likely to be driven off quite quickly if foods are cooked, or reheated, in open vessels.

Control Options

Too little is currently known about the formation, occurrence and potential risk of furan in foods for any valid control options to have been developed.

The FDA has published an Action Plan for furan in food. The goals of this plan are to develop reliable analytical methods, gather more data on dietary exposure to furan, learn more about the human toxicology of furan and produce sufficient data to undertake a full risk assessment. In the EU, EFSA has begun a similar programme of data collection and risk assessment.

Legislation

As yet there is no legislation limiting levels of furan in foods. Any future regulation will be based on the results of ongoing risk-analysis activities.

Sources of Further Information

Published

Yaylayan, V.A. Precursors, formation and determination of furan in food. *Journal für Verbraucherschutz und Lebensmittelsicherheit*, 2006 1(1), 5–9.

On the Web

FDA Furan documents. http://www.cfsan.fda.gov/~lrd/pestadd.html#furan
EFSA CONTAM Panel report on furan in food. http://www.efsa.europa.eu/EFSA/Scientific_Document/contam_furan_report7-11-051.pdf

2.2.1.5 Polycyclic Aromatic Hydrocarbons

Hazard Identification

What are Polycyclic Aromatic Hydrocarbons?

The polycyclic aromatic hydrocarbons (PAH) are a large group of stable, lipophilic organic chemical contaminants containing two or more fused aromatic rings. They can be produced during the partial combustion or pyrolysis of organic material and are common by-products of a number of industrial processes, including the processing and preparation of food. The presence of PAH in burnt and partially carbonised food was first reported over 40 years ago. PAH are potentially carcinogenic and their presence in food, even at low levels, is therefore undesirable.

Hundreds of PAH have been identified as by-products of incomplete combustion. However, by far the most studied PAH is benzo[*a*]pyrene (BaP, CAS No. 50-32-8). BaP is often used as a marker compound for all PAH in food, and also in environmental studies. Although the profile of PAH contamination in different foods varies, BaP has been shown to be valid marker compound for the most harmful group of higher molecular weight PAH compounds.

Occurrence in Foods

PAH are common environmental contaminants in water, air and soil, and so may contaminate many foods by this route. Vegetables are especially vulnerable to environmental PAH contamination, particularly when grown in areas where industrial pollution levels are high. Seafood, such as some shellfish and crustaceans, may also accumulate PAH from the water in which they are grown, but significant levels do not usually accumulate in the meat, milk, or eggs of food animals, because PAH are rapidly metabolised in these species.

However, the main source of PAH in the diet is generally considered to be food processing and preparation, especially foods processed at high temperatures. High levels (up to 130 μg/kg against a background level of <1 μg/kg) of individual PAH have been reported in grilled and barbecued meats. Smoked foods are also often contaminated, with levels of up to 200 μg/kg being reported in both smoked meat and fish. However, reported levels of PAH in smoked foods vary widely, and are probably dependent on the nature of the smoking process, with traditional methods generally producing higher levels than newer processes. Smoke flavourings may also be contaminated with PAH.

Vegetable oils, including olive pomace oils, are an important source of dietary PAH, which is usually present as a consequence of direct seed-drying methods where the product comes in contact with combustion gases. Reported levels in oils vary widely. Both roasted coffee beans and dried tea leaves may also contain high PAH levels – up to 1400 μg/kg in one report – but high levels have not been found in coffee or tea drinks as consumed. Dried fruits and nuts have also been reported to contain high levels of PAH on occasion.

Dietary intake of PAH across six EU countries has been estimated to be in the range 0.05 to 0.29 μg of BaP/day. Similar estimates have been produced in the USA. Food is thought to be the main source of PAH exposure in non-smokers.

Hazard Characterisation

Effects on Health

Little is known about the potential for acute toxicity of PAH, but it is extremely unlikely that this could occur through consumption of contaminated food, and it is the effect of low doses over a long time that is of most concern from a food safety point of view. A number of PAH, including BaP, have been shown to be both carcinogenic and genotoxic in experimental animals and are therefore potential human carcinogens. For example, BaP has been shown to cause tumours in the gastrointestinal tract, liver, lungs and mammary glands of rodents.

Individual PAH have also been found to produce other, non-carcinogenic effects in animals, including liver toxicity, reproductive and developmental toxicity and suppression of the immune system.

Because some PAH are likely to be both genotoxic and carcinogenic, the EU Scientific Committee on Food has recommended that no tolerable daily intake (TDI) be set for PAH. Instead the Committee recommended that levels in food should be as low as is reasonably achievable. However, it also noted that maximum dietary intakes are 5-6 times lower than the levels causing tumours in animals.

Sources

The main sources for PAH in foods are air-, soil-, or waterborne environmental contamination and food processing involving high temperatures. However, humans are also exposed to PAH in the air – from industrial and traffic pollution and from tobacco smoke.

The mechanism for PAH production during smoking, drying and cooking processes are not fully understood, but it is likely that more than one mechanism is involved. For example, when fat from cooking meat drips onto a heat source, it undergoes pyrolysis and PAH may be produced and deposited on the food itself. Meat heated to temperatures above 200 °C may also undergo pyrolysis, producing PAH on the surface. PAH production in grilled meats has been shown to be dependent on fat content and the time and temperatures used during cooking. In dried products, PAH contamination is most likely to come from exposure to partially burnt combustion gases in direct flame dryers.

Stability in Foods

PAH are generally very stable compounds, although photo-degradation does occur. They are highly lipophilic and are particularly stable in oils and fats. They also readily adhere to particles in the soil and in foods.

Control Options

The control of PAH in foods focuses on limiting their production during processing.

Processing

Although the mechanisms for PAH production in foods are still uncertain, it is known that processing conditions can have a dramatic effect on the levels present. It has therefore been possible to produce a number of recommendations for effective measures to reduce PAH production in a number of food types. For example:

- select leaner meat and fish for grilling and barbecuing;
- do not allow fat to come in contact with the heat source during cooking (*e.g.* by using vertical barbecues and grills);
- reduce cooking temperatures and do not brown food excessively;
- replace traditional direct smoking processes with indirect smoking, or use smoke flavouring;
- avoid direct contact of oil seeds and cereals with combustion gases during drying;
- wash, or peel, fruit and vegetables that have a waxy coating.

Product Use

Similar advice on safer barbecuing and grilling of meat and fish in the domestic environment may help consumers to reduce levels of PAH in their diet.

Legislation

In the EU, European Commission Regulation (EC) 1881/2006 sets permitted levels of BaP (as a marker for PAH) in a number of food products, including oils and fats, infant foods, and smoked meat and fish products. The maximum levels permitted in these products range from 1.0 μg/kg wet weight in baby foods to 10.0 μg/kg in bivalve molluscs.

Sources of Further Information

Published

Studer, A., Blank, I. and Stadler, R.H. Thermal processing contaminants in foodstuffs and potential control strategies. *Czech Journal of Food Science*, 2004, 22 (special issue), 1–10.

Phillips, D.H. Polycyclic aromatic hydrocarbons in the diet. *Mutation Research*, 1999, 15;443(1–2), 139–47.

On the Web

EFSA report on polycyclic aromatic hydrocarbons in food (2007). http://www.efsa.europa.eu/EFSA/Scientific_Document/datex_report_pah,0.pdf

JECFA monograph on benzo[*a*]pyrene (2006). http://whqlibdoc.who.int/publications/2006/9241660554_PAH_eng.pdf

2.2.2 CONTAMINANTS FROM FOOD-CONTACT MATERIALS

2.2.2.1 Bisphenol A

Hazard Identification

What is Bisphenol A?

Bisphenol A (BPA) is a phenolic compound ($C_{15}H_{16}O_2$, CAS No. 80-05-7), also referred to as 2,2-*bis*(4-hydroxyphenyl)propane. It was first synthesised over a hundred years ago and is an important industrial chemical used in manufacturing processes. BPA is a major component of rigid polycarbonate plastics and epoxy-resin coatings. Polycarbonate is commonly used in the food industry for water and soft-drink bottles, and epoxy resins are used as protective linings for metal food cans, wine storage vats and other liquid containers, and as coatings on metal lids used for glass bottles and jars. In addition, polycarbonate plastic containers and tableware are widely used by consumers and the material is also used to manufacture infant feeding bottles.

Although materials containing BPA have been used in packaging and storage vessels for food and beverages for over 50 years, some scientific studies have shown that under certain conditions BPA can migrate into food products. This is of concern because BPA is known to cause adverse health effects in animals at high levels.

Occurrence in Foods

Detection of BPA has been reported in various canned food and drink products, including canned fruit, vegetables, coffee, tea, infant formula concentrate and sake. A survey of 62 canned food and drink products by the UK Food Standards Agency (FSA) published in 2001 found detectable levels of BPA in fruits and vegetable products, stout, fish, soups, dairy products, meat products and pasta in tomato sauce. However, the Independent Committee on Toxicology of Chemicals in Food, Consumer Products and the Environment (COT) concluded that the levels of BPA found during the FSA survey were unlikely to be a concern for human health.

Estimates of dietary exposure to BPA vary widely and can be based on different methods of calculation.

Using migration figures from food-contact materials, levels of BPA found in foods and the amount of food consumed, a recent conservative estimate published in 2006 by the European Food Safety Authority (EFSA) Scientific Panel of Food Additives, Flavourings, Processing Aids and Materials in Contact with Food, gives values ranging from 0.2 μg/kg bodyweight (bw)/day for a breast-fed 3-month infant to 13 μg/kg bw/day for a 6-month infant fed formula from a polycarbonate bottle and consuming commercial foods and beverages. This highest value falls to 1.5 μg/kg bw/day for an adult consuming commercial foods and beverages.

Hazard Characterisation

Effects on Health

Based on studies in mice and rats, it is widely accepted that exposure to BPA (from the environment as well as from food) at high levels is potentially detrimental to human health. It is an endocrine disruptor and may have an effect on fertility. It has weak oestrogenic activity and has been shown to reduce sperm count and sperm activity. Studies indicate that it could affect development, and some research suggests that BPA may be carcinogenic, possibly leading to the precursors of breast cancer. Some reports indicate that it has liver toxicity and may even be linked to obesity by triggering fat-cell activity.

The effect of low level exposure to BPA on human health is far less clear. Some researchers believe that there is evidence in the literature demonstrating that animals exposed to very low doses of BPA suffer adverse affects. However, expert panels asked to review the data generally consider that there is not enough evidence from animal studies to suggest that low levels of BPA adversely affect humans.

Currently, both the US Environmental Protection Agency's (EPA) maximum acceptable or oral reference dose (RfD; established in 1993) and the European Food Safety Authority's (EFSA) tolerable daily intake (TDI; established in November 2006) are 0.05 mg/kg bodyweight (bw)/day. These values are considerably greater than the highest estimates of dietary intake.

Sources

BPA can be present in foods as a result of migration from the epoxy-resin coatings used to line metallic food cans and on metal closures for glass jars and bottles. The other main source is the polycarbonate plastic bottles and containers used to package a wide range of products, such as water, soft drinks and milk. BPA in food may also originate from epoxy coatings and polycarbonate plastic used in tanks and containers in the processing environment.

Another potential source of BPA in food is polycarbonate tableware used to store foods in the domestic environment. BPA may migrate from tableware to foodstuffs, either from residual BPA in the material, or because various extreme conditions, including repeated cleaning, exposure to heat and contact with acid foods, results in the polycarbonate breaking down to produce BPA, which subsequently migrates into the food.

BPA is also found in a wide variety of non-food sources, such as drinking-water storage tanks and water pipes, electrical equipment and various household appliances.

Stability in Foods

Bisphenol A appears to be readily biodegradable and after a short period of adaptation (3–8 days), levels in natural water environments rapidly decrease (100% removal in 2–17 days). Levels of bisphenol A also reduce rapidly during wastewater treatment.

Studies in fish indicate that bisphenol A has low potential for bioaccumulation.

Bisphenol A is very heat stable. It has a melting point of 155–157 °C and polycarbonate plastics can be used up to temperatures of around 145 °C.

Control Options

It is generally agreed that the levels of ingested bisphenol A should be as low as possible because of the uncertainties that exist about its potential adverse effects on human health.

Processing

The food industry is being encouraged to implement techniques and procedures to reduce the migration of bisphenol A into foodstuffs and to source can and container coatings that contain lower levels of bisphenol A, or are bisphenol A-free. It is important to note that for canned food products, alternatives should not permit bacterial or metallic contamination of the contents, and should not give rise to other safety concerns. The use of alternatives can also reduce the final shelf life of a canned product, because the resistance of the alternative is lower than that of an epoxy-resin-based lining.

Product Use

Alternatives to bisphenol-A-containing plastics can be used for feeding infants and for storing and serving food.

Legislation

At present there are no restrictions on the amount of bisphenol A that can be present in a final plastic product, but the tendency of bisphenol A to migrate from food-contact materials has been acknowledged in European Union food law. In 2002, EU legislation was introduced setting a Specific Migration Limit (SML) of 3 mg bisphenol A per kg food. This was amended in 2004 to set a SML(T) of 0.6 mg bisphenol A per kg food.

The migration limit in Japan allows a maximum of 2.5 ppm. There is no SML in the USA at present.

Sources of Further Information

Published

Kamrin, M.A. Bisphenol A: A Scientific Evaluation. *Medscape General Medicine*. 2004. 6(3) 7.

On the Web

Opinion of the Scientific Panel on Food Additives, Flavourings, Processing Aids and Materials in Contact with Food on a request from the Commission related to 2,2-*bis*(4-hydroxyphenyl)propane (bisphenol A). European Food Safety Authority (November 2006). http://www.efsa.europa.eu/etc/medialib/efsa/science/afc/afc_opinions/bisphenol_a.Par.0001.File.dat/afc_op_ej428_bpa_op_en.pdf

Bisphenol A (BPA) Risk Assessment Report. Japanese Research Center for Risk Management (January 2006). http://unit.aist.go.jp/crm/mainmenu/e_1-10.html

European Union Risk Assessment Report. 4,4′-isopropylidenediphenol (bisphenol-A) European Chemicals Bureau (2003). http://ecb.jrc.it/DOCUMENTS/Existing-Chemicals/RISK_ASSESSMENT/REPORT/bisphenolareport325.pdf

Opinion of the Scientific Committee on Food on bisphenol A. European Commission. (April 2002). http://ec.europa.eu/food/fs/sc/scf/out128_en.pdf

Chemical Study on bisphenol A. Dutch National Institute for Coastal and Marine Management (July 2001). http://www.rikz.nl/thema/ikc/rapport2001/rikz2001027.pdf

2.2.2.2 Phthalates

Hazard Identification

What are Pthalates?

The phthalates (also known as phthalic acid diesters) are a group of related organic chemicals commonly used in the plastics industry as plasticisers. Plasticisers are routinely added to other materials, particularly polyvinyl chloride (PVC) and other polymers such as rubber and styrene, to make them more pliable and elastic.

The five phthalates most commonly used by industry are di-(2-ethylhexyl) phthalate (DEHP), dibutyl phthalate (DBP), di-isononyl phthalate (DINP), di-isodecyl phthalate (DIDP) and benzyl butyl phthalate (BBP).

Since the early 1980s there have been concerns about the effect that phthalates have on human health. Phthalates are able to leach from the materials, to which they have been added and there is known to be widespread environmental exposure to the chemicals. Food products can become contaminated with phthalates from a wide variety of sources, but there has been particular concern over migration from food packaging. Phthalates can be present in some food-packaging materials, including printing inks used on flexible food packaging, adhesives used for paper, board and plastics, regenerated cellulose film (cellophane), aluminium foil-paper laminate and closure seals in bottles. It should be noted that many PVC "clingfilm" food wraps are no longer made with phthalates, but are now manufactured using other plasticisers.

Occurrence in Foods

Food can become contaminated with phthalates during processing, handling, transportation, and by migration from packaging, as well as from food-storage containers used in the home. Phthalates are fat soluble and have been found in many high-fat products, such as dairy products, meat and poultry, eggs, fish, fats and oils. High levels of phthalates have been found in some olive-oil samples.

Phthalates have also been found in a variety of other foods, such as infant formula, ready to use baby foods, bakery products, gravy granules, confectionery, pasta and cereal products, flour, sugar, vegetable burger mix and vegetables. They also occur in drinking water and in breast milk. In a UK survey of phthalates in foods from animal sources collected in 1993, DEHP was the most abundant individual phthalate found in each sample.

A total diet study conducted in the UK on samples collected during 1993 estimated that the total phthalate intake for an average or high-level consumer is 0.013 and 0.027 mg/kg bodyweight/day, respectively. Later Danish studies have suggested that this figure may be an underestimate of phthalate dietary intake, because the UK figures were based on foods from animal sources and did not take into account the high contribution that vegetables can make to phthalate intake.

Hazard Characterisation

Effects on Health

Most of the data on the health effects of phthalates comes from experiments exposing rats and mice to high levels of the chemicals for prolonged periods. Long-term health effects of phthalates may include changes in sperm production, adverse effects on fertility and birth defects. They have also been reported to cause kidney and liver damage. Phthalates may be potential carcinogens and also endocrine disruptors, and as such could affect reproductive development.

Individuals exposed to very high levels of DEHP for relatively short periods may experience mild gastrointestinal disturbances, vertigo and nausea.

There is no group tolerable daily intake (TDI) figure for phthalates, but TDIs have been set for some individual phthalates. For the five most commonly used phthalates, the European Food Safety Authority's (EFSA) Scientific Panel on Food Additives, Flavourings, Processing Aids and Materials in Contact with Foods (AFC) set TDI figures in 2005. The TDI for DEHP is 0.05 mg/kg bodyweight (bw); for BBP it is 0.5 mg/kg bw; for DBP it is 0.01 mg/kg bw; and for both DIDP and DINP it is 0.15 mg/kg bw. The US Environmental Protection Agency (EPA) has set an oral reference dose (RfD) of 0.02 mg/kg bw day for DEHP, 0.20 mg/kg bw day for BBP and 0.10 mg/kg bw day for DBP.

It is generally considered that the levels of individual phthalates currently found in foods are not a significant concern for human health.

Sources

Phthalates may be naturally produced by some animals and plants, and are also released into the environment during the manufacturing, use and disposal of products that contain them. As a result, consumers are exposed to these chemicals from a wide variety of environmental sources including air, drinking water and their physical surroundings. Phthalates are found in many plastics, cosmetics, medical devices, paints, lubricants, flooring materials, cleaning products, adhesives, inks, clothing, pesticides and toys. As a result, materials containing phthalates can be found everywhere in the domestic environment, but they are also used by the food industry and can be found in packaging materials, and also in manufacturing equipment such as conveyor belts and plastic hoses and tubing.

A number of food-packaging materials can contain phthalates, including PVC and other plastics, printing inks used on flexible food packaging, adhesives used for paper and board, regenerated cellulose film (cellophane), aluminium-foil–paper laminates and closure seals in bottles. Phthalates are known to migrate from packaging into foods, especially high-fat products and oils, and the rate of migration into food from packaging rises with increasing temperature.

Food surveys have determined that, although packaging contributes to the presence of phthalates in food products, it is not the only source of the chemicals. A UK survey published in 1995 found that measured levels of DEHP and DBP in some products were higher than would be expected if all the DEHP and DBP in the packaging had migrated into the foods concerned. In addition, the level of the phthalate at the core of products was equal to, or higher than, the level at the surface where the product was in contact with the packaging. These results may indicate that environmental sources contribute, at least in some part, to the presence of phthalates in foods.

Stability in Foods

Although phthalates are widespread in the environment, levels tend to be low because phthalates do not generally persist for extended periods when exposed to photochemical and biological breakdown.

DEHP in its gas form is broken down in the atmosphere by other chemicals in 1–2 days and solid particles are removed by various natural mechanisms in about 2–3 weeks. The chemical is broken down in surface soils by micro-organisms into harmless components, but the rate of degradation is temperature dependent and is slower at cooler temperatures. However, DEHP persists for much longer in deep soil or at the bottom of lakes and rivers because anaerobic degradation is considerably slower than aerobic breakdown. The contaminant is found in plants and fish, but bioaccumulation is limited and animals higher up in the food chain can break the chemical down so that tissue levels tend to be low.

DBP persists in air for about 1.5 days, and in water environments for 2–20 days. As with DEHP, aerobic degradation is more efficient than anaerobic breakdown.

Control Options

Food campaign groups have raised consumer awareness of the possible health effects associated with soft PVC plastics and other materials containing phthalates. In the EU, there have been bans on the use of phthalates in some toys and cosmetics. Some measures designed to reduce the levels of phthalates in the environment and in foods have been introduced to address these concerns.

Processing

Reducing the levels of phthalates in the food-processing environment and in food packaging can have a direct effect on the level of phthalates in food products. Where possible, soft PVC equipment parts containing phthalates, such as hoses, can be replaced with non-plastic parts, with other soft materials that do not contain plasticiser, or with plastics containing non-toxic plasticisers.

Manufacturers have developed glues and inks that do not contain phthalates to reduce levels in food packaging. PVC-free plastic food-wrap materials have also been introduced as replacements for older "clingfilm" type food wraps. Products vulnerable to phthalate contamination, especially fatty foods, can be packaged in materials that do not contain phthalates.

Product Use

Advice for consumers on the safe use of plastic containers and food wraps in the home has been issued by a number of food safety authorities, including the UK Food Standards Agency.

Legislation

The European Community now has legislation that limits the use of phthalates in food plastics, and where use is permitted, it limits the migration of these chemicals into foods by setting specific migration limits (SML). EU directive 2007/19/EC was adopted on 30 March 2007, and the manufacture, or import, of products that do not comply with this legislation are prohibited from 1 June 2008.

US regulations treat phthalates that migrate into foodstuffs from food-contact materials as indirect additives. In the USA, indirect food additives are defined as additives "that become part of the food in trace amounts due to its packaging, storage or other handling." The onus is on the food-packaging manufacturers to prove to the U.S. Food and Drug Administration (FDA) that food-contact materials are safe.

Sources of Further Information

Published

Jarosova, A. Phthalic acid esters (PAEs) in the food chain. *Czech Journal of Food Sciences*, 2006. 24, 223–231.

Mikula, P., Syobodova, Z. and Smutna, M. Phthalates: toxicology and food safety – a review. *Czech Journal of Food Sciences*, 2005. 23, 217–223.

On the Web

Opinion of the Scientific Panel on food additives, flavourings, processing aids and materials in contact with food (AFC) related to di-Butylphthalate (DBP) for use in food-contact materials. European Food Safety Authority (June 2005). http://www.efsa.europa.eu/EFSA/efsa_locale-1178620753812_1178620770694.htm

Opinion of the Scientific Panel on food additives, flavourings, processing aids and materials in contact with food (AFC) related to Bis(2-ethylhexyl)phthalate (DEHP) for use in food-contact materials. European Food Safety Authority (June 2005). http://www.efsa.europa.eu/EFSA/efsa_locale-1178620753812_1178620770530.htm

Opinion of the Scientific Panel on food additives, flavourings, processing aids and materials in contact with food (AFC) related to Butylbenzylphthalate (BBP) for use in food-contact materials. European Food Safety Authority (June 2005). http://www.efsa.europa.eu/EFSA/efsa_locale-1178620753812_1178620770710.htm

Opinion of the Scientific Panel on food additives, flavourings, processing aids and materials in contact with food (AFC) related to Di-isodecylphthalate (DIDP) for use in food-contact materials. European Food Safety Authority (July 2005). http://www.efsa.europa.eu/EFSA/efsa_locale-1178620753812_1178620770412.htm

Opinion of the Scientific Panel on food additives, flavourings, processing aids and materials in contact with food (AFC) related to Di-isononylphthalate (DINP) for use in food-contact materials. European Food Safety Authority (July 2005). http://www.efsa.europa.eu/EFSA/efsa_locale-1178620753812_1178620770396.htm

2.2.2.3 Semicarbazide

Hazard Identification

What is Semicarbazide?

Semicarbazide (SEM) is a chemical contaminant that has been found in a number of food products – probably originating from several different sources. It is of concern from a food safety point of view because it has been shown to be a weak carcinogen in laboratory animals.

SEM is a member of a group of chemicals known as the hydrazines. It is a small molecule with the chemical formula $H_2N\text{-}NH\text{-}CO\text{-}NH_2$ (CAS No. 57-56-7). It was first detected in foods in 2003, when it was identified as a contaminant in foods packed in glass jars and bottles with sealed lids.

Occurrence in Foods

A number of European studies were conducted during 2003 and 2004 to determine levels of SEM in foods. Baby foods in sealed glass jars contained the highest reported levels of SEM, ranging from not detectable to 140 μg/kg. Levels were similar in all European countries reporting data, and the average level of SEM found in 385 samples of baby foods was 13 μg/kg. The average SEM levels found in 121 samples of other food types (fruit, fish, vegetables, jams, pickles and sauces) included in these studies was 1.0 μg/kg.

Using figures derived from the studies of foods packaged in glass and jars and bottles, the European Food Safety Authority (EFSA) estimated daily intakes of SEM. "Reasonable worst case estimates" of daily intakes of SEM for infants fed on products packed in glass jars and bottles ranged from 0.35 to 1.4 μg/kg bodyweight/day. For adults, the estimates of SEM exposure from this source were much lower at 0.02 μg/kg bw/day, but these figures do not account for exposure to SEM from other dietary sources.

Canadian tests have found levels of SEM of up to 28 μg/kg in bread, with most SEM being found in the crust. Frozen breaded chicken or fish products can contain SEM in the breadcrumb coating, possibly at levels up to 5 μg/kg of product. SEM has also been detected in egg-white powder and in some types of carrageenan (particularly processed Euchema seaweed, E407a). SEM also appears to occur naturally in some foods, but may also originate from currently unidentified sources. For example, wild crayfish caught in Finland during autumn 2004 were found to contain SEM at levels of up to 18 μg/kg.

Hazard Characterisation

Effects on Health

Many of the hydrazine group of chemicals are known to cause cancer in laboratory animals. However, SEM is one of the least carcinogenic hydrazines. In 2005 the European Food Safety Authority's (EFSA) Scientific Panel of Food Additives, Flavourings, Processing Aids and Materials in Contact with Food concluded that evidence indicates that SEM is a weak non-genotoxic carcinogen. Data on the potential developmental and reproductive toxicity of SEM is limited.

Based on recent studies, and the fact that efforts are being made in Europe to reduce the levels of SEM from its main food-related source, products in glass jars and bottles, European experts consider that the risk, if any, to human health from SEM is very small, not only for adults, but also for infants.

Sources

There are thought to be several sources of SEM in foods, but by far the most significant source is considered to be migration into foods from sealing gaskets fitted to the lids of glass jars and bottles. SEM from this source arises as a by-product of the breakdown of azodicarbonamide used as a "blowing agent" in the formulation of PVC gaskets found on the inside of metal lids. Blowing agents change the texture of the gaskets and help to produce a better airtight seal. Azodicarbonamide has been used to help seal metal "twist" caps on glass jars used for a wide range of products including baby foods, fruit juices, conserves, pickles, mustard, mayonnaise and ketchups. Levels of SEM in the gaskets themselves have been found to vary from 1–7 mg per kg of gasket material.

In some countries, although not in the EU, azodicarbonamide is also approved as a food additive. It is used as a dough improver, and as a bleaching agent in cereal flour. SEM has been found in products made using flour to which azodicarbonamide has been added.

SEM is a metabolite of the veterinary drug nitrofurazone, and is used as a marker for the use of this drug in foods of animal origin. Nitrofurazone is not permitted for use in food-producing animals in the European Union and so SEM from this source should not be detected in foods. However, this may be a source of dietary exposure in other countries where nitrofuran drugs are not illegal.

SEM is also formed during some manufacturing processes used to produce egg-white powder and some types of carrageenan, particularly processed Euchema seaweed. SEM is thought to be produced as a by-product of these processes as a result of a reaction between hypochlorite bleach and organic substances.

SEM may also occur naturally in the environment, and it is thought that there may be some still unidentified sources of the contaminant in foods.

Stability in Foods

There is no available data on the persistence, or bioaccumulation of SEM in the environment. A study has shown that concentrations of SEM in pig muscle and liver did not drop significantly during storage for 8 months at –20°C, and that working standard solutions prepared in methanol stored at 4°C for 10 months were generally stable.

The melting point of SEM is around 173–177°C. A study concluded that SEM is largely resistant to conventional cooking techniques such as frying, microwaving, grilling and roasting.

Control Options

Processing

The World Health Organization has said "that the presence of SEM in baby foods is considered particularly undesirable". Therefore as a precaution, efforts should be made to reduce levels, or eliminate SEM from foods, particularly baby foods, and these efforts should focus at avoiding processes that produce the chemical.

In order to eliminate SEM from the gaskets used for metal twist caps, food manufacturers have been encouraged to develop alternative materials so that azodicarbonamide is no longer used in food packaging. Care should be taken to choose alternative types of sealing for bottles and jars that do not compromise the microbiological safety of the contents.

In the EU, the use of azodicarbonamide in food-contact materials was prohibited from August 2005, and once existing stocks of products have been used, the dietary intake of SEM derived from gaskets should have been eliminated.

The use of flour containing azodicarbonamide as an additive should be avoided to prevent the formation of SEM in baked foods, and in products with crumb coatings.

Legislation

Azodicarbonamide is not permitted as a flour-treatment agent in the European Union. At the time of writing it is permitted in some countries (*e.g.* the USA, Canada and Brazil), and can be used at levels up to 45 mg/kg flour.

The use of azodicarbonamide as a blowing agent has been prohibited in the European Union since 2 August 2005. Products filled before this date could continue to be placed on the market provided that the date of filling, or a mark indicating when it was filled, appeared on the product. At the time of writing the use of azodicarbonamide for food-contact materials is still permitted in some other countries, including the USA.

Sources of Further Information

Published

de la Calle, M.B. and Anklam, E. Semicarbazide: occurrence in food products and state-of-the-art in analytical methods used for its determination. *Analytical and Bioanalytical Chemistry*. 2005. 382, 968–977.

On the Web

Opinion of the Scientific Panel on Food Additives, Flavourings, Processing Aids and Materials in Contact with Food on a request from the Commission related to Semicarbazide in Food. European Food Safety Authority. 2005. http://www.efsa.europa.eu/EFSA/Scientific_Opinion/afc_op_ej219_semicarbazide_en2.pdf

Semicarbazide. World Health Organization. http://www.who.int/foodsafety/chem/sem/en/index.html

2.2.3 ENVIRONMENTAL CONTAMINANTS

2.2.3.1 Dioxins and PCBs

Hazard Identification

What are Dioxins and PCBs?

The term dioxins refers to a group of compounds with similar chemical and physical properties and structures. Dioxins are colourless, odourless organic compounds containing carbon, hydrogen, oxygen and chlorine. There are many different dioxins, of which 17 are known to be toxic to humans. The most toxic known dioxin is 2,3,7,8-tetrachlorodibenzo-p-dioxin (2,3,7,8-TCDD), and significant concentrations of this compound can be measured in parts per trillion (PPT).

Dioxins are ubiquitous environmental contaminants, having been found in soil, surface water, sediment, plants, and animal tissue worldwide. They are highly persistent in the environment with half-lives ranging from months to years. They have low water solubility and low volatility, meaning that they remain in soil and sediments that serve as environmental reservoirs from which the dioxins may be released over many years.

PCBs, or polychlorinated biphenyls, are chlorinated aromatic hydrocarbons produced by the direct chlorination of biphenyls. There are about 209 related PCBs, known as congeners of PCBs, of which 20 reportedly have toxicological effects. Some of the PCBs have toxicological properties similar to those of dioxins and are therefore often referred to as "dioxin-like PCBs".

Like dioxins, PCBs are widespread environmental contaminants and are very persistent in soil and sediments. It has been suggested that highly contaminated bottom sediments in sewers and receiving streams may represent a reservoir for the continued release of PCBs into the environment.

Occurrence in Foods

Dioxins and PCBs enter the food chain through a variety of routes. Grazing animals and growing vegetables may be exposed directly, or indirectly, to these contaminants in the soil. Leafy vegetables, pasture and roughage can also become contaminated through airborne transport of dioxins and PCBs. Dioxins in surface waters and sediments are accumulated by aquatic organisms and bioaccumulated through the food chain. The concentration of dioxins in fish may be hundreds to thousands of times higher than the concentrations found in surrounding water and sediments.

Because dioxins are not very soluble in water, they tend to accumulate in the fatty tissues of animals and fish. Theoretically, the longer the lifespan of the animal, the longer the time it has to accumulate dioxins and PCBs. Foods that are high in animal fat, such as milk, meat, fish, eggs and related products are the main source of dioxins and PCBs and contribute about 80% of the overall human exposure, although almost all foods will contain these contaminants at some (generally very low) level owing to their ubiquitous nature.

The main contributors to the average daily human intake of dioxins and PCBs have been found to be milk and dairy products, contributing between 16 and 39%; meat and meat products, contributing between 6 and 32%; and fish and fish products, contributing between 11 and 63%. Other foods, mainly vegetables and cereals, contributed 6–26% in the countries for which data was available (Codex Alimentarius Commission, 2001).

Human milk can contain elevated levels of dioxins, some of which can pass to the infant during lactation. However, the intake of babies from their mothers is limited to a relatively short period of their lives.

It is estimated that the average dietary intake of dioxins and dioxin-like PCBs has fallen amongst adults in the UK from 1.8 picograms World Health Organization toxic equivalents (WHO-TEQ)/kg[i] of bodyweight per day in 1997 to 0.9 picograms WHO-TEQ/kg bodyweight per day in 2001. Similar decreases have been reported in other countries. In November 2001, the Independent Committee on Toxicity recommended a TDI (tolerable daily intake) of 2 picograms WHO-TEQ/kg of bodyweight per day.

Hazard Characterisation

Effects on Health

Humans accumulate dioxins in fatty tissue mostly by eating dioxin-contaminated foods. The toxicity of dioxins is related to the amount accumulated in the body during the lifetime. Dioxins and PCBs have a broad range of toxic and biochemical effects, and some are classified as human carcinogens. In animal testing, dioxins have been implicated in causing damage to the immune and reproductive systems, developmental effects and neuro-behavioural effects.

Despite the variety of adverse effects observed in animals exposed to dioxins, documented adverse health effects in humans have generally been limited to highly exposed populations in industrial environments, or following accidental chemical contamination.

The most commonly observed adverse health effect in humans following acute over-exposure to dioxins and PCBs is the skin disease chloracne, a particularly severe and prolonged acne-like skin disorder. The accidental contamination of edible rice bran oil with PCBs in Japan in 1968 led to a poisoning epidemic amongst those who consumed the oil. The poisoning caused chloracne, liver disturbances, abdominal pain, headaches, skin discolouration, and the birth of abnormally small babies to mothers who had consumed the oil. A more recent example of dioxin contamination happened in Belgium in 1999, when PCB-contaminated feeds were fed to farm animals. The contamination was discovered as a result of the direct biological effects of dioxins observed in poultry.

[i] *Note*: the TEQ is a weighted toxicity value designed to take into account the variable toxicity of different dioxins and dioxin-like PCBs in comparison with the most toxic dioxins, and give a comparable overall measure of dioxin and PCB levels.

Sources

Dioxins are often man-made contaminants and are formed as unwanted by-products of industrial chemical processes, such as the manufacture of paints, steel, pesticides and other synthetic chemicals, wood pulp and paper bleaching, and also in emissions from vehicle exhausts and incineration. Dioxins are also produced naturally during volcanic eruptions and forest fires. Most industrial releases of dioxins are strictly controlled under pollution prevention and control regulations. Currently, the major environmental source of dioxins is incineration.

PCBs have been used in manufacturing industry since the early 1930s, mainly as cooling and insulating fluids in electrical equipment. The manufacture and general use of PCBs was banned in the 1970s because of environmental and health concerns. However, some PCBs remain in use, sealed inside older electrical equipment, although the use of this equipment must be phased out, and the PCBs removed and destroyed.

Stability in Foods

Dioxins and PCBs are highly stable with reportedly long half-lives. In animals, they accumulate in fat and in the liver and are only very slowly metabolised by oxidation or reductive dechlorination and conjugation. They are therefore likely to persist in animal tissues, especially fatty tissue, for long periods. They are not generally affected significantly by food processing such as heat treatments, or fermentation.

Control Options

There is very little scope for the removal of dioxins and PCBs from foods once they have entered the food chain. It is generally agreed that the best means for preventing dioxins and PCBs from entering the food chain is to control their release into the environment.

The overall goal of European policy is to reduce the contamination levels of dioxins and PCBs in the environment, and in food and feed. The EU has prohibited the use of most PCBs from 1978 and for certain applications from 1986. A deadline of 2010 has been set for removing all PCB-containing equipment from service. Dioxins, on the other hand, cannot be banned owing to their formation as unwanted by-products of many industrial processes. The amounts of dioxins and PCBs ingested in food are similar in the EU and the USA. Intakes are falling and have reduced by 85% since 1982, demonstrating some international success in controlling environmental contamination by these compounds.

Product Use

While studies suggest that there is no cause for alarm from potential health issues concerning dioxins in the diet, choosing leaner cuts of meat, removing the skin from chicken or trimming the fat off meat may help to minimise any potential exposure of consumers to dioxins in food. Similarly, drinking reduced- or low-fat milk may also help to reduce exposure slightly, as may the washing of fruit and vegetables to remove any airborne dioxin-contaminated dust particles that might have been deposited on produce in fields.

Legislation

EU

New EU regulations on contaminant levels in foods have recently been introduced (March 2007). These new regulations will require tougher safety controls in food-manufacturing plants. The regulations aim to ensure a harmonised approach to the enforcement of permitted contaminant levels across the EU.

Regulation (EC) 1881/2006 sets maximum levels for certain contaminants, including dioxins and dioxin-like PCBs in foods.

The limits for dioxins and PCBs as set out in this Regulation are as follows:

Foodstuff	*Maximum Levels (Sum of dioxins)*	*Maximum Levels (Sum of dioxins and dioxin-like PCBs)*
Meat and meat products (excluding edible offal) of the following animals:		
Bovine animals and sheep	3.0 pg/g fat	4.5 pg/g fat
Poultry	2.0 pg/g fat	4.0 pg/g fat
Pigs	1.0 pg/g fat	1.5 pg/g fat
Liver of terrestrial animals above and derived products thereof	6.0 pg/g fat	12.0 pg/g fat
Muscle meat of fish and fishery products and products thereof, excluding eel. The maximum level applies to crustaceans, excluding the brown meat of crab and excluding head and thorax meat of lobster and similar large crustaceans (*Nephropidae* and *Palinuridae*)	4.0 pg/g wet weight	8.0 pg/g wet weight
Muscle meat of eel (Anguilla anguilla) and products thereof	4.0 pg/g wet weight	12.0 pg/g wet weight
Raw milk and dairy products including butterfat	3.0 pg/g fat	6.0 pg/g fat
Hen eggs and egg products	3.0 pg/g fat	6.0 pg/g fat
Bovine and sheep fat	3.0 pg/g fat	4.5 pg/g fat
Poultry fat	2.0 pg/g fat	4.0 pg/g fat
Pig fat	1.0 pg/g fat	1.5 pg/g fat
Vegetable oils and fats	0.75 pg/g fat	1.5 pg/g fat
Marine oils (fish-body oil, fish-liver oil and oils of other marine organisms intended for human consumption)	2.0 pg/g fat	10.0 pg/g fat.

Methods of Sampling for Dioxins

Regulation (EC) 1883/2006 lays down the methods for sampling and analysis for the official control of levels of dioxins and dioxin-like PCBs in certain foodstuffs.

USA

There are no tolerances or other administrative levels for dioxins in food or feed in the USA and the FDA considers all detectable levels to be of concern. Action levels have been set for PCBs in red meat and fish. Temporary tolerances have also been set for animal feeds and paper packaging. These are published in the Federal Register.

Sources of Further Information

Published

Codex Alimentarius Commission, 2001 Position paper on Dioxins and Dioxin-like PCBs. CX/FAC 01/29.

Joint FAO/WHO Food Standard programme. Codex Committee on Food Additives and Contaminants, 33rd Session, The Hague, The Netherlands, 12–16 March 2001.

Institute of Food Science and Technology, UK (IFST) Position Statement. Dioxins and PCBs in Food. *Food Science and Technology Today*, 1998, 12, 177179.

On the Web

EPA Dioxin Homepage http://www.ejnet.org/dioxin/

JECFA evaluation of the safety of some dioxins and PCBs. http://www.inchem.org/documents/jecfa/jecmono/v48je20.htm

Food contaminants, dioxins and PCBs: http://ec.europa.eu/food/food/chemicalsafety/contaminants/dioxins_en.htm

USDA Dioxin resources page. http://www.fsis.usda.gov/Fact_Sheets/Dioxin_Resources/index.aspOurFood Database: http://www.ourfood.com/Dioxin.html#SECTION00800070000000000000

Dioxinfacts.org: http://www.dioxinfacts.org/dioxin_health/dioxin_tissues/bio_techreport.html

2.2.3.2 Heavy Metals

Hazard Identification

What are Heavy Metals?

The term "heavy metal" refers to any relatively high-density metallic element that is toxic or poisonous even at low concentrations. Heavy metals are natural components of the earth's crust and cannot be destroyed. Although there are many elements that are classified as heavy metals, the ones of most concern, with respect to their biotoxic effects and presence in food, are arsenic, cadmium, lead, and mercury, and it is primarily these that are dealt with here. These elements have no known bio-importance in human biochemistry and physiology, and consumption, even at very low concentrations, can cause toxic effects, because they tend to accumulate in the human body over time.

Because of their potential toxicity, regulatory bodies throughout the world have set a limit on the acceptable amounts of these contaminants in certain foods. In the EU, limits have been set on the amounts of the heavy metal tin in foods as well as on cadmium, lead and mercury. For this reason, tin has also been included in this section.

Occurrence in Foods

A major study was carried out in March 2004, at the EU Directorate-General, Health and Consumer Protection, in order to assess the dietary intake of arsenic, cadmium, lead and mercury of the population of the EU Member States (the reference to the full report is given below). The report collected data on the occurrence, consumption and intake calculations for the populations of Belgium, Denmark, Finland, France, Germany, Ireland, Italy, the Netherlands, Norway, Spain, Portugal, and the UK. Some of the results from this report are briefly summarised below.

Arsenic

The major source of arsenic in the diet is from fish and other seafood, although the daily intake is estimated to be less than 0.35 mg. The marine environment has a great impact on arsenic levels as sea fish have arsenic levels about 10 times higher than freshwater fish. Children have a lower intake of arsenic than adults, and young children have the lowest intake.

Cadmium

None of the most commonly consumed foods were found to be high in cadmium. Cereals, fruit and vegetables are the main source of cadmium in the diet, making up about 66% of the mean cadmium intake. The other sources include meat and fish, with liver, kidney, crustaceans, molluscs and cephalopods containing comparatively higher cadmium levels. The PTWI (permitted tolerable weekly intake) is 0.49 mg for a person weighing 70 kg, and the mean intake for most EU Member States is less than 30% of the PTWI. Children have a lower intake of cadmium than adults, and young children have the lowest intake.

Lead

None of the most commonly consumed foods were found to be high in lead, although some Member States reported high lead levels in wine, game, meat and fish. The PTWI for lead in the EU is 0.025 mg/kg bodyweight, which is equivalent to 1.75 mg for a person weighing 70 kg. The average intake of lead was less than 25% of the PTWI in most Member States. Children have a lower lead intake than adults.

Mercury

The main source of mercury in the diet is fish, followed by fruit and vegetables. In fish and shellfish, mercury is present in the form of methylmercury, while in most other food groups it is present in its inorganic form. Methylmercury is formed from inorganic mercury by the action of micro-organisms in marine and freshwater sediments. Predatory species of fish at the top of the food chain, such as tuna and swordfish, generally contain higher levels of mercury, but their contribution to total mercury intake is small as consumption levels are low. Fruit, dried fruit, mushrooms and vegetables are other sources of mercury in the diet.

The PTWI for mercury is 0.35 mg for a person weighing 70 kg. The mean intake for total mercury within the Member States is less than 30% of the PTWI. The PTWI for methylmercury is 0.112 mg/week for a person weighing 70 kg (1.6 μg/kg bodyweight). The mean intake of methylmercury is less than 30% of the PTWI. However, for people who consume a lot of fish, such as some groups in Norway, the PTWI may be exceeded. Although children have a lower total intake of mercury than adults, they also have a lower bodyweight and so, potentially, a relatively larger intake/kg bodyweight. It is possible that, in some cases, the PTWI for methylmercury may be exceeded.

Hazard Characterisation

Effects on Health

Arsenic

Arsenic is one of the most toxic elements found, and is present in foods in organic or inorganic forms, with the latter being considered to be far more toxic than the former. Additionally, inorganic As^{3+} salts are more toxic than As^{5+} salts. Illnesses associated with excessive inorganic arsenic intake include skin, lung and heart conditions, gastrointestinal diseases and possible carcinogenic effects. As^{3+} compounds are bound by red blood cells and affect the activity of many enzymes, particularly those involved in the respiratory process. 100 mg of arsenic oxide is considered to be lethal.

Organic arsenic does not cause cancer, nor is it thought to damage DNA, but exposure to high doses may cause nerve injury and stomach problems.

The levels of arsenic in most foods are very low, with the exception of seafood. However, the majority of arsenic in seafood is present in the organic, less toxic form, and during digestion of such compounds, the arsenic is not released, or is released only very slowly. This explains why very few cases of arsenic poisoning are associated with seafood consumption, despite the high levels observed.

Cadmium

Human intake of cadmium occurs mostly through food or through smoking. In humans, long-term exposure may lead to kidney damage, as cadmium tends to accumulate in the kidneys. Other adverse health effects include diarrhoea, stomach pains and sickness, bone defects, immune-system damage, possible infertility, possible damage to DNA and carcinogenic effects.

Cases of foodborne cadmium poisonings were reported in the 1940s in England, France, the US, Russia, New Zealand and other countries, caused by consumption of lemonade, coffee, wine and other products that had been prepared or stored in cadmium-coated containers or in refrigerators with cadmium-coated freezers.

Lead

Lead enters the human body via food, water and air. It is very damaging to health, particularly for infants, children and the developing foetus. Its adverse effects include disruption of haemoglobin synthesis, kidney damage, increased blood pressure, miscarriage, nervous-system disruption, reduced fertility, and learning disabilities and behavioural problems in children. Lead can cross the placenta and may damage the nervous system and brain of the developing foetus.

Symptoms of chronic lead poisoning occur following daily ingestion of 2 to 4 mg for several few months, whilst acute poisoning will occur after daily doses of 8 to 10 mg for a few weeks.

Mercury

Mercury is present in foods such as vegetables, mushrooms and, particularly, fish. It is highly toxic and can cause disruption of the nervous system, brain damage, damage to DNA and chromosomes, allergic reactions and adverse reproductive effects.

The first reported outbreak of food poisoning attributed to mercury ingestion was in 1953 in Japan. This outbreak was caused by consumption of fish containing significant amounts of methylmercury and affected people living in Minamata Bay, leading to the term Minamata disease, now often used to describe any form of foodborne mercury poisoning. Severe outbreaks of mercury-borne food poisoning also occurred in Iraq between 1955 and 1960. Over 8000 people were affected as a result of consumption of bread made from grain treated with methylmercury.

Tin

Tin has been used since the Bronze Age and is still used widely today. It is used in the production of plastics, pesticides, wood preservatives and as a coating for metal food cans. In some countries, inorganic tin compounds are added to preserve the colour of vegetable preserves packed in glass jars. Tin can also enter foods via the use of tin-containing organo-pesticides.

Inorganic tin salts are poorly absorbed and generally almost completely excreted from the body via the stools. Organic tin compounds are thought to be more toxic. Long-term exposure to organic tin compounds can lead to nervous-system disorders and sex-gland atrophy. The average daily intake of tin is around 4 mg, but it is not accumulated in the body.

Sources

Heavy metals can be present in food either naturally, or as a result of human activities, such as mining, irrigation, energy extraction, agricultural practices, incineration, industrial emissions and car exhausts. They may also originate from contamination during manufacturing, processing and storage, or from direct addition.

Plants grown in contaminated soil can accumulate heavy metals, particularly lead and cadmium. Arsenic and cadmium are concentrated in coal ash, from which they can leach into surface waters and accumulate in fish and other aquatic organisms. Mercury tends to accumulate in birds, mammals and fish. Drinking water is another possible source of heavy metals.

Stability in Foods

Heavy metals are stable elements and persist for long periods in the environment. There is no evidence to suggest that levels of heavy metals in foods are changed significantly by processing. For example, methylmercury can be found in canned fish that has undergone a severe thermal process.

Control Options

Control of heavy metal levels in foods relies largely on avoiding those food commodities that are likely to have been exposed to large concentrations of metal contaminants in the primary production environment. Examples include vegetables and produce grown in soils contaminated naturally, or by industrial activity, and large predatory fish. Many health and food safety authorities advise that children under sixteen, pregnant women, and women hoping to become pregnant should avoid shark, marlin and swordfish, and limit the amounts of tuna consumed, because of the possibility of high levels of mercury.

It is also important to ensure that heavy-metal contamination cannot arise from the use of inappropriate food processing equipment. Manufacturers must ensure that all equipment is constructed from "food grade" materials that meet the required standards.

Regulations in many countries set maximum levels for heavy-metal contaminants in certain foodstuffs. It is the responsibility of manufacturers to ensure that these limits are observed, and that ingredients are sourced from reputable suppliers. It is also important to ensure that all processing water is sourced from potable supplies that are not contaminated with heavy metals.

Legislation

EU

New EU regulations on contaminant levels in foods have recently been introduced (March 2007). These new regulations will require tougher safety controls in food-manufacturing plants, and aim to ensure a harmonised approach to the enforcement of contaminant levels across the EU.

For the heavy metals cadmium, lead, mercury and tin, maximum levels in certain foods have been established by Commission Regulation (EC) No 1881/2006, which replaces Commission Regulation (EC) 466/2001 setting maximum levels for certain contaminants in food.

The limits for heavy metals as set out in this Regulation are as follows:

Lead

Foodstuffs	*Maximum levels (mg/kg wet weight)*
Raw milk, heat-treated milk and milk for manufacture of milk-based products	0.020
Infant formulae and follow-on formulae	0.020
Meat (excluding offal) of bovine animals, sheep, pig and poultry	0.10
Offal of bovine animals, sheep, pig and poultry	0.50
Muscle meat of fish	0.30
Crustaceans, excluding brown meat of crab and excluding head and thorax meat of lobster and similar large crustaceans.	0.50
Bivalve molluscs	1.50
Cephalopods (without viscera)	1.00
Cereals, legumes and pulses	0.20
Vegetables, excluding brassica vegetables, leaf vegetables, fresh herbs and fungi. For potatoes, the maximum level applies to peeled potatoes	0.10
Brassica vegetables, leaf vegetables and cultivated fungi	0.30
Fruit, excluding berries and small fruit	0.10
Berries and small fruit	0.20
Fats and oils, including milk fat	0.10
Fruit juices, concentrated fruit juices as reconstituted and fruit nectars	0.050
Wine (including sparkling wine, excluding liqueur wine), cider, perry and fruit wine	0.20
Aromatised wine, aromatised wine-based drinks and aromatized wine-product cocktails	0.20

Cadmium

Foodstuffs	Maximum levels (mg/kg wet weight)
Meat (excluding offal) of bovine animals, sheep, pig and poultry	0.050
Horsemeat, excluding offal	0.20
Liver of bovine animals, sheep, pig, poultry and horse	0.50
Kidney of bovine animals, sheep, pig, poultry and horse	1.0
Muscle meat of fish (excluding the species mentioned in the 2 rows below)	0.050
Muscle meat of the following fish: anchovy, bonito, common two-banded seabream, eel, grey mullet, horse mackerel or scad, louver or luvar, sardine, sardinops, tuna, wedge sole	0.10
Muscle meat of swordfish	0.30
Crustaceans, excluding brown meat of crab and excluding head and thorax meat of lobster and similar large crustaceans	0.50
Bivalve molluscs	1.0
Cephalopods (without viscera)	1.0
Cereals excluding bran, germ, wheat and rice	0.10
Bran, germ, wheat and rice	0.20
Soybeans	0.20
Vegetables and fruit, excluding leaf vegetables, fresh herbs, fungi, stem vegetables, pine nuts, root vegetables and potatoes	0.050
Leaf vegetables, fresh herbs, cultivated fungi and celeriac	0.20
Stem vegetables, root vegetables and potatoes, excluding celeriac. For potatoes, the maximum level applies to peeled potatoes	0.10

Mercury

Foodstuffs	Maximum levels (mg/kg wet weight)
Fishery products and muscle meat of fish, excluding species listed in the row below. The maximum level applies to crustaceans, excluding the brown meat of crab, and excluding the head and thorax meat of lobsters and similar large crustaceans	0.50
Muscle meat of the following fish: anglerfish; atlantic catfish; bonito; eel; emperor, orange roughy, rosy soldierfish; grenadier; halibut; marlin; megrim; mullet; pike; plain bonito; poor cod; Portuguese dogfish; rays; redfish; sail fish; scabbard fish; seabream, Pandora; shark (all species); snake mackerel or butterfish; sturgeon; swordfish; tuna	1.0

Tin (inorganic)

Foodstuffs	Maximum levels (mg/kg wet weight)
Canned foods other than beverages	200
Canned beverages, including fruit juices and vegetable juices	100
Canned baby foods and processed cereal-based foods for infants and young children, excluding dried and powdered products	50
Canned infant formulae and follow-on formulae (including infant milk and follow-on milk), excluding dried and powdered products	50
Canned dietary foods for special medical purposes intended specifically for infants, excluding dried and powdered products	50

Arsenic

Arsenic is not covered in this Regulation, but there are maximum limits for arsenic in food in the UK, as set down in the UK Arsenic in Food Regulations (as amended) 1959.

Heavy-Metal Analysis

Provisions for methods of sampling and analysis for the official control of lead, cadmium, mercury, inorganic tin in foodstuffs are laid down in Commission Regulation (EC) No 333/2007.

US Regulatory Information

The US Food and Drug Administration (FDA) publishes a booklet giving action levels established for poisonous or deleterious substances in human food and animal feed. These action levels for poisonous or deleterious substances are established to control levels of contaminants in human food and animal feed. The booklet provides action levels for the heavy-metal contaminants cadmium, lead and mercury in certain foods and commodities. It was published in August 2000 and any new action levels published since then are published in the Federal Register.

The document can be accessed on the FDA web site at: http://www.cfsan.fda.gov/~lrd/fdaact.html

Sources of Further Information

Published

Duruibe, J.O., Ogwuegbu, M.O.C. and Egwurugwu, J.N. Heavy metal pollution and human biotoxic effects. *International Journal of Physical Sciences*, 2007, 2 (5), 112–118.

Jarup L. Hazards of heavy-metal contamination. *British Medical Bulletin*, 2003, 68, 167–182.

On the Web

SCOOP Report on heavy Metals in Food – March 2004. http://ec.europa.eu/food/food/chemicalsafety/contaminants/scoop_3-2-11_heavy_metals_report_en.pdf

Food Standards Agency Survey on Heavy metals in Foods. http://www.foodstandards.gov.uk/news/newsarchive/2007/jan/heavymetals

EFSA's scientific opinion and summary on mercury and methylmercury http://www.efsa.europa.eu/en/science/contam/contam_opinions/259.html

Heavy metals – information sheet. http://www.lenntech.com/heavy-metals.htm

FDA Total Diet Study Results for Arsenic, Cadmium, Lead, Mercury and other Elements, December 2006. http://www.cfsan.fda.gov/~comm/tds-res.html

Mercury in Fish. http://www.fda.gov/oc/opacom/hottopics/mercury/backgrounder.html

Mercury levels in commercial fish and shellfish. http://www.cfsan.fda.gov/~frf/sea-mehg.html

Mercury in fish – your questions answered. http://www.food.gov.uk/multimedia/faq/mercuryfish/

2.2.3.3 Perchlorate

Hazard Identification

What is Perchlorate?

Perchlorate is a chemical that occurs naturally and is also manufactured. The perchlorate anion consists of a chlorine atom surrounded by four oxygen atoms, and it is a very strong oxidising agent.

Perchlorate is very soluble in water, stable under most environmental conditions and very mobile in most media. Perchlorate has been recognised in the United States as an emerging contaminant, mainly associated with industrial activity and space exploration. Owing to this, there has been increasing interest in the levels of perchlorate in soil, groundwater, drinking water, irrigation water and food.

Occurrence in Foods

During 2004, the US FDA (Food and Drug Administration) conducted an initial survey investigating the perchlorate levels in a variety of products, including bottled water, milk, lettuce, tomatoes, carrots, spinach, and melons. Produce samples were collected particularly from regions known to have perchlorate-contaminated water supplies, such as Southern California and Arizona. Bottled water and milk samples were collected across the entire USA.

A further study conducted in 2005, extended the scope of the investigation to include additional samples of previously examined produce, together with fruits, such as apples, grapes and oranges, and their juices, vegetables such as cucumbers, green beans and greens, and seafood. In addition, grain products such as wheat flour, cornmeal and rice were sampled. On this occasion, the samples were obtained from a broader range of locations.

The results from these studies indicated that perchlorate was present in almost all samples of milk tested, at levels from 1.91 to 11.3 ppb. Perchlorate was found in varying amounts in lettuce, tomatoes, carrots, spinach, and melon, and in oatmeal, whole-wheat flour and a single sample of cornmeal. Although some samples were found to contain relatively high perchlorate levels, they were not deemed by the FDA to represent a risk to public health.

The full results of the studies can be found on the FDA web-site at: http://www.cfsan.fda.gov/~dms/clo4data.html

Hazard Characterisation

Effects on Health

Exposure to high doses of perchlorate has been found to interfere with iodine uptake into the thyroid gland. Perchlorate appears to remove an iodine ion from a protein that transports the iodine to the thyroid, leading to iodine deficiency. This, in turn, disrupts thyroid development and function, and may lead to a reduction in thyroid production. The thyroid plays an essential role in regulating metabolism, and in the developing foetus and in infants, thyroid hormones are essential for normal growth and development of the nervous system. Pregnant women and their unborn children are therefore at the greatest risk of iodine deficiency.

Although no studies have indicated that perchlorate-induced changes to thyroid function occur, even at doses as high as 0.5 mg/kg bodyweight/day, a recent report has suggested that a significant association might be present between perchlorate exposure and reduced thyroid function in women with low urinary iodine levels.

A report to assess the health implications of perchlorate, published by the NAS in 2005 (see link below), recommended a reference dose for perchlorate (RfD) of 0.7 μg/kg bodyweight/day. Inhibition of iodine uptake, the precursor to hypothyroidism, was used to derive the reference dose, which has now also been adopted by the EPA. The RfD has been set so that it protects those most at risk from perchlorate; namely, the foetuses of pregnant women who might have hypothyroidism or iodine deficiency. The RfD is equivalent to a level of 24.5 ppb of perchlorate in drinking water, based on a daily consumption of 2 litres.

As yet, there is no established standard for perchlorate in bottled water.

Sources

Naturally occurring perchlorate is found in the soil, particularly in dry areas, in nitrate fertiliser deposits in Chile (Chile saltpetre), and in potash in the USA and Canada. Ammonium perchlorate is also manufactured in the USA, where it is used as an oxidising agent in missile and rocket fuel. The compound is also used in fireworks and airbag inflators. The highest levels of perchlorate contamination are found in water and soil near military installations and around the industrial plants where the chemical is manufactured.

Perchlorate is thought to enter plants when they are irrigated with perchlorate-containing water, or when they are cultivated in soil containing natural perchlorate or perchlorate-containing fertilisers or water.

Stability in Foods

Perchlorate is very soluble in water, stable under most environmental conditions and very mobile in most media. Because of its high water solubility and stability, it tends to accumulate in foods that have a high water content, such as cucumbers, melons and tomatoes, when they are grown in soils contaminated with perchlorate or irrigated with perchlorate-contaminated water.

Control Options

Control is currently centred on reducing contamination of soil and water with perchlorate. Biological remediation appears to have the most promise for dealing with contaminated sites. Some bacteria possess perchlorate reductase enzymes, which could possibly be used to treat contaminated water, although, currently, systems involving the use of these micro-organisms have not been commercialised and are not used by US water authorities. Commercial anion-exchange systems also offer promise for treating perchlorate-contaminated water.

Legislation

The US Environmental Protection Agency has recommended a safe level for perchlorate in drinking water of 24.5 μg/litre, based on a reference dose (RfD) of 0.7 μg/kg bodyweight/day (The RfD is an estimate of daily oral exposure that is unlikely to cause any deleterious effects over a lifetime), but suggests that a safe level for babies should be 4.0 μg/litre.

The FDA has not established a standard for perchlorate levels in bottled water, and current legislation does not require bottled water manufacturers to test for perchlorate.

Sources of Further Information

Published

Blount, B.C., Pirkle, J.L., Osterloh, J.D., Valentin-Blasini and L., Caldwell, K.L. Perchlorate and Thyroid Hormone Levels in Adolescent and Adult Men and Women Living in the United States. *Environmental Health Perspectives*, December 2006, 114, (12), 1865–1871.

On the Web

Perchlorate – questions and answers. http://www.cfsan.fda.gov/~dms/clo4qa.html

Health Implications of Perchlorate Ingestion (2005) http://www.nap.edu/openbook.php?isbn=0309095689

Perchlorate as an environmental contaminant – review article. http://www.clu-in.org/download/contaminantfocus/perchlorate/ESPR_9_187_192.pdf

State of California Department of Toxic Substances Control – perchlorate page. http://www.dtsc.ca.gov/hazardouswaste/perchlorate/

2.2.4 VETERINARY RESIDUES

2.2.4.1 Antibiotics

Hazard Identification

What are Antibiotics?

The term "antibiotics" is now used to describe a broad and diverse range of chemical compounds that destroy, or limit, the growth of micro-organisms. Antibiotics may have activity against bacteria, fungi, or protozoa, though not viruses, and are used widely as veterinary drugs in food animals by the farming industry. There are many classes of compound with antibiotic properties, but some of the major groups in use are the β-lactams (including the penicillins), macrolides, ionophores, quinolones, lincosamides and tetracyclines.

Antibiotics may be administered to food animals for two reasons. They may be used, at relatively high doses, as therapeutic agents to treat clinical infections, or they may be administered at low, subtherapeutic doses as "growth promoters". The use of antibiotic growth promoters in intensive livestock farming has been shown to be an effective means of increasing the growth rate of food animals and improving the quality of meat by raising the protein content. It is not entirely clear how this effect is achieved, but it seems likely that antibiotic growth promoters in animal feed suppress some of the bacteria in the gut and allow more of the energy in the feed to be diverted to the growth of the animal. A further benefit of antibiotic growth promoters is said to be improved control of disease caused by bacterial pathogens, including *Salmonella* and *Campylobacter*, in intensively reared livestock.

The use of antibiotics in food animals has both direct and indirect implications for food safety. Some antibiotics and their metabolites may be toxic to humans, or may cause serious reactions in sensitive individuals (*e.g.* penicillins). Therefore antibiotic and antibiotic metabolite residues in meat, milk and other animal products may be a direct risk to human health. However, many experts currently consider that the development of antibiotic resistance in pathogenic bacteria that can cause disease in animals and humans (zoonoses) is a much more serious potential threat to human health, and the use of antibiotic growth promoters is widely thought to have contributed to reported increases in the prevalence of antibiotic resistance. The farming industry is a significant consumer of antibiotics, and it has been estimated that as much as 60–80% of antibiotics produced in the United States are administered in feed to healthy livestock at non-therapeutic levels. Many of these antibiotics are closely related to compounds that are administered to humans in clinical settings, and include tetracyclines, macrolides, streptogramins, and fluoroquinolones.

Occurrence in Foods

Antibiotic residues are most likely to be found in foods of animal origin, such as meat, poultry, fish, eggs and honey. They are usually present as a result of the use of therapeutic veterinary drugs to control infection and disease in food animals. Antibiotics are frequently used to treat mastitis in cows, and therefore antibiotic residues may be present in milk. Antibiotic residues in milk can pose significant problems to the dairy industry, as many of the antibiotics used may inhibit the starter cultures used in cheese and yoghurt production.

The use of antibiotic growth promoters in animals is unlikely to give rise to detectable residues in meat and other animal products unless they have been administered at levels much higher than are permitted.

The use of veterinary drugs for therapeutic use is highly regulated within the EU and in the US, and only certain drugs that have met stringent safety requirements are permitted (see Control Options). However, residues of antibiotics not authorised for food use may sometimes be found in certain foods. An example of this is the occasional detection of chloramphenicol residues in honey imported from China. Chloramphenicol is suspected of involvement in a form of anaemia in humans and is banned from food-animal use worldwide. Nitrofurans are also banned from food use in most of the world, but have been regularly detected in poultry and farmed crustaceans imported from East Asia and South America.

It is difficult to estimate current dietary intake of antibiotic residues from animal-derived products, but it is likely to be very low.

Hazard Characterisation

Effects on Health

The control of veterinary medicines in the EU and the USA is sufficiently strict that potentially toxic antibiotic residues are now very unlikely to be found in commercially produced animal products. Furthermore, most of the permitted antibiotics used are not considered to present a risk to human health at the levels likely to be found in meat, fish, milk, or eggs. However, there are still some concerns over the possible presence of penicillin and its derivatives. A number of individuals are sensitive to penicillins, and exhibit an immunopathogenic response that can be life threatening. This makes it essential that MRLs for this class of drugs are strictly adhered to. In addition, some hypersensitive individuals may develop a reaction to low levels of tetracyclines, also used in veterinary medicine.

Of much more concern is the possible role of antibiotic growth promoters in the development of antibiotic resistance in zoonotic bacterial pathogens. There is now considerable evidence that the use of medically important antibiotics as growth promoters in food animals may have contributed significantly to a reported rise in antibiotic resistance in several pathogenic bacterial species that cause zoonotic infections, notably *Salmonella enterica* serotypes,

Campylobacter jejuni, *Escherichia coli* and enterococci. For example, *Salmonella* Typhimurium definitive phage type (DT) 104 is a strain first isolated in the UK in 1988. At that time it already showed resistance to ampicillin, tetracycline and other antibiotics, but since 1988 it has spread all over the world and is often isolated from food animals. Many isolates are now resistant to other antibiotics, including fluoroquinolones, some of which have been used as growth promoters. Human infections caused by these bacteria now have very limited treatment options. The prevalence of fluoroquinolone-resistant *Campylobacter* in poultry is also increasing, especially in countries that permit the use of these antibiotics as growth promoters. The incidence of human infections caused by these pathogens is reported to be rising, especially in the USA.

The increasing prevalence of antibiotic resistance in zoonotic pathogens is now a global problem and many experts believe that the practice of using antibiotic growth promoters in food animals must be banned worldwide as it is in the EU. There are fears that, unless action is taken, antibiotics will soon no longer be effective as a treatment for many bacterial infections in animals and humans.

Sources

It is now thought that all antibiotic residues found in food are present as the result of being administered to animals for therapeutic reasons, or as growth promoters. There is little or no evidence to support suggestions that some antibiotics, such as chloramphenicol, can be produced naturally by microbial action in the soil.

Stability in Foods

Many studies have been carried out investigating the effects of processing on the stability of antibiotic residues in food, with very variable results reflecting the wide range of chemical compounds concerned. For example, the penicillins and tetracyclines are known to be heat sensitive and may degrade during cooking or canning processes, although the degree of degradation is variable and depends often on the nature of the food containing the residues. In addition, the implications of this to food safety are uncertain, since the nature of the degradation products is unknown in most cases. It is possible that some degradation products may be more toxic than the antibiotic from which they are derived.

Control Options

Control of antibiotic residues in food is focused on the strict regulation of the veterinary medicines administered to food animals.

Primary Production

To safeguard human health, maximum residue limits (MRLs) at the time of slaughter can be determined for veterinary medicines in order to set permissible limits for antibiotic residues in foods. The limits depend on the toxicity of the drug in question. Establishing an MRL also requires the setting of a minimum withdrawal period. This is the time that passes between the last dose administered to the animal and the time when the level of residues in the tissues, milk or eggs are lower than, or equal to, the MRL. Neither the animal, nor its products can be used for human consumption until the withdrawal period has elapsed. The withdrawal period is set out in the data sheet for the medicine and on the product packaging instructions. In the EU, only those drugs with established MRLs are permitted for use in food animals. MRLs are set with very large safety margins. For example, the calculation of the MRL value is based on the acceptable daily intake (ADI) for the drug in question. The calculation of the ADI includes an extremely large safety factor, and the MRL calculation assumes an average daily intake of 500 g of meat, 1.5 litres of milk, 2 eggs and 20 g of honey.

A full list of all permitted medicines and their established MRLs can be accessed on the European Medicines Agency web site at: http://www.emea.europa.eu/htms/vet/mrls/mrlop.htm

The use of subtherapeutic doses of antibiotics as growth promoters was banned in the EU from 1 January 2006 (Regulation (EC) No.1831/2003). It is still permissible to add coccidiostats and histomonostats (used to control protozoa), but their use as feed additives will be banned in the EU by 2009. The addition of subtherapeutic levels of antibiotics to animal feeds is currently still permitted in the United States and in other important meat-producing countries.

The effectiveness of all these controls is closely monitored in the EU by the use of extensive surveillance programmes.

Alternatives to Antibiotic Growth Promoters

A number of alternatives to the use of antibiotics as growth promoters in food animals have been suggested. These include the addition of digestive enzymes to animal feed to help break down certain feed components, the addition of probiotic microbes to animal feed, and the introduction of more effective infection controls, such as improved biosecurity measures.

In Sweden, where the use of antibiotic growth promoters was banned as long ago as 1985, it has been demonstrated that antibiotics are not necessary to produce healthy food animals in modern farming systems if accommodation, husbandry practices and feed quality are of a sufficiently high standard. However, Swedish production costs are still higher than those of other countries.

Legislation

EU

Information on legislation regarding MRLs for antibiotic residues and residues of other medicinally acceptable veterinary drugs for food-producing animals can be found on the European Medicines Agency web site at: http://emea.europa.eu/htms/vet/mrls/mrlfaq.htm

Specific legislation that relates to the establishment of MRLs in the European Union is laid out in the following Regulations and amendments:

- Council Regulation (EEC) No 2377/90 of 26 June 1990 laying down a Community procedure for the establishment of maximum residue limits of veterinary medicinal products in foodstuffs of animal origin.

The Regulation has subsequently been amended by:

- Commission Regulation (EEC) No 762/92
- Council Regulation (EC) 434/97
- Council Regulation (EC) No 1308/1999

Substances with established MRLs are listed in the annexes to Regulation 2377/90 where the relevant MRL values and target animal species are identified. These annexes are updated by Commission Regulations published regularly in the L-series of the Official Journal of the European Communities.

US Legislation

Maximum tolerance levels for residues of animal drugs in food have also been laid down by the United States Food and Drug Administration. These levels can be accessed on the Internet at the following link: http://www.access.gpo.gov/nara/cfr/waisidx_02/21cfr556_02.html

Sources of Further Information

Published

Moats, W.A. The effects of processing on veterinary residues in foods. *Advances in Experimental Medicine and Biology*, 1999, 459, 233–241.

Long A.R. and Barker, S.A. Antibiotics in foods of animal origin. *Encyclopaedia of Food Science and Technology*, (1991), Vol. 1 (Y.H. Hui, ed.) Wiley Intersciences, John Wiley and Sons, p. 59.

On the Web

The UK Veterinary Medicines Directorate http://www.vmd.gov.uk/

The European Medicines Agency http://emea.europa.eu/

FDA Center for Veterinary Medicine http://www.fda.gov/cvm/antimicrobial.html

Antibiotic growth promoters in food animals http://www.fao.org/docrep/article/agrippa/555_en.htm

2.2.4.2 Hormones

Hazard Identification

What are Hormones?

Hormones are naturally produced chemicals that occur in the bodies of all animals, including humans. They can be proteins or steroids, and they act as chemical messengers. They are produced in specific hormone-producing organs or glands (the endocrine system) and circulate around the body until they reach the sites where they exert their effects. Although only produced in small amounts, they control essential body functions such as growth, development and reproduction.

Although normally produced naturally, hormones are sometimes used therapeutically. For example, insulin is a protein hormone that is administered to control Type-1 diabetes in humans. Certain hormones are also used as growth promoters to make young livestock develop and gain weight more rapidly and to increase yields. Hormone administration to cattle and sheep increases their growth rate and reduces the amount of feed needed before an animal is ready for slaughter. In dairy cattle, hormones can also be used to increase milk production. Thus, hormones are administered to animals mainly for economic purposes.

The use of hormones in food animals is controversial and there are concerns that the practice may have implications for human health.

Occurrence in Foods

Hormones are not permitted for use in meat- or milk-producing animals in the EU. However, they are permitted in the USA, where they can be used in cattle and sheep.

There are currently six different kinds of hormones, all steroids, approved for use in food production in the USA. These hormones are estradiol, progesterone, testosterone, zeranol, trenbolone acetate and melengestrol acetate. Estradiol and progesterone are natural female sex hormones, testosterone is a natural male sex hormone and zeranol, trenbolone acetate and melengesterol acetate are synthetic hormone-like chemicals that make animals gain weight faster. These hormones are permitted for use in cattle and sheep, but not in poultry or pigs.

The use of recombinant bovine growth hormone (rbGH) is also permitted in the USA for use only in dairy cattle. RbGH, also known as recombinant bovine somatotropin, is a protein hormone used to increase milk production in dairy cows. This hormone is not permitted for use in the EU. As long as hormones are used as directed and correct treatment and withdrawal times are adhered to, the likelihood of unwanted hormone residues in meat and milk is low.

There are also reported to be significant levels of certain natural hormones in some plant-based foods. For example, potatoes and wheat have been reported to contain progesterone, and testosterone has been found at detectable levels in wheat and oils.

Hazard Characterisation

Effects on Health

The main concern over the use of steroid hormones for promoting growth in meat-producing animals is whether these hormones present any risk to human health. Lifetime exposure to oestrogen is associated with an increased risk of breast cancer, and excess exposure to anabolic steroids may result in a precocious puberty effect. Steroid hormones in food were suspected of causing early puberty in girls in some studies. However, exposure to higher than natural levels of steroid hormones through hormone-treated meat has not been documented. Studies have suggested that if correct treatment and slaughter practices are followed, the levels of these hormones may be slightly higher in treated animals, but still within the normal range of natural variation known to occur in untreated animals. Given the increased levels of other endocrine-disrupters in the environment, it is very difficult to attribute any increase in hormone-related cancers solely to hormone residues in meat.

With respect to milk from rbGH-treated dairy cows, scientists at the FDA Center for Veterinary Medicine have concluded that drinking milk with slightly higher levels of rbGH has no effect on human health, as the amount of rbGH present is insignificant compared with the amount of growth hormone produced naturally in the human body. Furthermore, because rbGH is a protein hormone, it is likely to be broken down during digestion.

There are, however, slight concerns over the effects of rbGH on the treated animal. The growth hormone acts by triggering cells to produce growth factors that cause an increase in growth rate and milk production. Milk from rbGH-treated cows has been found to contain slightly elevated levels of insulin-dependent growth factor-1 (IGF-1). Studies have indicated that higher levels of IGF-1 than normal are present in the blood of women with breast cancer, but it is unclear whether the higher levels are associated with increased breast cancer risk. Scientists at the FDA have concluded that IGF-1 in milk is unlikely to present any human food safety concern, particularly as it is a protein likely to be digested in the stomach.

There are also concerns that, because of increased milking, rbGH-treated cows may become more prone to mastitis, an infection of the udder. Growth-hormone treatment has also been shown to cause increased lameness and injection-site reactions in cattle. It has also been noted that there is a possible association between hormone use in large-scale beef-cattle production and undesirable effects in wild fish species living in rivers exposed to waste water from these farms.

Sources

The source of natural hormones in meat may be endogenous production by the endocrine system of the animal itself, or administration as a growth promoter. Synthetic and recombinant hormones can only originate from the latter source.

Stability in Foods

Some steroid hormones, including trenbolone and melengestrol acetate, have been shown to persist to some extent in animal dung, soil and water and so may cause environmental contamination. There are few documented reports on the stability of hormones in foods, but steroids are generally quite heat stable. For example, progesterone has been reported to survive heating at 56 °C for 30 min. It is therefore possible that steroid hormones might not be completely inactivated by typical meat-cooking processes or milk pasteurisation. It has been reported that pasteurisation destroys approximately 90% of residues of the protein hormone rbGH in bovine milk.

Control Options

Effective control of hormone residues in meat and milk depends on the careful administration of hormone preparations on the farm.

Primary Production

It is essential that hormones are used as directed, and that correct treatment and withdrawal times are adhered to. With these controls in place the likelihood of unwanted hormone residues in meat and milk is low. Steroid hormones are generally administered in the form of a pellet that is implanted beneath the skin of the ear. The ears of animals are then discarded at slaughter. Improper use of hormone-containing pellets, for example implantation into muscle tissue, results in higher levels of hormone residues in edible meat cuts. FDA regulations prohibit their use in this manner. Melengestrol acetate can also be added to animal feeds.

Recombinant growth hormone is administered as an injection beneath the skin of the animal. The hormone is available in single-dose packages to reduce the risk of accidental overdose.

Legislation

EU Legislation

The use of substances having a hormonal action for growth promotion in farm animals was prohibited in 1981 in the EU (Directive 81/602/EEC). This prohibition applies to Member States and imports from third countries. The legal instrument in force is Directive 96/22/EC as amended by Directive 2003/74/EC.

Recently, the European Food Safety Authority (EFSA) was asked by The European Commission to perform a review of scientific data on potential risks to human health from hormone residues in bovine meat and meat products. In accordance with the request, the Panel on Contaminants in the Food Chain reviewed the scientific literature between 2002 and 2007 before drafting an opinion, which was published in July 2007 and concluded that there were no grounds to call for revision of previous risk assessments.

The full text of the opinion can be found at the following web link: http://www.efsa.europa.eu/en/science/contam/contam_opinions/ej510_hormone.html

US Legislation

The FDA permits the use of the following hormones and synthetic hormone-like growth promoters in food production in the USA: estradiol, progesterone, testosterone, melengestrol acetate, trenbolone acetate, and zeranol. These substances are permitted for use in cattle and sheep, but not in pigs or poultry. Maximum tolerance levels for hormone residues in food have been laid down by the FDA.

These can be accessed at the following web link: http://www.access.gpo.gov/nara/cfr/waisidx_02/21cfr556_02.html

Meat from animals is regularly monitored for residues of synthetic hormones by the Food Safety Inspection Service (FSIS) of the US Department of Agriculture (USDA). Meat is also monitored for the presence of the illegal synthetic oestrogen, diethylstilbestrol.

Estradiol, progesterone and testosterone are all sex hormones produced naturally by animals and no regulatory monitoring of these hormones is possible, as it is difficult to differentiate administered hormones from those produced naturally in the body of the animal. Therefore, for naturally occurring hormones, the permitted residue levels are quoted in terms of an amount above the concentrations of hormone naturally present in untreated animals.

Use of recombinant bovine growth hormone (bovine somatotropin) is permitted in the USA, but only in dairy cattle.

World Trade Organization Dispute

The use of hormones in meat-producing animals has been a major source of contention between the EU and the USA. The import of hormone-raised beef into the EU was first banned during the 1980s. The USA, and later Canada, took the case to the World Trade Organization (WTO) for settlement of the dispute. The WTO ruled that the US and Canada could fine the EU for not abiding by world trade rules. Retaliatory trade restrictions and duties were then imposed by the US and Canada on the EU. The EU responded by issuing a new Directive on 22 September 2003, based on a full scientific risk assessment conducted between 1999 and 2002. The new Directive supported the continuation of the ban (Directive 2003/74/EC).

Sources of Further Information

Published

Andersson, A.M. *et al.* Exposure to exogenous estrogens in food: possible impact on human development and health. *European Journal of Endocrinology*, 1999, 140 (6), 477–485.

Collins, S.S. *et al.* The EEC ban against growth promoting hormones. *Nutrition Reviews*, 1989, 47 (8), 238–246.

Karg, H. The present situation and evaluation of the risk of using hormonal preparations in animals produced for food. *Monatsschrift fur Kinderheilkunde*, 1990, 138 (1), 2–5.

On the Web

Europa Food Safety – Hormones in Meat. http://ec.europa.eu/food/food/chemicalsafety/contaminants/hormones/index_en.htm

Hormones in Bovine Meat – Background and History of WTO Dispute. http://ec.europa.eu/dgs/health_consumer/library/press/press57_en.pdf

Human Safety of Hormone Implants used to Promote Growth in Cattle – A Review of the Scientific Literature. http://www.wisc.edu/fri/briefs/hormone.pdf

Section 3: Allergens

CHAPTER 3.1

Food Allergy

Hazard Identification

What is Food Allergy?

Food allergy can be defined as an adverse, immune-mediated reaction to food. Often, people will refer to any adverse reaction to food as an "allergy." However, it is important to remember that true food allergies involve the immune system and are almost invariably mediated through immunoglobulin E (IgE).

The majority of food allergies are caused by proteins, which sensitise and then elicit an allergic reaction in sensitive individuals. Food allergy needs to be differentiated from food intolerance, a condition that has no immune-system involvement and includes reactions to certain food components, such as lactose, amines and histamine. Adverse reactions that lack an immunological mechanism are sometimes referred to as non-allergic food hypersensitivity reactions. Food intolerances can sometimes be controlled by limiting the amount of a particular food eaten, but with food allergies, much stricter avoidance of the food is necessary. Only food allergy, and not food intolerance, can lead to the potentially fatal reaction of anaphylaxis.

Gluten intolerance or coeliac disease is also not to be confused with gluten or wheat allergy, even though the symptoms may be similar. Although coeliac disease is an immune system response, it is not mediated through immunoglobulin E, as all other true food allergies are. Unlike wheat allergy, coeliac disease is mediated through immunoglobulin A (IgA) and immunoglobulin G (IgG), and sufferers will develop gliadin-specific IgA and IgG antibodies. Coeliac disease does not cause the potentially fatal anaphylaxis associated with true food allergies if gluten is eaten.

Allergy-like food poisoning has also been confused, in some cases, with food allergy. The reaction occurs as a result of ingestion of histamine from products such as spoiled tuna, mackerel, other fish and occasionally cheese. Histamine is one of the primary mediators of allergic reactions and is released from the cells

The Food Safety Hazard Guidebook
By Richard Lawley, Laurie Curtis & Judy Davis

of the body during a true allergic reaction. In the case of allergy-like food poisoning, the histamine is ingested and then elicits the allergy-like symptoms.

Mechanism of Allergenicity

Immunoglobulins, such as IgE, are produced by the body's immune system as a defence against invading micro-organisms. Sometimes, the body also mounts an IgE response against certain agents, such as pollen, dust, house mites and food, and it is this response that gives rise to allergic reactions such as hay fever and food allergy.

There are two stages to the development of IgE-mediated allergies. The first is the sensitisation stage, in which an individual on first exposure to an antigen (usually a protein) will undergo a series of metabolic reactions resulting in the production of specific IgE (an antibody normally only produced in response to parasitic infections such as malaria).

The second stage involves elicitation of an allergic reaction. IgE becomes associated with specific receptors on the surface of special blood cells packed with inflammatory mediators, such as histamine. On the next exposure to the specific antigen, the cell-bound IgE reacts with the antigen, causing the cells to release the inflammatory mediators, which then trigger the symptoms associated with the allergic response, such as difficulty in breathing, gastrointestinal upsets and skin itchiness, *etc.* These symptoms normally occur within a very short time following exposure to the antigen.

The majority of food allergens are proteins. Sensitisation can occur through ingestion of the allergen, or through inhalation of certain allergens such as birch or grass pollen. Owing to the similarities between certain allergens, cross-reactions can occur in some unfortunate individuals, who might find themselves allergic to more than one type of allergen. Cross-reactions are particularly common between pollen or latex and some fruits and vegetables, giving rise to the syndrome known as pollen-fruit or latex-fruit syndrome.

Another subset of food allergies is known as "exercise-induced allergy". In this case, the allergic response occurs only when the specific food is eaten just before or after exercise.

Prevalence

The overall and worldwide prevalence of IgE-mediated food allergies is not precisely known. About 1–2% of adults and between 5 and 7% of children are believed to suffer from some type of food allergy, and it is believed that these numbers are increasing. The prevalence is higher amongst children who often grow out of allergies, such as cow's milk or egg allergy. Prevalence also depends on country, for example, peanut allergy is particularly common in the United States, where peanut butter is a very widely consumed food. Mustard allergy is particularly common in France, and celery allergy is very common in Switzerland, Germany and France.

Currently, legislation in the EU requires that the following allergens must be declared on food labels: cereals containing gluten, crustaceans, milk, eggs, fish, peanuts, soya beans, tree nuts, celery, mustard, sesame seeds and all their products, and sulfur dioxide. Legislation in the US requires that the following eight types of allergen be declared: cow's milk, eggs, peanuts, tree nuts, wheat, soya, fish and shellfish. (For more detail, please see Allergen Legislation).

This section of the Food Safety Hazard Guide covers the twelve major food allergens currently designated by EU legislation, although it is clear that allergies can be caused by many more foods than these.

Allergen Nomenclature

An allergen is termed "major" if it is recognised by IgE from at least 50% of a cohort of allergic individuals, but does not carry any connotation of allergenic strength; otherwise, allergens are termed "minor". The allergen designation is based on the Latin name of the species it originates from, and is made up of the first three letters of the genus followed by the first letter of the species finishing with an Arabic number, *e.g.* Ara h 1 is an allergen from peanuts (*Arachis hypogea*), and Gly m 1 is an allergen from soya (Glycine max.).

Hazard Characterisation

Effects on Health

The main symptoms of IgE-mediated food allergy are:

Gastrointestinal	Nausea, vomiting, abdominal cramping, diarrhoea
Respiratory	Wheezing, asthma, rhinitis
Cutaneous	Itching, urticaria (hives), eczema, atopic dermatitis, angioedema, rash
Other	Hypertension, increased heart rate, tongue swelling, anaphylactic shock, oral allergy syndrome, laryngeal oedema

Dose-Response

The amount of allergen required to elicit an allergic response varies tremendously between individuals and between allergens. In some cases, the dose required to elicit a response can be minute (measured in micrograms), and even kissing someone known to have eaten the allergen is sometimes enough to cause a reaction. Inhalation of vapours from cooking of the allergen can also cause life-threatening reactions for some individuals. For this reason, people with food allergies are generally advised to avoid the offending food completely.

Management of Food Allergy

Typically, the prevention of IgE-mediated food allergy involves avoidance of the offending food and strict observance of food labels. For management of specific food allergies, please refer to the relevant sections.

Sources of Further Information

Published

Bush R.K. and Hefle, S.L. Food Allergens, *Critical Reviews in Food Science and Nutrition*, 1996, 36, Suppl: S119–163.

On the Web

IFST Information Statement *Food Allergy*. www.ifst.org/uploadedfiles/cms/store/ATTACHMENTS/allergy.pdf

CHAPTER 3.2

Specific Allergens

3.2.1 CELERY ALLERGY

Hazard Identification

Celery (*Apeum graveolens*) grows wild in Europe, around the Mediterranean and in Asia west of the Himalayas. It is also widely cultivated as a vegetable, which is consumed raw, cooked or dried in spice mixtures. Celery is grown for its wide, fleshy stalks as well as its large, edible tuber, known as celeriac. Celery stalks are commonly used in soups, stews and in salads, and celeriac is used mainly as a cooked vegetable, but is becoming increasingly popular grated into raw salads. Celery is also grown for its seeds, which contain a valuable essential oil used in the flavouring, perfumery and pharmaceutical industries. Celery seeds are used as a flavouring, either whole or ground into a powder, which is mixed with salt to form celery salt. Celery salt is also sometimes made from celeriac.

Celery is one of the most common foods to cause oral allergy syndrome in adults in countries such as Switzerland, Germany and France.

Allergenicity

Allergy to celery root (celeriac) is more common than allergy to celery stalks. The principal allergen in celery is designated Api g1, and it appears to be resistant to heat, so that its allergenicity is retained even after extensive thermal treatment. Cooking, therefore, does not reduce the allergenicity of celery or its products. Celery spice and raw celery are equally allergenic.

Allergy to celery is often associated with allergy to tree and grass pollen. Individuals who develop allergy to birch pollen tend to be allergic to the birch-pollen allergen, designated Bet v 1. Proteins related to Bet v 1 are found in other plants and in the edible tissues of a number of fruits and vegetables, including

The Food Safety Hazard Guidebook
By Richard Lawley, Laurie Curtis & Judy Davis

celery. When people who have a Bet v 1-type allergy eat certain fruits and vegetables, such as celery, they often experience a reaction confined to the mouth, known as oral allergy syndrome. Because allergy to celery is frequently associated with birch and/or mugwort pollinosis, the term birch-mugwort-celery syndrome has been established.

Allergy to other vegetables, such as carrots and bell peppers, is also associated with celery allergy, as is allergy to certain other members of the *Apiaceae* family, such as parsley, aniseed, cumin and coriander.

Prevalence

Allergy to celery is particularly common in European countries, such as Switzerland, Germany and France. It is the most common pollen-related food allergy in Switzerland, where about 40% of patients with food allergy are allergic to celery root, and severe anaphylactic reactions have been observed. In France, about 30% of severe allergic reactions to food were thought to be caused by celery.

There is evidence that birch pollen and celery allergy are highly related in Central Europe, while in Southern Europe, celery allergy is most frequently related to mugwort pollen.

Hazard Characterisation

Effects on Health

The most common symptom associated with celery allergy is the oral allergy syndrome. During challenge testing with celery, 50% of patients developed local reactions in the mouth and 50% developed systemic reactions. Other symptoms include:

- itchiness and redness of the skin and skin swelling;
- stomach cramps and nausea;
- wheeziness, asthma and tightness of the chest;
- anaphylactic shock.

The symptoms associated with celery allergy are frequently more severe compared with allergic reactions associated with other fresh vegetables.

Dose-Response

The threshold dose needed to elicit an allergic reaction has not yet been established; however, in a study of patients undergoing oral challenge with celery, almost a half developed symptoms of allergy at a dose of 700 mg.

Management of Celery Allergy

Avoidance of celery, celeriac and all foods containing celery is the best way to manage the condition. The main difficulty arises in the extensive use of celery extracts in spices. The dried powder from celeriac is used as a flavouring ingredient in numerous processed foods, such as soups, stews, salad dressings and spice mixtures. Care should be taken when reading food labels. Owing to its high allergenic potential, celery has now been included as one of the major allergens that have to be labelled in pre-packed foods sold in the EU. This is not currently the case in the United States.

Sources of Further Information

Published

Ballmer-Weber, B. *et al.* Allergen Data Collection: Celery (Apium graveolens). *Internet Symposium on food allergens*, 2000, 2 (3), 145–167.

Ballmer-Weber, B. *et al.* Celery allergy confirmed by double-blind, placebo-controlled food challenge: A clinical study in 32 subjects with a history of adverse reactions to celery root. *Journal of Allergy and Clinical Immunology*, 2000, 106 (2), 373–378.

On the Web

Internet Symposium on Food Allergens http://www.food-allergens.de/
The InformAll Database http://foodallergens.ifr.ac.uk/

3.2.2 HEN'S EGG ALLERGY

Hazard Identification

Hen's egg allergy is one of the commonest immediate food allergies in children in Europe and America, but it also affects some adults too. It is caused by the proteins found in hen's eggs. Hen's eggs cannot be replaced by other eggs, such as those from ducks, turkeys, geese or quail as these are also known to cause allergic reactions in people who are sensitive. The correct name for the chicken is *Gallus gallus domesticus*, and therefore, the designated allergen names all start with the letters Gal.

Allergenicity

Eggs are made up of about 60% egg white and 35% egg yolk. The egg white appears to be slightly more allergenic than the egg yolk. Over 50% of the egg white is made up of the protein ovalbumin, the rest is made up from ovotransferrin, ovomucoid, ovomucin and lysozyme. Other minor proteins include ovoflavoprotein, ovodin, ovomacroglobulin and cystatin. The major egg-white allergens are ovomucoid, with the designated allergen name of Gal d 1, and ovalbumin, designated allergen name Gal d 2.

The proteins found in egg yolk include lipovittelin, phosvitin, egg yolk specific lipoprotein and apovittelin I and IV. It has been proposed that egg allergy in children is caused by egg-white proteins and in adults by livetins in the egg yolk.

Both of the major egg-white allergens, ovomucoid and ovalbumin are resistant to denaturation and enzymic digestion, but cooked egg appears to be less allergenic than raw egg.

It is thought that sensitisation occurs through ingestion of egg proteins in the diet. Even minute amounts of egg protein in human milk are sufficient to sensitise an infant, with a reaction occurring when the child eats food that contains egg. Consumption of poultry meat rarely causes a reaction. However, inhalation of allergenic proteins, which sometimes occurs in people who keep birds as pets, can cause sensitisation.

Prevalence

Egg allergy is one of the commonest allergies found in children, with a prevalence of about 2%. The majority outgrow their allergy before adulthood leaving less than 1% of the adult population allergic to hen's eggs. Early sensitisation to hen's eggs, however, may predispose some children to later development of asthma.

Hazard Characterisation

Effects on Health

In sensitised individuals, ingestion of egg or egg-white proteins will elicit an immediate response. The following symptoms have been observed:

- itching of the mouth and pharynx;
- eczema, pruritis and dermatitis, and urticaria;
- nausea and vomiting;
- rhinoconjunctivitis;
- in very rare cases – anaphylaxis.

Dose-Response

The minimum dose required to elicit an allergic reaction has been reported as 1 mg of liquid egg. The majority of those sensitive to egg allergy will respond to doses in the milligram to gram range. Reportedly, 5% will respond to doses below 5 mg, whereas about 50% will require doses of about 100 mg before symptoms are observed. As with most allergens, the threshold dose varies for each individual.

Management of Egg Allergy

Avoidance of eggs and all egg-derived products is the recommended way to treat this allergy. As the threshold dose varies so greatly between individuals, some may not need to avoid egg derivatives used as only very minor ingredients in foods, such as egg-yolk lecithin.

All pre-packed products containing eggs or egg-derived ingredients must now be labelled as such in the UK, EU and the United States. Egg-derived ingredients to look out for include albumin, ovalbumin, vitellin, globulin, and ovomucoid, *etc*. Prepared foods commonly containing eggs or egg derivatives include cakes, desserts, pasta, biscuits, mayonnaise, sauces and chocolate. Some childhood vaccines are also prepared in egg yolks and parents of very sensitive children need to be aware of this.

Sources of Further Information

Published

Poulsen, L.K., Hansen, T.K., Norgaard, A., Vestergaard, H., Skov, P.S. and Bindslev-Jensen, C. Allergens from fish and egg. Allergy (*European Journal of Allergy and Clinical Immunology*), 2001, (56), supplement 67, 39–42.

On the Web

Protall Information sheet http://www.ifr.ac.uk/protall/infosheet.html
The InformAll Database http://foodallergens.ifr.ac.uk/

3.2.3 FISH ALLERGY

Hazard Identification

Fin fish is one of the commonest causes of food allergy. It is a real food allergy resulting in IgE-mediated symptoms, which is not to be confused with the toxic reactions that occur after histamine ingestion from spoiled fish (which will usually cause a reaction in everyone who has eaten the fish).

The allergy is caused by ingestion of almost all fish because it involves a protein found in the muscle of the majority of fish species. Although not complete, the list of fish causing allergy includes cod, mackerel, herring, sardine, anchovy, bass, haddock, hake, plaice, sole, salmon, tuna, trout, Alaska pollock, eel, catfish, perch, and carp. Although fin fish and shellfish allergies are not linked by a common allergen, individuals may be allergic to both types of seafood.

Allergenicity

The major fish allergen is parvalbumin, a protein that is conserved across all species of fish. As the parvalbumins are similar in all species, individuals allergic to one type of fish are likely to be allergic to all others. Parvalbumin is heat stable and therefore, cooking is unlikely to remove the allergenicity from fish. In addition, other proteins in fish, apart from parvalbumin, have been shown to be allergenic. The designated allergen name for parvalbumin from cod is Gad c 1 (from the latin name for cod, *Gadus callarias*), and the designated allergen name for the allergen from salmon is Sal s 1 (from the official name *Salmo salar*). A few people who are allergic to fish also react to frog, as frog muscle also contains the protein parvalbumin.

Allergy to cartilaginous fishes also exists, but it is possible that there may be differences between these allergies and allergy to bony fish. The cartilaginous fishes include sharks, rays, dogfish and skate.

Prevalence

The prevalence of fish allergy varies, but it is generally thought to affect between 0.1 and 0.2% of the population. Both children and adults are affected, and fish allergy generally persists throughout the lifetime of an individual. Fish allergy is more prevalent in countries and parts of the world where fish constitutes a major part of the diet.

Hazard Characterisation

Effects on Health

As with most allergens the severity of the reactions varies depending upon the sensitivity of the subject and on how much of the allergen is consumed.

The first symptoms are generally itchiness and sensitivity of the mouth and throat, which can be followed by other reactions, such as:

- nausea, vomiting, stomach pains and diarrhoea;
- hives, itching, swelling and reddening of the skin;
- eczema, asthma and hay fever, accompanied by runny and itchy eyes and nose;
- swelling of the airways;
- anaphylactic shock.

Dose-Response

Doses as low as 5 mg of cod have been reported to elicit an allergic reaction. Allergic reactions to fish have also been reported after inhalation of allergens in the steam from cooking fish, and after kissing someone who had previously consumed fish. Cross-contamination from frying oil containing minute amounts of fish protein is also a problem. Manual handling of fish can also cause eczema or asthma in sensitive individuals.

Management of Fish Allergy

Once a diagnosis of fish allergy has been confirmed, the only way to successfully manage the allergy is by complete avoidance of fish in any form, and fish-derived ingredients. As one of the recognised major allergens, fish should always be labelled on pre-packaged foods in the EU and the US.

The following foods may contain hidden fish: surimi, pâté, Worcestershire sauce, Caesar salad dressing, oyster sauce, tapenade, pizza toppings, kedgeree, caponata, bouillabaisse, gumbo, paella, fruits de mer, frito misto (mixed fried-fish dish), fish sauce (Nuoc Mam and Nam Pla), gentleman's relish, sushi, and animal fat. Some fish or animal oils may also contain minute amounts of fish protein. Gelatine obtained from fish skin and bones and used in foods is not considered a problem for fish-allergic consumers.

Special care should be taken by people allergic to fish when they eat out in restaurants, as cross-contamination of foods can easily occur, for example, from the frying oil.

Sources of Further Information

Published

Bush, R.K. and Hefle, S.L. Food allergens. *Critical Reviews in Food Science and Nutrition*, 1996, 36 (Suppl. S), 119–163.

Poulsen, L.K., Hansen, T.K., Norgaard, A., Vestergaard, H., Skov, P.S. and Bindslev-Jensen, C. Allergens from fish and egg. Allergy (*European Journal of Allergy and Clinical Immunology*), 2001, (56), supplement 67, 39–42.

On the Web

The InformAll Database http://foodallergens.ifr.ac.uk/food.lasso?selected_food=5020

The Anaphylaxis Campaign http://www.anaphylaxis.org.uk/information/print_common_food_al.html

3.2.4 COW'S MILK ALLERGY

Hazard Identification

Hippocrates first observed and wrote about negative reactions to cow's milk around 370 BC, since when, the prevalence, awareness and understanding of this allergy has increased. Milk allergy is one of the major allergies in infants and is caused by the proteins present in cow's milk. Contrary to popular opinion, goats or sheep's milk cannot generally replace cow's milk for those who are sensitive. This is because of the similarity between the casein and whey proteins in cow's milk and those in the milk of goats and sheep.

Allergenicity

Most milk proteins are potential allergens and milk contains about 30–35 g protein/litre. The major allergens recognised in milk are casein, beta-lactoglobulin (a protein that is absent from human milk), alpha-lactalbumin and alpha-lactoglobulin. Although it may be reduced, the allergenicity of milk cannot be removed by simple thermal processing. Low heat treatment, like pasteurisation at 75 °C for 15 s, ensures the microbial safety of milk, but does not cause significant reduction in its allergenicity. Strong heat treatment (121 °C for 20 min) largely destroys the allergenicity of the whey proteins, but it only reduces the allergenicity of the caseins. Homogenisation has no effect on the allergenicity of milk proteins.

Casein appears to be the most potent allergen when it comes to skin tests, and beta-lactoglobulin appears to be the most potent in oral challenges.

The blood proteins present in cow's milk are also present in meat (beef). These proteins are not the most important allergens of milk, but for around 10% of milk-allergic patients, allergy to milk goes together with allergy to beef. Some of these people may tolerate well-cooked beef.

Prevalence

There are no definitive data on the prevalence of allergy to milk. However, in Western countries, it is believed to affect about 2–3% of children under the age of 2 years. In general, children lose this sensitivity as they grow up, with 90% losing it by the age of three. In a very few cases, milk allergy may persist and occur in adults. It is interesting to note that the pattern of sensitisation to milk proteins is not the same now as it was in 1990. For example, the prevalence of sensitisation to casein has dramatically increased, possibly in line with the much wider use of casein as a food ingredient.

Hazard Characterisation

Effects on Health

Cow's milk allergy differs from most other allergies, such as allergy to nuts or crustacea, in that the allergy generally develops before the age of three, and the majority of sufferers become tolerant to milk within a few years. Thus, the distribution of symptoms tends to be different from that of other allergies, with more cases of atopic dermatitis associated with milk allergy.

The majority of milk-allergic children demonstrate two or more types of symptoms in at least two different organs. Up to three quarters have skin symptoms, such as atopic dermatitis, eczema, and urticaria. Just over half have gastrointestinal symptoms such as vomiting, diarrhoea, constipation, and abdominal pain. About 20–30% have symptoms associated with breathing problems, such as hay-fever-like symptoms from the nose and eyes, and recurrent wheezing.

Systemic symptoms, such as anaphylactic shock may occur in up to 10% of subjects. In infants with cow's milk allergy, who are exclusively breast-fed, severe atopic eczema is the predominant symptom.

Symptoms can occur within a few minutes and up to an hour after milk exposure. These reactions are called immediate reactions. Reactions occurring after one hour are called delayed reactions. In some cases, symptoms even occur after a few days have passed. These late reactions are generally limited to atopic eczema and gastrointestinal disorders like constipation.

Dose-Response

The lowest dose of milk protein capable of eliciting an allergic reaction during challenge studies has been reported to be in the range of 0.6 mg to 180 mg. The minimum amount of milk reported to cause an allergic reaction is 0.02 ml cow's milk.

Management of Milk Allergy

Giving cow's milk formula as a first feed to babies with a family history of atopy may possibly lead to development of cow's milk allergy. Mothers in this situation should be advised accordingly.

Complete avoidance of cow's milk protein is the best way to manage the allergy. For babies and young infants, a hypoallergenic formula, *i.e.* one that has been extensively hydrolysed, is recommended if breast-feeding is not possible. Hydrolysis degrades the large allergenic milk proteins into smaller peptides that have lost their allergenicity. In rare cases, an amino-acid-based formula may be required (amino acids are the building blocks of proteins and peptides). Partially hydrolysed formulas are not well tolerated, as large protein fragments may still be allergenic. In older children, soya milk or soy-milk formula may offer an alternative. However, it has been shown that about 25% of individuals allergic to cow's milk will also be allergic to soya milk. Advice of a clinical dietician may help to ensure an adequate diet and in order to avoid "hidden" cow's milk proteins in commercial foods.

Casein and caseinates are widely used as extenders and tenderisers in foods such as sausages, soups and stews. Both casein and whey are used in high-protein powdered drinks. Other ingredients to look out for that may indicate the presence of milk include, butter, butterfat, butter oil, ghee, cheese, yoghurt and ice cream. Foods that may contain "hidden" milk proteins are so numerous it would be difficult to list them all, therefore, strict observance of food-package labels is essential, as pre-packed foods containing cow's milk and its derivatives have to be labelled by law.

Sources of Further Information

Published

Eigenmann, P.A. Anaphylaxis to cow's milk and beef meat proteins, *Annals of Allergy Asthma and Immunology*, 2002, 89(6 Suppl 1), 61–4.

Wal, J.M. Cow's milk proteins/allergens. *Annals of Allergy Asthma and Immunology*, 2002, 89(6 Suppl 1):3–10.

On the Web

Protall Information sheet http://www.ifr.ac.uk/protall/infosheet.html

Anaphylaxis Campaign factsheet http://www.anaphylaxis.org.uk/information/factsheets/23552%20milk%20allergy.pdf

LabSpec fact sheet http://www.labspec.co.za/l_milk.htm

3.2.5 MUSTARD ALLERGY

Hazard Identification

There are various varieties of mustard belonging to the *Brassicaceae* family. Mustard powder is typically a mixture of *Sinapsis alba* (white mustard) and *Brassica juncea* (oriental mustard). The mustard seeds are ground to form a powder that is used as a condiment and as flavouring in numerous dishes. The whole seeds are often used in pickling solutions to add flavour, and mustard oil is occasionally used in cooking. Because of its use as a flavouring, mustard can often act as a masked allergen, giving rise to serious allergic reactions. France is the largest European producer of mustard and also the biggest consumer, ahead of Germany and Great Britain. This explains the high prevalence of mustard allergy in France. In addition, the mustard varieties *Brassica nigra* and *Brassica juncea* are extensively cultivated in India.

Allergenicity

The major allergen of white mustard is designated Sin a 1, and that of oriental mustard Bra j 1. These allergens are heat stable and resistant to digestion by proteolytic enzymes, such as trypsin and proteases. Therefore, roasting mustard seeds has little effect on their allergenicity. Also, their resistance to proteolytic enzymes means that they have a high resistance to digestion in the stomach and will pass unchanged into the GI tract.

Numerous members of the *Brassicaceae* family are used as food plants, including cabbage, cauliflower, broccoli, watercress, horseradish and turnips. However, cross-reactions involving clinical symptoms between mustard and other *Brassicaceae* family members are rare. Cross-reactions with ragweed pollen have been reported.

Prevalence

As an emerging allergen, the prevalence of allergic reactions to mustard is on the rise. In Europe, it is particularly common in France, the largest producer and consumer of mustard, and most published research has been conducted by French researchers. Regional differences in prevalence have been reported. In the eastern part of France, a prevalence of 0.8% to 1% of food allergies is attributed to mustard; in the centre of France it is 3% and in the south of France, 8.9%. In Spain, 1.5% of food allergies are attributed to mustard.

India is another country where production and consumption of mustard is high. Prevalence of allergy to mustard is also very high in India. Because mustard is introduced into the diet at an early age, prevalence of mustard allergy is high in infants and children. There are no data indicating whether the allergy is outgrown.

Hazard Characterisation

Effects on Health

The initial clinical features are atopic dermatitis, urticaria and/or angioedema. Other typical symptoms include:

- asthma and wheeziness;
- abdominal pain and diarrhoea;
- dizziness, low blood pressure, and anaphylactic shock.

Contact dermatitis has also been reported in workers involved in salad production, and contact urticaria for workers in food factories.

Many incidents of anaphylactic shock to mustard have been documented, indicating the seriousness of this allergy; however, no deaths have been recorded.

Dose-Response

The dose of mustard required to elicit an allergic response is unclear. In studies, individuals have been shown to react to between 40 mg and 440 mg of a mustard condiment containing about 33% of seeds. Based on these findings, the smallest dose of mustard needed to elicit a response is approximately 14 mg.

Management of Mustard Allergy

As with all other food allergies, the best way to manage this allergy is by avoidance of all food products containing mustard. Because of its use as a seasoning and condiment, this is not always easy.

Foods to avoid include spicy sauces, curry sauces, mayonnaise, vinaigrette, crackers, flours, dried soups, and some baby foods. The whole seeds are used in pickling spices, so products such as baby gherkins and some pickled onions may be contaminated with mustard. Care should be taken when eating out in restaurants and at fast-food stands. Hot dogs are likely to be contaminated, as the individual preparing and serving the product will probably have handled mustard at some point.

All pre-packed food containing mustard must be labelled in the EU under the provisions of recent allergen legislation. This is not the case in the United States.

Sources of Further Information

Published

Rancé, F. Mustard allergy as a new food allergy. *Allergy*, 2003, 58, 287–288.
Montreal, P. *et al*. Mustard allergy. Two anaphylactic reactions to ingestion of mustard sauce. *Annals of Allergy*, 1992, 69, 317–320.

On the Web

UK Food Standards Agency information sheet http://www.eatwell.gov.uk/healthissues/foodintolerance/foodintolerancetypes/mustardallergy/
The InformAll Database http://foodallergens.ifr.ac.uk/

3.2.6 PEANUT ALLERGY

Hazard Identification

Peanuts are unrelated to tree nuts such as almonds and hazelnuts, and actually develop in a seed-pod below ground, which explains their alternative name – groundnuts. They are also sometimes called monkey nuts. Botanically, peanuts are a member of the legume family, which includes peas, soya beans and lentils.

Peanuts are one of the most common causes of food allergy and can cause severe reactions, including anaphylaxis. Very tiny amounts of peanut can cause a reaction in people who are sensitive. An adverse reaction to peanuts is a true food-allergy response, involving an over-reaction of the immune system and production of IgE antibodies.

Allergenicity

Peanuts are harvested as shelled products containing the fruit surrounded by a skin and formed into two halves. Peanut proteins make up about 25% of the fruit, and it is these proteins that are responsible for peanut allergenicity. Peanut proteins are thought to contain numerous allergenic fractions, many of which still remain unidentified and uncharacterised. Neither roasting nor other heat treatment of peanuts seems to reduce the allergenic response. In fact, roasting peanuts may actually increase their potential allergenicity. This is quite unusual, as most allergenic proteins can be made less allergenic, or non-allergenic, by heat treatment. On the other hand, when peanuts are boiled in water, their allergenicity is reduced. This is because some of the allergenic proteins leach out into the cooking water.

Prevalence

Peanuts are a common cause of food allergy in the USA, where consumption of peanuts is very high. Peanut allergy is also becoming increasingly common in the UK in line with the increasing popularity of peanut products. Although exact numbers are unknown, some studies suggest that one person in 200 might be affected to some degree, although a recent study in children, carried out in 2002, indicated that as many as one in 70 children across the UK was allergic to peanuts. At one time, it was thought that peanut allergy was life-long in all cases, but recently it has been shown that about 20% of young children outgrow their peanut allergy.

It is thought that the increased incidence of peanut allergy is the result of increased dietary exposure to peanuts at an earlier age than previously occurred. Susceptible infants can probably become sensitised through breast-feeding, via certain ointments used for skin lesions, or via the respiratory system following exposure to peanut allergen. Sensitisation may even occur *in utero*. Atopic individuals with asthma seem to be more at risk of developing food allergies.

Hazard Characterisation

Effects on Health

The symptoms of peanut allergy can vary tremendously, from very mild to severe. The most common mild symptoms include:

- tingling in the mouth and lips and facial swelling;
- nausea and colicky pain, accompanied by a feeling of tightness in the throat;
- urticaria or nettle rash.

Severe reactions, exhibited by those more sensitive to peanuts include:

- swelling of the airways and obstructed breathing;
- sudden drop in blood pressure;
- collapse and unconsciousness.

These symptoms result from the widespread release of pre-formed histamine and other inflammatory mediators from mast cells and basophil cells. The more severe reactions are classified as anaphylaxis and require immediate medical attention. The onset of anaphylactic reactions is generally extremely rapid and can proceed very quickly to unconsciousness.

A recent analysis was carried out of 32 fatal cases of food-related anaphylaxis reported to a national registry, established by the American Academy of Allergy, Asthma, and Immunology, with the assistance of the Food Allergy and Anaphylaxis Network. The 32 individuals could be divided into 2 groups. Group 1 had sufficient data to identify peanut as the responsible food in 14 (67%), and tree nuts in 7 (33%) of the cases. In group 2 subjects, 6 (55%) of the fatalities were probably due to peanut, 3 (27%) to tree nuts, and the other 2 cases were probably due to milk and fish. The sexes were equally affected; most victims were adolescents or young adults, and all but one subject were known to have a food allergy before the fatal event. In those subjects for whom data were available, all but one was known to have asthma, and most of these individuals did not have epinephrine available at the time of their fatal reaction. In this series, peanuts and tree nuts accounted for more than 90% of the fatalities.

Dose-Response

The amount of peanuts required to elicit an allergic reaction has not been extensively studied, although sensitive individuals can react to minute amounts (100 μg–50 mg). Some case studies report reactions to extremely low doses of peanut. For example, children have been reported to exhibit symptoms after contact with a table, reportedly wiped clean of all visible peanut butter; other cases have been documented as being caused by kissing someone who had previously eaten peanuts, or by sharing drinks. Symptoms were even reported by a patient when a jar of peanut butter was opened in their presence. Even being close to someone eating peanuts can be sufficient to cause a reaction in some individuals.

Management of Peanut Allergy

Complete avoidance of peanuts and all peanut products is the best way to manage peanut allergy, although this may not be straightforward. The presence of "hidden" peanut products in processed foods is always a risk for sensitised individuals. Food labels must always be read carefully as peanuts and their products may appear under different names, such as groundnuts, monkey nuts, earth nuts, mixed nuts, peanut butter, peanut oil, groundnut oil and arachis oil. Products such as cakes, biscuits, desserts, ice cream, cereal bars, satay sauces, breakfast cereals, ready meals (particularly Thai, Indonesian, Chinese and Indian meals), curry sauces, salad dressings, marzipan and praline and vegetarian products such as veggie-burgers, *etc.* may all contain hidden peanut products.

Eating out in catering establishments and buying unwrapped foods also pose a risk, as no labelling laws exist to cover these situations. Care is needed in preparation and storage of food to ensure that no cross-contamination occurs.

It is probably wise for children who are allergic to peanuts to avoid other nuts, sesame seeds, nut mixes and possibly other legumes to prevent further sensitisation.

Sources of Further Information

Published

Sampson, H.A. Clinical practice: Peanut allergy. *New England Journal of Medicine*, 2002, 346(17), 1294–1299.

On the Web

The Anaphylaxis Campaign www.anaphylaxis.org.uk/
British Nutrition Foundation fact sheet http://www.nutrition.org.uk/upload/Peanut%20Allergy.doc
The InformAll Database http://foodallergens.ifr.ac.uk/
Food Allergy Info web site (Institute of Food Research) www.foodallergens.info

3.2.7 SHELLFISH ALLERGY

Hazard Identification

Shellfish are biologically very different from finfish. They are aquatic invertebrates that can be divided into four main groups:

1. Crustaceans – crabs, prawns, shrimps, lobsters, crayfish.
2. Bivalve molluscs – mussels, oysters, clams, scallops.
3. Gastropod molluscs – winkles, whelks, periwinkles, limpets, snails.
4. Cephalopod molluscs – octopus, squid, cuttlefish.

As far as allergen legislation currently stands in the EU, it is the crustacean shellfish that have to be declared on food labels, and not (as yet) molluscan shellfish, although both types of shellfish are associated with allergic reactions. Allergic reactions reported from consumption of crustacean shellfish tend to be far more frequent and more severe than those reported from consumption of molluscan shellfish.

Allergenicity

Crustacean shellfish allergy is relatively common, and is thought to be caused by a protein known as tropomyosin, which is very similar in the majority of crustaceans. Thus, a person who is allergic to one type of crustacean shellfish is most likely to be allergic to others. Tropomyosin is also found in certain insects, such as cockroaches, dust mites and chronomid (used as fish food), and people allergic to crustacean shellfish may also be allergic to these. Cooking does not destroy the allergenicity, although some allergens may leach out into the cooking water, making this allergenic too.

Allergy to molluscan shellfish is not as common as allergy to crustacean shellfish, and people who are allergic to crustacean shellfish are not necessarily allergic to molluscan shellfish, although a small proportion may be. Allergy to one type of mollusc most likely pre-supposes allergy to others, as they are all very similar, although cross-reactions are most likely to occur within a specific group of molluscs. Thus someone allergic to squid may well also be allergic to octopus, whilst someone allergic to mussels may well be allergic to other bivalve molluscs, such as oysters or clams.

Prevalence

Allergy to crustacean shellfish is the third most common allergy after peanuts and tree nuts. It is thought that about 1% of the population may be affected, although the frequency varies tremendously throughout the world. Scandinavian countries, for example, appear to have higher rates of allergy to crustacean shellfish than other northern European countries. It has been estimated that approximately three-quarters of people allergic to one type of crustacean shellfish are also allergic to others.

Allergy to molluscan shellfish is not quite so common, and this may be why molluscan shellfish are not yet included in the list of major allergens that must be declared on food labels in the EU and US. It has been reported that allergy to molluscan shellfish is less frequent than allergy to crustacean shellfish by a factor of about three.

Food allergy to both types of shellfish has been reported in both children and adults, and, although little is known about the persistence of shellfish allergies, evidence suggests that they are not outgrown.

Hazard Characterisation

Effects on Health

As with most allergies, symptoms vary depending upon the sensitivity of the individual. Common symptoms include:

- itching of the lips, mouth and throat;
- swelling of the lips, tongue, throat and palate;
- urticaria, itchy skin, and swelling beneath the skin;
- nausea, vomiting and diarrhoea;
- asthma, difficulty breathing, wheeziness, and sore and runny eyes;
- anaphylaxis.

Shellfish are the third most common cause of anaphylaxis after peanuts and tree nuts.

Symptoms can occur after ingestion of shellfish, when shellfish are handled, or even by inhalation of steam from cooking shellfish.

Dose-Response

There is very little evidence in the literature relating to the minimum amount of shellfish needed to cause an allergic reaction, although it is likely to be very small, as inhalation of shellfish allergens in the steam from cooking water has been known to elicit an allergic reaction in some people.

Management of Crustacean Shellfish Allergy

Once a diagnosis of crustacean shellfish allergy has been confirmed, the only way to successfully manage the allergy is by complete avoidance of crustacean shellfish in any form, or of crustacean shellfish-derived ingredients. As one of the recognised major allergens, crustacean shellfish should always be labelled on pre-packaged foods in the EU and the US, although this is not as yet the case for molluscan shellfish, such as mussels and clams.

As crustacean shellfish is a relatively expensive ingredient, it is rarely undeclared on the label, or an unexpected ingredient. Stocks and soups may contain shellfish extract to enhance flavour, and surimi may contain shellfish extract. Dishes to avoid include paella and many South East Asian dishes. People with crustacean shellfish allergy are also advised to avoid the food supplement, glucosamine, as this is made from the shells and skeletons of shellfish.

People with crustacean shellfish allergy need to be especially careful when eating out, as very sensitive individuals have been known to suffer anaphylactic shock from breathing in airborne particles of crustacean shellfish originating from cooking fumes. For the same reason, sensitive individuals should avoid open fish markets.

Sources of Further Information

Published

Bush, R.K. and Hefle, S.L. Food allergens. *Critical Reviews in Food Science and Nutrition*, 1996, 36 (Suppl. S), 119–163.

On the Web

The InformAll Database http://foodallergens.ifr.ac.uk/food.lasso?selected_food=5012#summary

The Anaphylaxis Campaign http://www.anaphylaxis.org.uk/

3.2.8 SOYA ALLERGY

Hazard Identification

Soya beans (*glycine max.*) are one of the most common causes of food-related allergic reactions. It is the protein fraction of soya that causes the reaction and, unfortunately, this protein fraction is found in many soya products, including soya flour, soya milk, soya meal, soya protein isolate, soya protein concentrate, tofu, miso, textured vegetable protein and many more. Soya derivatives are very commonly used as food ingredients in numerous processed foods. For example, soya products are widely used as texturisers, emulsifiers and protein fillers. Soya bean lecithin is also used as an emulsifier (E322).

Allergenicity

Soya-bean allergy appears to occur in both infants and adults, but it is generally accepted that it is less severe and less frequent than peanut allergy. As with all the other food allergies, soya allergy does not appear on first exposure to the allergen, symptoms only occur upon re-exposure to soya. The first contact only sensitises the individual to soya. It is still unclear exactly which components of soya are responsible for allergenicity, but so far, at least fifteen different allergenic proteins have been found in soya. People who are allergic to soya are frequently also sensitive to tree pollen, such as birch.

Some fermented soya foods appear to be less allergenic than the unfermented soya products, most likely because fermentation may cause the degradation of allergenic proteins.

The major known allergens in soya are the 7S seed storage globulin, the 11S seed storage globulins, the Betv 1 homologue and an inactive papain-related thiol protease. Some of the designated allergen names of soya allergens, as given by the Allergen Nomenclature subcommittee of the International Union of Immunological Societies are Gly m 1 (hydrophobic soya bean protein), Gly m 2 (disease response protein), Gly m 3 (a profilin), Gly m 4 and Gly m Bd 30K. The Kunitz-trypsin inhibitor has also been recognised as an important allergen in people suffering with baker's asthma. However, this is a respiratory rather than a food allergen.

Processing of soya beans may alter their allergenicity. For example, the Bet v 1 allergen is found in textured soya protein but is absent from roasted soya beans and fermented soya products, such as soya sauce.

Prevalence

Epidemiological data on soya allergy are poor and the data relating to identity of soya-bean allergens are inconsistent. Studies suggest that the prevalence of this allergy is between 0.3 and 1.0%, with a slightly higher prevalence in children than in adults. The higher prevalence in children is most likely the result of infant exposure to soya-bean-based infant formula, or to pre-sensitisation in the womb. Many infants outgrow soya allergy, so the prevalence is therefore lower in adults.

Hazard Characterisation

Effects on Health

The symptoms of soya allergy range from relatively mild symptoms to severe symptoms that require emergency treatment. Soya is considered one of the most important food allergies and it elicits a true food-allergy response involving over-reaction of the immune system and production of IgE antibodies.

There are significant differences in the reported reactions to the molecular allergens of soya in different parts of the world. It appears that different allergens are involved in Japan compared with those in North America and Europe, although the basis for these differences remains unclear.

Symptoms range from mild, including the oral allergy syndrome (mouth tingling, *etc.*), nausea and vomiting, diarrhoea, urticaria and itchy skin, to severe reactions requiring treatment, such as a sudden drop in blood pressure, asthma, breathing difficulties and anaphylaxis.

There are numerous reports of incidents in which soya has been implicated in causing allergic reactions. For example, in Sweden, researchers examined cases that came to light after a young girl suffered an asthma attack and died after eating a hamburger that contained only 2.2% soya protein. The researchers evaluated 61 cases of severe reactions to food, of which five were fatal, and found that peanut, soya and tree nuts caused 45 of the 61 reactions. Of the five deaths that occurred, four were attributed to soya. The four children who died from soya had known allergies to peanuts but not to soya. The amount of soya eaten ranged from 1–10 g, which is typical of the levels found when soya protein is used as a meat extender in ready-made foods such as hamburgers, meatballs, spaghetti sauces, kebabs and sausages or as an extender in breads and pastries.

Dose-Response

There is very little information concerning the threshold dose of soya required to elicit an allergic response, but one report suggested it was in the region of 1 g of soya bean in dry matter, far higher than the threshold level reported for peanut. There have been a number of reports describing asthmatic symptoms suffered by workers handling soya flour, suggesting that powder inhalation can also elicit allergic reactions.

Management of Soya Allergy

The best way to manage soya allergy is by employing an exclusion diet and vigilant avoidance of foods that may contain soya ingredients. As soya is recognised as one of the major allergens, both in the EU and in the US, any pre-packed food products containing soya should be labelled as such. Strict observance of all food labels is therefore recommended.

Foods that may contain soya include bakery products, breakfast cereals, ice cream, margarine, chocolate, pasta, processed meats, ready meals, vegetarian convenience foods, tofu, tempeh, miso, and soya-protein concentrates and isolates. Food additives that may contain soya include hydrolysed vegetable protein, certain flavourings and lecithin (E322). Studies indicate that most individuals allergic to soya protein are able to consume refined soya oil safely, as virtually all traces of protein are removed during the refining process.

Eating out in catering establishments and buying unwrapped foods also pose a risk as no labelling laws exist to cover these situations. Care is needed in preparation and storage of food to ensure that no cross-contamination occurs.

Sources of Further Information

Published

Arshad, S.H. and Venter, C. Allergens in food. *Reviews in Food and Nutrition Toxicity*, volume 1. 2003, 129–157 Taylor and Francis.

On the Web

InformAll Database http://foodallergens.ifr.ac.uk/
Food Allergy Info (Institute of Food Research) www.foodallergens.info
Allergy UK http://www.allergyuk.org/
Soy Foods Association of North America http://www.soyfoods.org/top/technical/sana-allergen-guidelines-for-processing-dairy-like-soy-foods/

3.2.9 SESAME ALLERGY

Hazard Identification

Sesame (*Sesamum indicum*) is an oilseed plant originating in India and cultivated in Africa, Asia, the Middle East, the Balkans, Latin America and the United States. It belongs to the *Pedaliaceae* family. Sesame seeds have an oil content of between 50% and 60%. In contrast to other vegetable oils, such as sunflower or groundnut oils, sesame-seed oil for food use is always cold pressed to preserve its delicate flavour. The production and consumption of sesame seeds have increased dramatically over the past few years, in line with the increasing prevalence of sesame-seed allergy.

Sesame seeds are used whole or can be crushed to form a paste used as an ingredient in many foods. The oil is used for cooking and in salad dressings. Sesame-seed oil is also often used in cosmetic and pharmaceutical products.

Allergenicity

The major allergens in sesame belong to the seed-storage proteins and are very resistant to processing and proteolysis. At least four proteins in sesame are thought to be responsible for the allergenicity. These are, a 7S vicilin-type globulin, two seed-storage proteins of sesame (Ses i 3, and Ses i 2) and a 2S albumin.

Homology between Ses i 3 and the peanut allergen Ara h 1 has been found. Allergy to poppy seed and/or sesame seed has also been reported to occur with simultaneous sensitisation to nuts and flour. Common allergenic structures have also been identified in sesame, poppy seed, hazelnut and rye. In patients with sesame allergy, associated allergy to almond, Brazil nut, walnut and pistachio has also been reported.

Sesame oil has reportedly been the cause of a number of incidents of anaphylactic shock. This is probably because, for culinary purposes, sesame oil is used unrefined, to retain its delicate flavour and aroma. Therefore, tiny traces of allergenic proteins are likely to remain in the oil.

Prevalence

Sesame allergy was almost unheard of twenty years ago, but today it is increasingly common. In Australia, the prevalence amongst children was reported to be 0.42%, and in the UK, a figure of 0.04% amongst adults has been suggested, although it is likely to be much higher. In fact, the first survey of the Allergy Vigilance Network, launched in 2000, indicated that 4% of life-threatening food allergies were caused by sesame seeds.

Sesame allergy is far more common in Japan and China, the main global producers of sesame, and where sesame seed is a common constituent of the diet. The prevalence is increasing dramatically in countries such as Australia and France, and particularly in Israel, where sesame seed pastes in the form of tahini, hummus and halva are common snack foods. Sesame was found to be a major cause of IgE- mediated food allergy in Israel, and it is second only to cow's milk as a cause of anaphylaxis in that country. The increasing use of sesame in food products, including food preparations for infants, may also explain the increase in sesame allergy in extremely young children.

Sesame allergy is also a cause of occupational allergy in people involved in the production of speciality breads and pastries containing sesame. Many people with sesame allergy are also allergic to nuts.

The natural course of sesame allergy is unknown; however, it is reported that only 15% of infants diagnosed at the age of 10–12 months outgrew their allergy within 2 years. In adults, there are no examples of recovery from allergy to sesame.

Hazard Characterisation

Effects on Health

The predominant clinical features of sesame allergy in children are asthma and atopic dermatitis. About half of affected adults have been reported to experience anaphylactic shock, with loss of consciousness in some cases. In general, the principal symptoms are:

- skin rash, urticaria, hives, itchiness, angioedema and skin swelling;
- hayfever, asthma, coughing, wheeziness and tightness of the chest;
- oral allergy syndrome, nausea, vomiting, diarrhoea, stomach cramps;
- dizziness, drowsiness, low blood pressure, collapse, anaphylaxis.

Symptoms generally occur within a few minutes to up to two hours after ingestion of sesame-containing products. The incidence of gastrointestinal symptoms with sesame allergy is low compared with other symptoms experienced.

Dose-Response

Doses of as little as 100 mg of sesame seeds or 3 ml of sesame-seed oil have been reported to elicit an allergic response in sensitive individuals. In general, however, the threshold dose for most people is around 2–10 g of sesame seeds or sesame-seed flour.

Management of Sesame Allergy

Complete avoidance of sesame seeds, flour and oil is the recommended course of action for anyone found to have an allergy to sesame. As the allergy appears to be particularly prevalent in individuals already known to be susceptible to allergies, such as those with eczema or other food allergies, it is recommended that sesame be excluded from the diet of infants with a history of atopic dermatitis or atopic family history. Because the incidence of sesame allergy has increased so dramatically in infants and young children, it has also been suggested that sesame be added to the list of allergenic foods to be avoided in the first year of life.

Sesame can be present as a hidden ingredient, especially in margarines and salad dressings, where the label merely states "vegetable oil". However, the requirements of recent EU labelling legislation are that it is mandatory to include sesame on the label of pre-packed foods that contain it as an ingredient. In the United States, sesame is not yet among the list of allergenic ingredients that have to be labelled by law.

Common foods containing sesame include sesame-topped burger buns, tahini, halva, salad dressings, sauces, falafel, Turkish cakes, Chinese foods, breads, muesli bars, and mixed seed products. Sesame oil is also commonly used in cosmetics, such as lipsticks and moisturising creams.

Sources of Further Information

Published

Spirito Perkins, M. Raising awareness of sesame allergy. *Pharmaceutical Journal*, 2001, 267 (24 November), 757–758.

Levy, Y. *et al.* Allergy to sesame seed in infants. *Allergy*, 2001, 56, 193–194.

On the Web

American College of Allergy, Asthma and Immunology (ACAAI) http://www.acaai.org/public/linkpages/Sesame_Allergy.htm

The Anaphylaxis Campaign information sheet http://www.anaphylaxis.org.uk/information/print_common_food_al.html

3.2.10 SULFITE ALLERGY

Hazard Identification

Sulfites are compounds containing the sulfite ion (sulfur and oxygen), most often in combination with sodium (sodium sulfite) or potassium (potassium sulfite). Sulfites release the irritant gas sulfur dioxide, which acts as a preservative and bleaching agent. As well as occurring naturally in some foods and in the human body, sulfites are added to certain foods to act as a preservative, as they inhibit microbial growth, maintain food colour and increase shelf life. Foods to which sulfites are commonly added include wines, beer, and dried fruit. They are also used to bleach food starches, such as potato starch, and are used in the production of some food-packaging materials such as cellophane.

Allergenicity

It is still unclear why sulfites elicit an allergic reaction in some people but not in others. Sulfur dioxide is an irritant gas and so reflex contraction of the airways has been proposed as one possible mechanism, as the majority of sulfite-allergic individuals exhibit asthma-like symptoms. IgE involvement has also been demonstrated in some subjects who exhibit a positive skin-prick allergy reaction to sulfites, and a few subjects have a partial deficiency of the enzyme sulfite oxidase that helps to degrade sulfur dioxide. Sulfite allergy is unlike other food allergies in that it is not triggered by a protein.

Prevalence

The true prevalence of sulfite allergy in the general population is unknown. Figures of prevalence amongst asthmatic individuals vary. Prevalence of sulfite allergy in steroid-dependent asthmatic children is estimated to be between 20 and 66%, whilst prevalence in steroid-dependent asthmatic adults is lower, and estimated at between 3.9 and 4.5%.

Hazard Characterisation

Effects on Health

The majority of sulfite-allergic individuals exhibit an asthma-like reaction with the following possible symptoms:

- trouble breathing, speaking or swallowing;
- wheezing.

A few will exhibit symptoms similar to anaphylaxis:

- flushing, fast heartbeat and dizziness;
- stomach upset and diarrhoea;
- collapse.

Dose-Response

There are no clear data on dose-response effects of sulfites; however, sensitive individuals have been known to react to the smallest amounts of sulfite used as an additive in products such as jam.

Management of Sulfite Allergy

As with all other food allergies, avoidance is the best way to manage sulfite allergy. In the EU, the following preservatives should be avoided by those with sulfite allergy:

E number	*Name*
E220	Sulfur dioxide
E221	Sodium sulfite
E222	Sodium hydrogen sulfite
E223	Sodium metabisulfite
E224	Potassium metabisulfite
E226	Calcium sulfite
E227	Calcium hydrogen sulfite
E228	Potassium hydrogen sulfite

Other additives also contain sulfites, but they are not used as preservatives, nor are they normally referred to as sulfites. These include:

E number	*Name*
E150b	Caustic sulfite caramel
E150d	Sulfite ammonia caramel

It is therefore essential to read all food labels properly to ensure that the food is free of these additives.

Foods that might contain sulfites include beer, cider and wine, bottled lemon or lime juice concentrate, canned vegetables, condiments, deli meats, sausages, dressings, dried fruits, dried herbs, fish, fresh grapes, lettuce, fruit fillings, jams, fruit juices, glacée fruits, processed potatoes, soya products, starches, sugar syrups, sugar and vinegar.

Sulfites can also occur naturally in foods. For example, wine-making yeasts generate sulfur dioxide in wines and some strains produce over 100 ppm. Sulfites are also generated naturally in the human body by metabolism of sulfur-based amino acids.

Sources of Further Information

Published

Gunnison, A.F. and Jacobesen, D.W. Sulfite hypersensitivity: A critical review. *CRC Critical Reviews in Toxicology*, 1987, 17 (3), 185–214.

Bush, R.K., Taylor, S.L., Holden, K., Nordlee, J. and Busse, W.W. Prevalence of sensitivity to sulfiting agenst in asthmatic patients. *The American Journal of Medicine*, 1986, 81, 816–820.

On the Web

Allergy Capital – Allergic reactions to sulfites (sulfite allergy) www.allergycapital.com.au/Pages/sulfites.html

3.2.11 TREE-NUT ALLERGY

Hazard Identification

Tree nuts are botanically defined as the edible kernels of the seeds of trees. Included in the category of tree nuts as potential allergens are almonds, Brazil nuts, cashew nuts, hazelnuts, Macadamia nuts, pecans, walnuts and Queensland nuts.

Allergenicity

Tree-nut allergies are common, potentially life-threatening, food allergies. The allergy frequently lasts throughout an individual's lifetime. Tree nuts may belong to different families that are unrelated to one another, and tree nuts are also not related to peanuts. Peanut allergic individuals can often eat tree nuts and those allergic to tree nuts can often tolerate peanuts. However, some allergic individuals may be allergic to both peanut and tree nuts. In addition, individuals can be allergic to some, but not all, tree nuts. Of all the common tree nuts, almond appears to cause the fewest cases of allergy.

An adverse reaction to tree nuts is a true food allergy, involving an overreaction of the immune system and production of IgE antibodies. The major allergens in tree nuts include the 2S albumin, the 7S storage globulins, the 11S seed storage globulins, non-specific lipid-transfer proteins and the Bet v 1 homologue. Some of the designated allergen names of tree-nut allergens, as given by the Allergen Nomenclature subcommittee of the International Union of Immunological Societies are: Brazil nuts – Ber e 1, Ber e 2; Walnuts – Jug r 1, Jug r 2, Jug r 3; Cashews – Ana o 1, Ana o 3; and Hazelnuts – Cor a 8, Cor a 11.

Prevalence

Food surveys suggest that tree-nut allergy affects about 1% of the population. It appears to be more common in the United States than in some parts of Europe, such as Spain, although it is unclear why this should be so. Genetic or environmental factors may play a part. Tree-nut allergy is not generally as common as peanut allergy, although in Germany, hazelnut allergy is more common than peanut allergy.

Hazard Characterisation

Effects on Health

Allergies to tree nuts tend to be of a more severe nature, causing life-threatening and occasionally fatal reactions. People with tree-nut allergies also often suffer from reactions triggered by a number of different types of nuts, even though they do not come from closely related plant species. In general, these allergies are triggered by the major proteins found in nuts and seeds, many of which are heat resistant.

There is also a milder form of tree-nut allergy, which is associated with birch-pollen allergy, where symptoms are confined largely to the mouth, causing a condition called "oral allergy syndrome" (OAS). This condition is triggered by molecules found in tree nuts, which are very similar to pollen allergens like the major birch-pollen allergen Bet v 1. These molecules tend to be destroyed by cooking, which can therefore reduce the allergenicity of nuts for some consumers.

The symptoms of tree-nut allergy can vary from mild to severe. The most common mild symptoms include:

- tingling in the mouth and lips and facial swelling;
- nausea and colicky pain, accompanied by a feeling of tightness in the throat.
- urticaria or nettle rash.

Severe reactions, exhibited by those more sensitive to tree nuts include:

- swelling of the airways and obstructed breathing;
- sudden drop in blood pressure;
- collapse and unconsciousness.

These symptoms result from the widespread release of pre-formed histamine and other inflammatory mediators from mast cells and basophil cells. The more severe reactions are classified as anaphylaxis and require immediate medical attention. The onset of anaphylactic reactions is generally extremely rapid and can proceed very quickly to unconsciousness.

Dose-Response

There is very little information concerning the dose required to elicit an allergic response to tree nuts. Sensitivity appears to be very variable and dependent on the particular individual.

Management of Tree-Nut Allergy

Complete avoidance of all tree nuts and their products is probably the best way to manage this allergy. Despite the fact that allergy to one type of tree nut does not necessarily pre-suppose allergy to other types of tree nut, this may not necessarily be the case. Those allergic to tree nuts would be best advised to avoid other tree nuts, unless their tolerance has been clearly proven by reliable tests.

The types of product likely to contain tree nuts include chocolate, candies, cookies, desserts, sweets, almond paste, doughnuts, ice cream, cereals, ready meals, granola bars, trail mixes, pesto sauce, muesli, vegetarian ready meals and products, and care should be taken when checking the labels.

Eating out in catering establishments and buying unwrapped foods also pose a risk, as no labelling laws exist to cover these situations. Care is needed in preparation and storage of food to ensure that no cross-contamination occurs.

Sources of Further Information

Published

Angus, F. Nut allergens. *Natural Toxicants in Foods*, Sheffield Academic Press, 1998, 84–104.

On the Web

The InformAll Database http://foodallergens.ifr.ac.uk/food.lasso?selected_food=53#summary

The Calgary Allergy Network http://www.calgaryallergy.ca/Articles/English/treenuthp.htm#chart

Food Allergy Info (Institute of Food Research) http://www.foodallergens.info

3.2.12 WHEAT ALLERGY

Hazard Identification

Wheat and wheat products are frequently implicated in food-allergy reactions in adults and children. As with other food allergies, it is the protein fractions that are responsible for causing the allergic reaction. The proteins found in wheat are similar to those found in related cereals such as rye, barley and spelt. Although less closely related, oats may also be a problem. Therefore, people allergic to wheat are likely to be allergic to some other cereals. Rice and maize (corn) do not appear to pose the same problems.

Allergenicity

Wheat allergy is an adverse reaction involving production of immunoglobulin E (IgE) antibodies in response to one or more of the protein fractions found in the wheat kernel. These include gliadin, glutenin (gluten), albumin, and globulin. The majority of allergic reactions to wheat are caused by the albumin and globulin fractions, although gliadin and gluten may also be responsible, though far less frequently. Allergic reactions to wheat are caused by ingestion of wheat-containing foods or by inhalation of flour containing wheat.

Individuals with wheat allergy will often also be allergic to related cereals, such as barley, rye and spelt, and possibly oats. Some wheat allergens are the same proteins as the allergens found in grass pollen.

Heating does not appear to reduce the allergenicity of wheat. In fact, it has been shown that the baking process actually increases the resistance of the allergens in wheat flour to proteolytic enzymes, allowing the allergenic proteins to reach the digestive tract undegraded, where they can elicit an immunological response. Therefore, baked bread appears to be potentially more allergenic than raw flour.

Wheat allergy is not to be confused with coeliac disease, although the symptoms may be similar. Coeliac disease, also known as gluten enteropathy, was, until recently, known as gluten intolerance. It is a hereditary disorder of the immune system, during the course of which, eating gluten causes damage to the lining of the small intestine. This results in malabsorption of nutrients and vitamins. Unlike wheat allergy, coeliac disease is mediated through immunoglobulin A (IgA) and immunoglobulin G (IgG), and sufferers will develop gliadin-specific IgA and IgG antibodies. Coeliac disease does not cause the potentially fatal anaphylaxis associated with true food allergies if gluten is eaten.

Allergy to wheat can occur in any individual, but coeliac disease is hereditary.

Prevalence

Wheat allergy occurs in both children and adults, although it is more likely that young children will outgrow it. Individuals who develop wheat allergy in later life are likely to retain the allergy. It is more common in children than in adults, but there are few data indicating exactly how prevalent wheat allergy actually is. It is probably less common than peanut, tree nut, shellfish, fish, milk, egg and soya allergies. A study in Australia suggested that there was a prevalence of about 0.25% amongst young adults.

In certain subgroups, wheat allergy may be more common. For example, in the baking industry, it is reported that wheat allergy is responsible for occupational allergy in up to 30% of individuals.

A specific type of allergy, known as wheat-dependent exercise-induced anaphylaxis is linked to physical exercise after consumption of wheat. This type of allergy is more often reported in adults with no previous history of wheat allergy in childhood.

Hazard Characterisation

Effects on Health

Allergic reactions to wheat generally start within minutes and up to a few hours of eating wheat (or inhaling it). The most common symptoms are:

- itching of the skin, hives, urticaria, eczema;
- angioedema (swelling of the skin, lips and throat);
- abdominal cramps, nausea, vomiting, diarrhoea;
- asthma, wheezing, allergic rhinitis.

In severe cases:

- blood pressure drop, collapse;
- anaphylactic shock;
- exercise-induced anaphylaxis.

Dose-Response

It is unclear how much wheat is needed to cause a reaction in sensitive individuals; however, a recently reported challenge protocol in Germany used doses of between 4 mg and 3.5 g of wheat flour, suggesting that only small quantities of wheat would be required to induce symptoms.

Management of Wheat Allergy

As with most food allergies, avoidance is the best way to treat allergy to wheat. As wheat is such a widely used ingredient in common foods, avoidance can be difficult. Wheat is frequently present as an invisible ingredient. However, to comply with recent allergen legislation, it is required that all pre-packed foods containing wheat be labelled as such in both the EU and in the United States.

Wheat-allergic individuals should also avoid other gluten-containing cereals such as rye, barley and oats, although rice- and maize-based foods may be suitable substitutes. Wheat is used for making bread, biscuits, crackers, pastry, breakfast cereals, pasta and thickening agents. It is also used to make alcoholic beverages such as beer, lager and whisky. Ingredients to look out for and avoid include breadcrumbs, bran, cereal extracts, gluten, couscous, semolina wheat, wheat germ, wheat malt, gelatinised starch, modified starch, soya sauce and vegetable gums and starches.

Sources of Further Information

Published

Bush, R.K. and Hefle, S.L. Food allergens. *Critical Reviews in Food Science and Nutrition*, 1996, 36 (Suppl. S), 119–163.

Simonato, B. *et al.* Food allergy to wheat products: the effect of bread baking and *in vitro* digestion of wheat allergenic proteins. *Journal of Agriculture and Food Chemistry*, 2001, 49, 5668–5673.

On the Web

Allergy Society of South Africa fact sheet. www.allergysa.org/wheat.htm

UK Food Standards Agency fact sheet. www.eatwell.gov.uk/healthissues/foodintolerance/foodintolerancetypes/wheatallergy/

CHAPTER 3.3

Allergen-Control Options

Manufacturing

It is the responsibility of food manufacturers to minimise the risks of their products to individuals with food allergies. The UK Institute of Food Science and Technology (IFST) advises that the following strategies should be adopted:

- Implementation of a HACCP plan to analyse the entire manufacturing process in relation to allergen hazards.
- In a multiproduct company, wherever possible, segregate manufacturing operations involving the allergen-containing food into a separate building.
- When possible, formulate foods that are free of all unnecessary major allergens as ingredients.
- Organise raw materials supplies, storage and handling, production schedules and cleaning procedures to prevent cross-contamination of products by "foreign" allergens.
- Ensure all personnel are fully trained to understand the necessary measures and the reasons for them.
- Comply with the relevant labelling legislation, ensuring that appropriate warnings are included on the product label warning the consumer of the presence of a major allergen.
- Have in place an appropriate recall system for any product found to contain a major allergen not indicated on the product label.

By following strict GMP, most problems can be avoided. Misformulation results from inattention or inadequate quality control. Cross-contamination stems from residues in shared equipment caused by inadequate cleaning, airborne dust, or even incorporation of rework without consideration of the allergen problem. Ideally, separate equipment should be used for products containing the specific allergen in question. For larger companies, designation of an allergen-only site is the most effective way to prevent any cross-contamination. If it is impossible to

The Food Safety Hazard Guidebook
By Richard Lawley, Laurie Curtis & Judy Davis

avoid sharing production equipment, then it is preferable to schedule the allergen-containing product at the end of the day, just before cleaning.

Allergen-Control Plan

In order to develop an effective allergen-control plan, every aspect of the manufacturing operation must be examined for the risk of allergens. The following is an example of a checklist providing the components of an allergen control strategy:

- Develop a list of all the raw materials used in your factory/production area, including all processing aids, additives, flavourings, *etc*. Specify which of them are allergens, or contain allergens. In the case of outside suppliers, ensure that they too have a documented allergen-control plan in place. Specify that any purchased ingredients are free of undeclared allergens and that a letter guaranteeing this be supplied with each shipment.
- Compile a list of all finished products, and state which ones are produced using allergenic ingredients.
- Deal with allergen-containing incoming ingredients appropriately. Allergens should be transported in clearly marked containers and must be separated physically from non-allergenic ingredients. All incoming containers should be checked for possible damage or spillage. Allergenic ingredients should ideally be kept in an area separate from non-allergenic ingredients. The different areas should be well marked and colour coded if possible. Allergenic materials should always be stored below non-allergenic materials.
- Where bulk tanks are used, try to dedicate them to allergenic or non-allergenic materials only. Where this is impossible, ensure an appropriate and thorough sanitation programme is carried out between shipments.
- If possible, dedicate processing equipment, production lines and personnel to allergenic products, to prevent cross-contamination. Where this is not feasible, the alternatives are to segregate production to different days of the week, and if not possible, run non-allergenic products before allergen-containing products; schedule long production runs of allergen-containing products to minimise changeover; and schedule cleaning to follow immediately after allergen-containing products have been run.
- In the case of rework, the ideal would be to advocate an "exact into exact" approach, *i.e.* rework should only be used in the same product from which it was generated. Containers for rework should be clearly labelled, for example, by using colour-coded tags.
- Ensure that the correct packaging materials are used. Discard all obsolete packaging materials immediately. Packaging materials should ideally be stored in a designated area, and the accuracy of labels should be thoroughly checked.
- Cleaning and sanitation are of prime importance, particularly where equipment is shared. Wet cleaning is generally preferred as allergenic proteins tend to be soluble in hot water and detergents can help in

removing proteins. Where wet cleaning is impossible, wipe downs are often needed and other approaches are available. Validation of sanitation practices on shared equipment is recommended. Various analytical kits are available, such as ELISA kits, and lateral flow devices (dipsticks), which can be used to validate sanitation practices.

Precautionary Labelling

Many manufacturers use precautionary labelling in cases where it is impossible to guarantee that the manufactured product is completely allergen-free. Precautionary statements such as "may contain" or "may contain traces of" are often used. However, these can often even further limit the allergic individual's choice of foods, with the result that some consumers choose to ignore precautionary labels putting their health at risk.

Sources of Further Information

Published

Hignett J. Controlling allergens in the manufacturing environment. *Food Allergy and Intolerance*, 2004, 5 (1), 5–13.

Taylor S.L., Hefle S.L. Allergen control. *Food Technology*, 2005, 59 (2), 40–43 + 75.

On the Web

IFST Information Statement *Food Allergy*. http://ww w.ifst.org/uploadedfiles/cms/store/ATTACHMENTS/allergy.pdf

CHAPTER 3.4

Allergen Legislation

Pre-packed Foods

EU Legislation

In recent years the food-labelling regulations have been amended to help people suffering from allergies. This legislation came into force in November 2004, and was fully implemented on 25th November 2005. Since that date, all pre-packed food and drink has to comply with the new labelling rules.

The major difference between current and previous legislation is that the so-called "25% rule" has now been abolished. Manufacturers have to list product ingredients in descending order of weight, but there was a previous exclusion for ingredients if they were part of a compound ingredient that constituted less than 25% of the product. For example, if sliced salami were included in the topping of a pizza, and the salami made up less than 25% of the whole product, then there was no legal requirement to list the ingredients of the salami. This meant that consumers with food allergies would not necessarily have all the information they needed to make an informed choice as to whether the food was suitable for them.

The European Directive 2003/89/EC (European Commission, 2003), which amends Directive 2000/13/EC, came into force in November 2004. This legislation gives a list of allergenic food ingredients that now have to be indicated on the label when they, or their derivatives, are used in food sold pre-packed in the EU. The legislation includes all food ingredients, including carry-over additives, additives used as processing aids, solvents and media for additives and flavours. It also applies to alcoholic beverages. The Directive can be found at the following web address: http://www.foodallergens.info/industry/fl_com 2003-89_en.pdf

In England, the equivalent legislation – The Food Labelling (Amendment) (England) (No. 2) Regulations 2004, also came into force on the 26th November 2004. Similar regulations apply to Scotland, Wales and Northern

The Food Safety Hazard Guidebook
By Richard Lawley, Laurie Curtis & Judy Davis

Ireland. The Regulations can be found at the following web address: http://www.legislation.hmso.gov.uk/si/si2004/20042824.htm

The new rules require that for all allergenic ingredients, the source must be indicated. Thus, if vegetable oil contains peanut oil, then this has to be declared on the label. If the source of a natural flavour is allergen-based, *e.g.* from nuts, then this must also be declared, rather than "natural flavour".

Currently, there are twelve allergenic foods on the list of those that must be declared:

- cereals containing gluten (*i.e.* wheat, rye, barley, oats, spelt or their hybridised strains) and products thereof;
- crustaceans and products thereof;
- fish and products thereof;
- egg and products thereof;
- peanuts and products thereof;
- soya beans and products thereof;
- milk and products thereof;
- tree nuts – almond, hazelnuts, walnuts, cashews, pecans, brazil nuts, pistachio nuts, Macadamia nuts, Queensland nuts and products thereof;
- celery and products thereof;
- mustard and products thereof;
- sesame seeds and products thereof;
- sulfur dioxide and sulfites at concentrations of more than 10 mg/kg or 10 mg/litre, expressed as sulfur dioxide.

Whenever any of these ingredients are used in the production of foods, they must be labelled. At the moment, many other allergens, which are less common, have been omitted from the list. However, this may change, as other allergenic foods can be added to the list on the advice of the European Food Safety Authority (EFSA). Because different people have different tolerances to allergens, it is impossible to define an acceptable threshold limit, as is the case with setting acceptable levels for other chemicals in food.

In some cases, processing removes the allergenic risk from ingredients derived from some of the foods on the list. The European Commission has recognised this and granted them a temporary labelling exemption through the Food Labelling (Amendment) (No.2) Regulations 2005.

The list of provisionally exempt derived ingredients has been published (Table 3.4.1) and a permanent list of exemptions is scheduled for publication in November 2007.

Fully refined peanut oil is not currently included on this list of exempt derived ingredients, because, on the basis of existing information, EFSA is of the opinion that it could possibly cause allergic reactions in sensitive individuals. However, this may change should further information in support of its exemption be submitted for evaluation by EFSA.

The Directive and the UK Regulations do not specify the format in which allergen declarations must appear, other than that they have to be included

Table 3.4.1

Ingredients	*Products thereof provisionally excluded*
Cereals containing gluten	• Wheat-based glucose syrups including dextrose • Wheat-based maltodextrins • Glucose syrups based on barley • Cereals used in distillates for spirits
Eggs	• Lysozyme (produced from egg) used in wine • Albumin (from egg) used as a clarifying agent in wine and cider
Fish	• Fish gelatine used as a carrier for vitamins and flavours • Fish gelatine or isinglass used as a fining agent in beer, cider and wine
Soya bean	• Fully refined soya bean oil and fat • Natural mixed tocopherols (E306), natural D-α-tocopheryl succinate from soya sources • Phytosterols and phytosterol esters derived from vegetable oils from soya-bean sources • Plant stanol esters produced from vegetable-oil sterols from soya-bean sources
Milk	• Whey used in distillates for spirits • Lactitol • Milk (casein) products used as fining agents in cider and wines
Nuts	• Nuts used in distillates for spirits • Nuts (almonds, walnuts) used as flavourings in spirits
Celery	• Celery leaf and seed oil • Celery seed oleoresin
Mustard	• Mustard oil • Mustard-seed oil • Mustard-seed oleoresin

somewhere in the list of ingredients. It has been suggested that allergen information on a label should be made more prominent, for example, by putting it in a box labelled "Allergen Information". Some manufacturers are currently doing this, but it is not yet required by law.

A detailed guidance document to the legislation has been produced by the UK Food Standards Agency and is available at the following web address: http://www.food.gov.uk/multimedia/pdfs/allergenukguidance.pdf

US Legislation

In the USA, regulation is by the Food Allergen Labelling and Consumer Protection Act 2004, which can be found at the following web address: http://www.cfsan.fda.gov/~dms/alrgact.html

The law in the USA requires that food manufacturers identify, in plain common language, the presence of any of eight major food allergens; namely, wheat, eggs, milk, fish, crustacean shellfish, peanuts, nuts and soya beans. The legislation states that the presence of the major food allergens in spices, flavourings, colourings and additives must be declared.

Non-pre-packed Foods

A wide number of establishments and organisations produce food for the general public that is not pre-packed. A number of food-allergy accidents have been attributed to food sold in this way, for example from restaurants, bakeries and other food-catering establishments. Many fatal allergic reactions have occurred when allergic consumers eat out. The UK Food Standards Agency has provided some information and guidance for the catering industry, which can be found at the following web address: http://www.food.gov.uk/safereating/allergyintol/

In the United States, guidance for caterers and retailers has been produced by the Hospitality Institute of Technology and Management, which can be found at: http://www.hi-tm.com/Documents2005/allergens-retail-and-list.pdf

Sources of Further Information

On the Web

UK Food Standards Agency Allergy Labelling Guide for Small Businesses. www.food.gov.uk/multimedia/pdfs/allergyleaflet.pdf

Chartered Institute of Environmental Health article on allergy labeling. http://www.cieh.org/ehp1/article.aspx?id=695

IFST Information Statement *Food Allergy*. http://www.ifst.org/uploadedfiles/cms/store/ATTACHMENTS/allergy.pdf

Section 4: Food Safety Legislation

CHAPTER 4.1

Food Safety Legislation

Introduction

The history of much modern food safety legislation can be traced back to Victorian England, when widespread adulteration of food was a serious problem. This was not only fraudulent, but was often dangerous. For example, toxic salts of lead and mercury were sometimes used to provide additional colour in sugar confectionery intended for children. The urgent need to curb these practices led to the introduction of the first Food Adulteration Act in 1860. Since then, food law has evolved steadily into the sophisticated framework of legislation that now exists to protect consumers in most parts of the world.

Food safety legislation is a very complex subject, and a detailed examination of the law as it relates to food safety hazards is beyond the scope of this book. Furthermore, the body of food safety legislation is constantly being added to and amended, so that any written work on the subject is almost certain to be out of date by the time it is published.

Note: Readers are strongly advised to consult a reputable specialist or legal adviser if they require more detailed information, or have specific questions on food safety legislation.

What follows, therefore, is a concise overview of food safety legislation in the EU and in the USA, with a brief mention of some of the international aspects of food law. It is intended to be neither detailed, nor exhaustive. The intention is to give an overall impression of the approach to food safety regulation and enforcement taken by the authorities in two of the worlds' most highly developed and complex food markets.

European Legislation

Much of the food safety legislation now in force in the countries of the European Union (EU) originates from the European Commission (EC), rather

The Food Safety Hazard Guidebook
By Richard Lawley, Laurie Curtis & Judy Davis

than from national authorities. There are two main legal instruments by which the Commission can introduce new food legislation. The first of these is the Directive, which sets out an objective, but allows national authorities to determine how that objective is to be achieved, and cannot be enforced in individual Member States until implemented into national legislation. The second instrument is the Regulation, which is "directly applicable" and becomes law in all Member States as soon as it comes into force, without the need to change national legislation. Both Directives and Regulations may be described as "horizontal", dealing with one aspect of food, such as hygiene, across all commodities, or "vertical", applying to particular foods.

Although the EC initiates new Directives and Regulations, an established path of consultation, amendment and review must be followed before proposed legislation can be formally adopted by the European Parliament and by the Council of Ministers. Finally, the new legislation is published in the Official Journal of the EU and then comes into force. This process can take years, especially if there are contentious issues involved. The development of new food safety and hygiene measures is now informed by the scientific analysis and evaluation of food safety hazards. It is usual for the EC to submit a request for a risk analysis to be undertaken by the European Food Safety Authority (EFSA), before legislative proposals are drawn up.

Until comparatively recently, food safety in the EU was largely regulated by a complicated system of horizontal and vertical food hygiene Directives that had evolved over many years. This system inevitably included some anomalies and duplication, and was not implemented uniformly in all Member States. The situation became increasingly unsatisfactory, particularly in view of the planned accession of a number of new member countries. Consequently, the Commission carried out a comprehensive review of the EU food hygiene legislation in the late 1990s. The result was the introduction of the "Food Hygiene Package" of EU legislation, which came into force on 1 January 2006.

The Food Hygiene Package

The Package consists of three main Regulations, which applied immediately throughout the EU. These are:

- Regulation (EC) 852/2004 on the hygiene of foodstuffs;
- Regulation (EC) 853/2004 setting out specific hygiene requirements for foods of animal origin;
- Regulation (EC) 854/2004 setting out specific requirements for organising official controls on products of animal origin intended for human consumption.

Regulation 852/2004 contains general hygiene requirements for all food businesses and covers a wide range of topics, including the general obligations of businesses with regard to food hygiene, the requirements for hazard-analysis critical control point- (HACCP) based food safety management procedures, hygiene requirements for premises and equipment, staff training and personal hygiene, heat processes and packaging. Regulation 853/2004 supplements 852/2004 by adding specific hygiene requirements for meat, milk, fish and egg production, as well as for by-products, such as gelatine. Regulation 854/2004 deals only with the organisation of the official controls needed for animal products in the human food chain.

The approach of the new Regulations is described as "farm to fork", in that it applies to all stages in the food supply chain, including farmers and growers involved in primary production – a sector not covered by previous food hygiene legislation. All food businesses must also register with the "competent authority", so that they can be clearly identified. The inclusion of HACCP in the Regulations is another key development, clearly signifying that this is now the preferred method of ensuring food safety.

The development of guidance documents on the new legislation in individual Member States has been encouraged, and a number of these have been produced by the EC and at national level, by authorities such as the UK Food Standards Agency, and by industry bodies and trade associations.

Other EU Legislation

While the 2006 Food Hygiene Regulations provide the current backbone of food safety legislation in the EU, they do not by any means include all of the food safety requirements that food businesses need to be aware of. For example, a large number of new "implementing regulations" have also been introduced to deal with specific topics and amendments to the Hygiene Regulations.

The Microbiological Criteria Regulation

One of the most important implementing regulations for all food businesses is Regulation (EC) 2073/2005 on microbiological criteria for foodstuffs, often referred to as the MCR, which came into force on 1 January 2006. This Regulation brought together microbiological criteria for specific foods that had previously been scattered across a number of vertical directives and presented them in a common format.

The MCR includes some of the criteria from previous legislation in unchanged form, but others have been removed and some new criteria have been introduced. The primary purpose of the criteria set out in the Regulation is the validation and verification of HACCP procedures, rather than as standalone food safety controls. It is important for all food businesses to be aware of the requirements of this Regulation.

Food Contaminants Regulations

On 1 March 2007, three new EU regulations came into force, dealing with a range of chemical contaminants in foods. The most important of these from a food industry point of view is Regulation (EC) 1881/2006, which replaces (EC) 466/2001 and sets maximum permitted levels for certain contaminants in foodstuffs. This Regulation covers a number of contaminants, including mycotoxins, heavy metals, chloropropanols, PAH, dioxins and PCBs.

US Legislation

The system of food safety legislation in the USA is quite different in structure from that of the EU. Despite this, the main objective of protecting the consumer from exposure to unsafe and unwholesome food products is much the same. The system is based on flexible and science-based federal and state laws and the basic responsibility of industry to produce safe foods. A risk-based, precautionary approach is built in to the legislative system.

Federal Legislation

The basic foundation of US food safety legislation is determined by Congress in the form of authorising statutes, which are designed to achieve specific food safety objectives and to establish the level of public protection. These are generally broad in scope, but also define the limits of regulation. Important statutes include the following:

- Federal Food, Drug and Cosmetic Act;
- Federal Meat Inspection Act;
- Poultry Products Inspection Act;
- Egg Products Inspection Act;
- Food Quality Protection Act;
- Public Health Service Act.

Implementation of these statutes is the responsibility of a number of executive agencies, and is accomplished by the development and enforcement of regulations. The main federal regulatory organisations concerned with food safety are the Food and Drug Administration (FDA) and the US Department of Agriculture (USDA) Food Safety and Inspection Service (FSIS). However, other agencies, including the Department of Health and Human Services (DHHS) and the Environmental Protection Agency (EPA), also play important regulatory roles.

Responsibility for food safety is divided largely between the FSIS and the FDA according to food sector. The FSIS is responsible for the safety of all meat, poultry and egg products, while the FDA assumes responsibility for all other foods. In addition, the EPA has a key role in protecting consumers from risks posed by pesticides in food.

Food safety regulations are developed using a risk-analysis approach in a transparent process that encourages the participation of industry and consumers. All significant comments must be addressed in the final regulation.

Once this has been published in the *Federal Register*, it can be enforced. Examples of regulations developed in this way include the HACCP regulations and the introduction of performance standards for pathogen reduction and control. All current regulations are listed in the *Code of Federal Regulations*.

State Legislation

In addition to the federal system of food safety legislation, there is an additional layer of regulation at the state level. States have their own legislative assemblies that are able to pass state laws and these may then be implemented as regulations by the local authorities for health and/or agriculture. Generally, state regulations should follow national food safety policy, but there may be differences in the detail, and some states, such as California, have passed state food safety laws. For example, the California state legislature recently passed a law requiring that meat from cloned animals be labelled as such, although this conflicts with current federal policy. Many states also have their own microbiological standards or guidelines for foods.

International Aspects of Food Safety Legislation

Although most countries have developed food legislation structures on a national, or regional basis, there has also been a degree of international cooperation. This has been achieved mainly through the activities of the Codex Alimentarius Commission, a body set up in 1963 by the World Health Organization and the Food and Agriculture Organization with the aim of promoting the coordination of food standards work carried out by national authorities and other bodies.

Since its inception, Codex has developed and agreed a series of food standards, codes of practice, guidelines and other recommendations intended to protect consumer health and ensure fair trade practices. Codex standards cover a range of topics, including maximum residue limits for pesticides, food contaminants and toxins. Codes of practice include food hygiene principles, HACCP and control of veterinary drug use. Codex has also published "principles" covering microbiological criteria and risk assessment.

Sources of Further Information

Published

Europe

MacMaolain C. EU food law: protecting consumers and health in a common market Oxford, Hart Publishing, 2007.

USA

Curtis P. A. A guide to food laws and regulations Oxford, Blackwell Publishing, 2005.

On the Web

Europe

European Commission – basic food hygiene legislation page http://ec.europa.eu/food/food/biosafety/hygienelegislation/comm_rules_en.htm

European Commission – basic food hygiene guidance documents http://ec.europa.eu/food/food/biosafety/hygienelegislation/guide_en.htm

EUR-Lex – Direct free access to European Union Law with full search facility http://eur-lex.europa.eu/en/index.htm

UK Food Standards Agency European legislation pages http://www.food.gov.uk/foodindustry/regulation/europeleg/

UK Food Standards Agency guidance on the 2006 food hygiene legislation http://www.food.gov.uk/foodindustry/guidancenotes/hygguid/fhlguidance/

USA

Code of Federal Regulations http://www.gpoaccess.gov/cfr/index.html

Information for FDA-regulated industry http://www.fda.gov/oc/industry/default.htm

USDA Food Safety and Inspection Service – Regulations & Policies page http://www.fsis.usda.gov/Regulations & Policies/index.asp

California Department of Health Services Food and Drug Branch – Sherman food, drug and cosmetic law http://www.dhs.ca.gov/fdb/HTML/General/Sheindex.htm

Other Useful Sites for Legislation

Codex Alimentarius Commission http://www.codexalimentarius.org/

Foodlaw Reading – Reading University site on EU and international food law maintained by Dr D. J. Jukes http://www.rdg.ac.uk/foodlaw/

Food Standards Australia New Zealand (FSANZ) – Food Standards Code pages (includes food safety standards for Australia) http://www.foodstandards.gov.au/thecode/

Japanese Food Safety Commission – pages in English http://www.fsc.go.jp/english/

Section 5: Sources of Further Information

CHAPTER 5.1

Sources of Further Information

Today's food safety professional can access an enormous amount of information on most of the topics covered in this book from a variety of sources. More information is freely available than ever before, but this availability brings its own problems. Identifying reliable and authoritative sources of technical and scientific information can be difficult and time consuming, especially when looking for material on-line. The following pages are intended to guide readers to some of the most reputable information resources available to them, predominantly on the Internet.

Traditional Publications

The number of published scientific journals and reference books containing information relevant to food safety is constantly increasing as the body of scientific knowledge supporting food safety practice grows.

Journals

Scientific journals provide an excellent means of keeping abreast of the latest food safety research and discovery. However, with a few exceptions (see below), most require a fairly substantial subscription for full access. Nevertheless, most journals now have dedicated web pages on the Internet that allow the visitor to browse the contents of each issue, and often to view abstracts and purchase individual articles on-line. Traditional library facilities also allow readers to search for specific articles and papers and obtain copies at a reasonable cost. Some scientific publishers are now beginning to adopt an "open access" approach to their journals through the Internet (see below), and this may mean that current papers are more freely accessible to individuals in the future.

The Food Safety Hazard Guidebook
By Richard Lawley, Laurie Curtis & Judy Davis

Reference Books

A very large number of reference books and textbooks relating to food safety have been published in recent years. Many of these are excellent sources of detailed information on specific food safety issues and the best examples are regularly updated with revised editions. Most food manufacturers will neither have the time nor the budget to assemble their own libraries of specialist food safety books, but it is worth seeking out some of the broader, more practical titles. Unfortunately, many reference texts have been written from an academic perspective and may be somewhat inaccessible for the non-specialist reader, but food safety books written from a practical viewpoint are also available. Useful reference titles can be found through library catalogues, or the web sites of publishers and on-line booksellers. It is often possible to find reviews of books prior to purchase.

The Internet

The Internet has developed in recent years into a very valuable and accessible information resource for the food safety professional. Most of the organisations concerned with food safety now have their own web sites, as do many scientific and professional bodies, scientific publishers and commercial organisations. The majority of these web sites contain information, or links to information, that will be of value to food businesses, but it is important to be aware of the following warning.

Note: Any individual can post information on the Internet, or set up a web site. It is therefore critical to ensure that any information used professionally within a commercial food business is obtained from a reputable and authoritative source, and is referenced accordingly. Ideally, the reader should cross-check information between at least two reliable sources wherever possible.

List of Useful Web Sites

The Internet is a great source of food safety information, but it is also an ever-changing resource. Web sites come and go and their addresses change. The following is a compendium of some of the most useful food safety information web sites and on-line resources. The web addresses, and descriptions, were correct at the time of writing, but are subject to change.

Libraries

The British Library (fully searchable catalogue). http://www.bl.uk/
The Library of Congress. http://www.loc.gov/index.html
New York State Library. http://www.nysl.nysed.gov/

Scientific Search Engines

In recent years a number of Internet search engines specifically designed to identify scientific papers and publications have been developed. Some are freely accessible, but are not specific to food science and technology. These search engines include:

Google Scholar
Google Scholar is a broad search engine that finds articles from peer-reviewed papers, theses, books, abstracts and articles, from academic publishers, professional societies, preprint repositories, universities and other scholarly organisations. http://scholar.google.co.uk/
Scirus
Scirus examines only science-based web pages for articles containing the search terms. It classifies results into "journal results", "preferred web results" and "other web results" enabling the user to view the search results by source preference. http://www.scirus.com/

Journals

The following is a list of some useful journals that provide some free access to papers relevant to food safety.

Applied and Environmental Microbiology
Published by the American Society for Microbiology. For the food technologist the journal is a useful source of research on food and industrial microbiology. Papers are freely available four months after publication for the primary research journals, and one year after publication for the review journals. http://aem.asm.org/
Comprehensive Reviews in Food Science and Food Safety
A journal produced by the Institute of Food Technologists through Blackwell publishing. Includes papers on risk management, food microbiology and food safety. All articles are freely available on-line. http://www.blackwell-synergy.com/loi/CRFS
Emerging Infectious Diseases
A publication produced by the US Centers for Disease Control and Prevention (CDC). It provides information on emerging and re-emerging infectious diseases worldwide including foodborne microbial pathogens. Access to all papers is free. http://www.cdc.gov/eid/
Eurosurveillance
An on-line journal published by the European Center for Disease, Prevention and Control. It is devoted to the epidemiology, surveillance, prevention and control of communicable diseases; including foodborne disease. It is published in 3 formats (weekly, monthly and quarterly) all of which are freely accessible. http://www.eurosurveillance.org/

Journal of Infectious Diseases
Produced by the University of Chicago Press. This journal specialises in papers on the pathogenesis, diagnosis, and treatment of infectious diseases. There are some useful microbiology papers and access is free for papers over 12 months old. http://www.journals.uchicago.edu/JID/

Morbidity & Mortality weekly report
A publication produced by the US Centers for Disease Control and Prevention (CDC). It provides epidemiological and statistical data on a range of health care issues. Access to all papers is free. http://www.cdc.gov/mmwr/

Government Agencies

Government agencies dealing with food safety issues usually have web sites that not only inform on current food safety issues but also have good archives of reports, and other documents discussing food safety issues. Some useful sites are listed below:

European Center for Disease Control (ECDC)
The ECDC is an EU agency set up to work with health authorities in EU member states to help develop policies on risks posed by current and emerging disease. The agency provides information on health issues including microbiological hazards associated with food. http://www.ecdc.eu.int/

European Commission Food Safety Site
The Food Safety site of the European Commission provides information on food safety and on European food safety legislation. This site includes the Rapid Alert System for Food and Feed (RASFF), a weekly overview of food or feed products associated with "risks", and the countries involved. http://ec.europa.eu/food/index_en.htm

European Food Safety Authority (EFSA)
EFSA is an independent agency funded by the European Union. It supplies and publishes independent scientific advice on existing and emerging risks associated with food and feed safety. The agency provides opinions and information on biological hazards, chemical contaminants, issues associated with food-contact materials allergens, pesticides, genetically modified organisms (GMOs) and animal health and welfare. http://www.efsa.europa.eu/

Food Safety Authority Ireland (FSAI)
The FSAI is an independent and science-based body funded by the Irish government. It provides information and advice in the area of food safety and hygiene. It is also responsible for the enforcement of food safety legislation in Ireland. The site contains some useful publications on HACCP, and foodborne pathogens, particularly reducing the risk of *Escherichia coli* O157. Most publications are freely accessible. http://www.fsai.ie/

Food Standards Australia New Zealand (FSANZ)
The FSANZ is an independent agency set up and funded by the Australian government. For Australia the agency is responsible for food safety standards. http://www.foodstandards.gov.au/

Health Canada
Health Canada is a government agency responsible for providing information and advice on food safety and nutrition. This site has good consumer information on allergens as well as other food safety information. http://www.hc-sc.gc.ca/

New Zealand Food Safety Authority (NZFSA)
The NZFSA is a government agency providing the government, consumers and the food industry in New Zealand with information, analysis and advice on food safety issues. The site includes some useful microbial hazard data sheets and risk profiles specific to a particular hazard/food combination. http://www.nzfsa.govt.nz/

UK Food Standards Agency (FSA)
The FSA is an independent organisation set up by the UK government to protect public health and consumer interests in relation to food. This site contains useful information of microbiological and chemical hazards associated with food. http://www.food.gov.uk/

UK Health Protection Agency (HPA)
The HPA provides advice on health issues including hazards associated with food. http://www.hpa.org.uk/

US Food and Drug Administration (FDA) Center for Food Safety and Applied Nutrition (CFSAN).
A useful and comprehensive web site providing information on many aspects of food safety. Includes: data on organisms associated with food poisoning via the "bad bug book"; heavy metals in food; pesticides; acrylamide; dioxins; furan; and natural toxins. This web site also includes microbiological methods and analytical methods for drugs and chemical residues. http://www.cfsan.fda.gov/

US Department of Agriculture's Food Safety and Inspection Service (FSIS)
The FSIS is the public agency responsible for the safety of meat, poultry and egg products in the US. The web site provides a wealth of information relating to safe production, cooking and storage of these commodities, including fact sheets and published risk assessments. http://www.fsis.usda.gov/home/index.asp

International Organisations

The following international organisations deal with global food safety issues.

Codex Alimentarius Commission
Develops international food standards and guidelines. http://www.codexalimentarius.org/

Food and Agriculture Organization of the United Nations (FAO)
This site contains useful risk assessments on microbial hazards and expert committee reports on food additives. http://www.fao.org/

International Programme on Chemical Safety (IPCS) INCHEM.
INCHEM is an intergovernmental chemical safety web site that gives access to peer reviewed information on chemicals including those found in food, such as pesticides and food additives. http://www.inchem.org/

World Health Organization (WHO)
The site contains information on microbiological and chemical risks associated with food. The web site has also been used as vehicle for publishing information on avian influenza H5N1. http://www.who.int/foodsafety/en/

Universities

Science faculties within some universities providing food safety courses have published very useful information on their web sites. Some examples are as follows.

FoodRisk.org
A University of Maryland and FDA joint project to provide a searchable database to support food safety risk analysis. http://www.foodrisk.org/index.cfm

The Food Safety Network
A Kansas State University based web site providing information and daily news on many food safety issues. http://www.foodsafety.ksu.edu/en/

Nottingham Trent University
General food microbiology information can be found on this web site. http://www.foodmicrobe.com/

University of California
The web site of this university includes the Seafood Information Network Center (Seafood NIC) giving access to seafood safety and quality information. http://seafood.ucdavis.edu/

University of Iowa
The web site provides useful food safety information easy to find and presented in a easily understandable form. http://www.extension.iastate.edu/foodsafety/

Research Institutes and Professional Bodies

American Food Safety Institute (AFSI)
An American Institute providing food safety training and certification. The web site contains some useful information for food processors on food biosecurity. http://www.americanfoodsafety.com/

Hospitality Institute of Technology and Management (HITM)
This web site contains publications relevant to the food service industry. http://www.hi-tm.com/index.html

International Life Sciences Institute (ILSI)
Provides information on food safety, toxicology and risk assessment. Many peer-reviewed publications are freely available on this web site. http://europe.ilsi.org/
Institute of Food Research (IFR)
A UK-based provider of scientific research and information on food. The site contains some useful fact sheets relating to food safety. http://www.ifr.ac.uk/
Institute of Food Science and Technology (IFST)
A UK-based professional body concerned with all aspects of food science and technology, including food safety. There is free information available on the web site including "Information Statements" on a range of topics, including *Campylobacter*, avian influenza and food, and acrylamide. http://www.ifst.org/
Institute of Food Technologists (IFT)
The US-based IFT web site provides free access to many expert reports, scientific summaries, research summits and policy comments relating to food science and food safety. http://www.ift.org/

Trade Associations

Web sites for trade associations often provide very useful sector-specific information on food safety to help with risk assessment. Some of these are listed here.

Chilled Food Association (CFA)
A UK-based association. The web site provides some free information on the principles of food safety when producing and storing chilled foods. http://www.chilledfood.org/
Food and Drink Federation
A UK-based association. Provides a focus for all food processing related issues including food safety. Some food safety information is available for public viewing on the web site. http://www.fdf.org.uk/
Grocery Manufacturers Association (GMA)
The GMA is a US-based trade association with a web site that provides some food safety and food science factsheets, although some are available to members only. http://www.gmabrands.com/index.cfm
Ice Cream Alliance
The web site includes some free information on the safe handling of ice cream. http://www.ice-cream.org/
The British Sandwich Association
Web site provides food safety information to subscribers only. http://www.sandwich.org.uk/
The Specialist Cheesemakers Association
Allows free access to a code of best practice. http://www.specialistcheesemakers.co.uk/

Organisations Publishing Official Standards for Methods of Food Analysis

Some standards organisations web sites have a facility to purchase copies of descriptive methods of analysis that have been developed and published for public scrutiny. These methods of analysis can be referenced and are usually recognised by manufacturers and retailers.

American National Standards Institute (ANSI) http://www.ansi.org
BSI Management Systems http://www.bsi-uk.com/
International Organization for Standardization (ISO) http://www.iso.org/

Organisations Providing Methods of Analysis On-line

There are many useful methods of analysis relating to food safety that are freely available on-line. Some useful web sites for methods are as follows:

AOAC International
Provides guidelines for the validation of microbiological methods of analysis. Provides a list of test kit methods (allergen, toxin, microbiology, biochemical, GM organisms and antibiotic) that have been successfully validated by AOAC. http://www.aoac.org/
Bacteriological Analytical Manual On-line (BAM)
Provides full free details of the US Food and Drug Administration's preferred laboratory methods of microbiological analysis for food and cosmetics. http://www.cfsan.fda.gov/~ebam/bam-toc.html
Compendium of Fish and Fishery Product Processes, Hazards, and Controls
From the University of California web site. http://seafood.ucdavis.edu/HACCP/Compendium/compend.htm
Compendium of methods for Chemical Analysis of Foods
Chemical methods of analysis from the Health Canada web site. http://www.hc-sc.gc.ca/fn-an/res-rech/analy meth/chem/index_e.html
Detection and Quantification of acrylamide in foods
A draft method published by the US Food and Drug Administration's Center for Food Safety and Applied Nutrition. http://www.cfsan.fda.gov/~dms/acrylami.html
Determination of furan in foods
A method published by the US Food and Drug Administration's Center for Food Safety and Applied Nutrition. http://www.cfsan.fda.gov/~dms/furan.html
Food Contact Materials
European legislation giving rules and specific guidance for migration testing of the constituents of plastic materials and articles intended to come into contact with foodstuffs. http://ec.europa.eu/food/food/chemicalsafety/foodcontact/legisl_list_en.htm

Mycotoxins Analytical Methods
A series of factsheets giving details of analytical methods for various mycotoxins produced by the European Mycotoxins Awareness Network, a project funded by the European Commission. http://www.mycotoxins.org/
Pesticide Analytical Manual (PAM)
Analytical methods used by the US Food and Drug Administration's laboratories to examine foods for pesticide residues. http://www.cfsan.fda.gov/~frf/pami1.html
Rapid Microbiology
A web site providing free information on rapid test kits and methods for micro-organisms. Also provides details of suppliers and testing laboratories by country. http://www.rapidmicrobiology.com/
Rapid Test Methods for Seafood Hazards
From the University of California web site. http://seafood.ucdavis.edu/organize/rapid.html
The Compendium of Food Allergen Methodologies
From the Health Canada web site. http://www.hc-sc.gc.ca/fn-an/res-rech/analy meth/allergen/index_e.html
The Compendium of Analytical Methods
Microbiological methods from the Health Canada web site. http://www.hc-sc.gc.ca/fn-an/res-rech/analy meth/microbio/index_e.html

Information on Pesticide Residues

Codex Alimentarius Commission – pesticide residues in food. http://www.codexalimentarius.net/mrls/pestdes/jsp/pest_q-e.jsp
UK Pesticides Residues Committee homepage. http://www.pesticides.gov.uk/prc_home.asp
US Environmental Protection Agency Pesticides page. http://www.epa.gov/pesticides/

Miscellaneous

ComBase
Combined database for predictive microbiology http://www.combase.cc/default.html
Food Safety Watch
An independent web site operated by Food Safety Info supplying food safety news and information. http://www.foodsafetywatch.com
International Food Information Council (IFIC) Foundation
US-based web site for disseminating scientific information on food safety and nutrition. http://ific.org/food/
ProMed-Mail
International reporting forum for outbreaks of infectious diseases and toxins, including food-poisoning outbreaks. http://www.promedmail.org

Subject Index